CURRICULUM FOR WALES

MATHEMATICS AND NUMERACY

Mastering Mathematics

FOR 11–14 YEARS

Linda Mason, Jonathan Agar, Laszlo Fedor

BOOK 2

The Publishers would like to thank the following for permission to reproduce copyright material.

Photo credits

p. 60 © Rawpixel - Fotolia.com; **p. 129** © butenkow - stock.adobe.com; **p. 160** © Christian Mueller/Shutterstock.com; **p. 183** © Luke Robinson; **p. 185**a © darkhriss - stock.adobe.com, b © arbalest - stock.adobe.com, c © Lynne Carpenter/stock.adobe.com, d © tnehala - stock.adobe.com, e © eldadcarin - stock.adobe.com, f © AlenKadr - stock.adobe.com, g © bsanchez - stock.adobe.com, h © akiyoko - stock.adobe.com, i © Luke Robinson; **p. 277**©francescodemarco - stock.adobe.com.

Acknowledgements

This book is based on material written for and published in *Key Stage 3 Mastering Mathematics: Book 2*, Second Edition (978 1 3983 0840 4) by Sophie Goldie, Luke Robinson and Andrew Ginty, with Series Editor Steve Cavill. The publisher would like to thank them for permission to re-use their work in the present volume.

Typical stopping distances data on page 38 © The Official Highway Code, 2021

This information is licensed under the Open Government Licence v3.0. To view this licence, visit www.nationalarchives.gov.uk/doc/open-government-licence/ **OGL**

Although every effort has been made to ensure that website addresses are correct at time of going to press, Hodder Education cannot be held responsible for the content of any website mentioned in this book. It is sometimes possible to find a relocated web page by typing in the address of the home page for a website in the URL window of your browser.

Hachette UK's policy is to use papers that are natural, renewable and recyclable products and made from wood grown in well-managed forests and other controlled sources. The logging and manufacturing processes are expected to conform to the environmental regulations of the country of origin.

Orders: please contact Hachette UK Distribution, Hely Hutchinson Centre, Milton Road, Didcot, Oxfordshire, OX11 7HH. Telephone: +44 (0)1235 827827. Email: education@hachette.co.uk Lines are open from 9 a.m. to 5 p.m., Monday to Friday. You can also order through our website: www.hoddereducation.co.uk

ISBN: 978 1 3983 4446 4

© Hodder & Stoughton Limited 2022

First published in 2022 by
Hodder Education,
An Hachette UK Company
Carmelite House
50 Victoria Embankment
London EC4Y 0DZ

www.hoddereducation.co.uk

Impression number   10 9 8 7 6 5 4 3 2 1

Year   2026 2025 2024 2023 2022

All rights reserved. Apart from any use permitted under UK copyright law, no part of this publication may be reproduced or transmitted in any form or by any means, electronic or mechanical, including photocopying and recording, or held within any information storage and retrieval system, without permission in writing from the publisher or under licence from the Copyright Licensing Agency Limited. Further details of such licences (for reprographic reproduction) may be obtained from the Copyright Licensing Agency Limited, www.cla.co.uk

Cover photo © juhrozian - stock.adobe.com

Typeset in India by Aptara Inc.

Printed in India.

A catalogue record for this title is available from the British Library.

Contents

The curriculum vi

How to use this book vii

1 Sequences 1
1.1 Function machines 1
1.2 Arithmetic sequences 6
Review exercise 14

2 Graphs 17
2.1 Straight line graphs 17
2.2 Real life graphs 26
Review exercise 35

Consolidation 1 38

3 Angles 42
3.1 Parallel lines 42
3.2 Polygons 49
Review exercise 54

4 Constructions 56
4.1 Bearings 56
4.2 Scale drawings 60
4.3 Constructions 63
Review exercise 67

5 Calculations 70
5.1 Calculations review 71
5.2 Multiplying decimals 76
5.3 Dividing decimals 79
Review exercise 82

Consolidation 2 85

6 Negative numbers 89
6.1 Negative numbers 89
6.2 Adding and subtracting negative numbers 92
6.3 Multiplying and dividing negative numbers 98
Review exercise 103

7 Fractions 105
   7.1 Fractions review 105
   7.2 Multiplying fractions 111
   7.3 Dividing fractions 115
   Review exercise 118

8 Expressions and formulas 120
   8.1 Working with letter symbols review 121
   8.2 Expanding brackets 124
   8.3 Factorising expressions 126
   8.4 Rearranging formulas 128
   Review exercise 131

Consolidation 3 133

9 Equations 137
   9.1 Solving equations review 137
   9.2 Solving equations with an unknown on both sides 143
   9.3 Solving equations with brackets 149
   Review exercise 154

10 Working with 2D shapes 157
   10.1 Types of quadrilateral 158
   10.2 Area 166
   Review exercise 174

Consolidation 4 178

11 Properties of 3D shapes 182
   11.1 Properties of 3D shapes 182
   11.2 Nets 188
   11.3 Surface area and volume of a cuboid 192
   Review exercise 199

12 Percentages 202
   12.1 Working with percentages 202
   12.2 Percentage increase and decrease 206
   12.3 Percentage change 210
   Review exercise 217

⑬ Multiplicative reasoning 219
  13.1 Ratio and proportion review 220
  13.2 Conversion graphs 224
  13.3 Best buys 231
  Review exercise 237

Consolidation 5 240

⑭ Working with data 244
  14.1 Frequency tables 244
  14.2 Pie charts 251
  Review exercise 258

⑮ Circles 261
  15.1 Circumference 262
  15.2 Area of a circle 268
  Review exercise 275

⑯ Pythagoras' theorem 278
  16.1 Investigating triangles 279
  16.2 Using Pythagoras' theorem 284
  Review exercise 289

Consolidation 6 292

Glossary 296

The curriculum

The Curriculum for Wales has been developed in Wales, by practitioners for practitioners, bringing together educational expertise with wider research and evidence. Our resources are designed to reflect the Welsh context and to help develop your identity as a citizen of Wales and the world.

We have worked in collaboration with University of Wales Press to produce this resource. They have reviewed it to make sure it is tailored to the new curriculum and explores Welsh culture and heritage in an authentic way. Find out more about University of Wales Press and their resources in Welsh and English languages by visiting their websites **www.uwp.co.uk** and **www.gwasgprifysgolcymru.org**

Our authors have a wealth of experience teaching, examining and working in education in Wales:

- **Laszlo Fedor** has 20 years' experience of working in state-funded schools in Wales. He teaches Mathematics to children of all abilities in secondary and sixth-form education.
- **Jonathan Agar** has been teaching in South Wales for over a decade, including time spent as head of department, assistant headteacher and principal examiner. He has a particular interest in pupils' misconceptions in mathematics and has completed a Research Masters in Mathematics Education.
- Series Editor **Linda Mason** has many years of experience as teacher, adviser, curriculum consultant and principal examiner in Wales, including work across North Wales and with WJEC.

We would also like to thank the teachers from schools across Wales who helped to plan and review this title, including:

- Llanishen High School, Cardiff
- Cefn Hengoed Community School, Swansea
- Bishop Hedley High School, Merthyr Tydfil.

How to use this book

▶ How to get the most from this book

Hodder Education's Mathematics resources support the learning and experience of Mathematics for years 7–9 and comprise:

- three books to support the Wales National Curriculum for ages 11–14
- Boost online content.

Our Book 2 material is split into 16 chapters, and each chapter comprises two, three or four units. In total there are 42 units in the book. The material across all three books, and the editable course planner, is designed to be used whenever the teacher feels it is appropriate for the class; for example, some content in Book 1 or Book 3 may be suitable for some teaching in year 8. Similarly, our scheme of work is designed to be flexible.

The book contains indication of five proficiencies: **conceptual understanding**, **communication using symbols**, **fluency**, **logical reasoning** and **strategic competence**.

- Fluency
- Logical reasoning
- Strategic competence
- Conceptual understanding
- Communication using symbols

These five proficiencies are intertwined, so no individual proficiency is developed in isolation. Consequently, in general, many of the proficiencies could be highlighted in activities, examples and exercises throughout the book.

As an aid, the best fit or a principal proficiency is flagged as guidance only, to raise awareness of a particular proficiency for the learner. We have chosen to highlight good examples of conceptual understanding and communication using symbols as relevant alongside various mathematical explanations, activities and worked examples. Fluency, logical reasoning and strategic competence are highlighted in relation to individual questions in our exercises, reviews and consolidation sections. All of these indicators are intended as guidance to aid the learner in understanding their own proficiency development.

In summary, the five proficiencies capture a learner's developing understanding of the multi-faceted nature of their learning.

- **Conceptual understanding** allows learners to develop their ability to connect ideas through increasing depth of knowledge. Understanding the way in which concepts are connected aids learner development.
- Through progression in **communication using symbols**, learners develop understanding of conventions and abstract representation.
- With experience, learners will develop **fluency** in remembering facts, relationships and techniques.
- Learners develop **logical reasoning**, including justification and proof, in understanding the relationship between concepts.
- In developing **strategic competence**, learners show independence in applying ideas within a problem, and recognise mathematical structure.

All exercise questions that relate to finance are indicated with this symbol. £

Each **chapter** includes:

- *Coming up* – a list of learning objectives that will be tackled in the chapter
- a *Starter* problem – either an activity or a puzzle – to engage the students with a new topic and designed to be used before the first lesson, or at the start of the first lesson in that topic

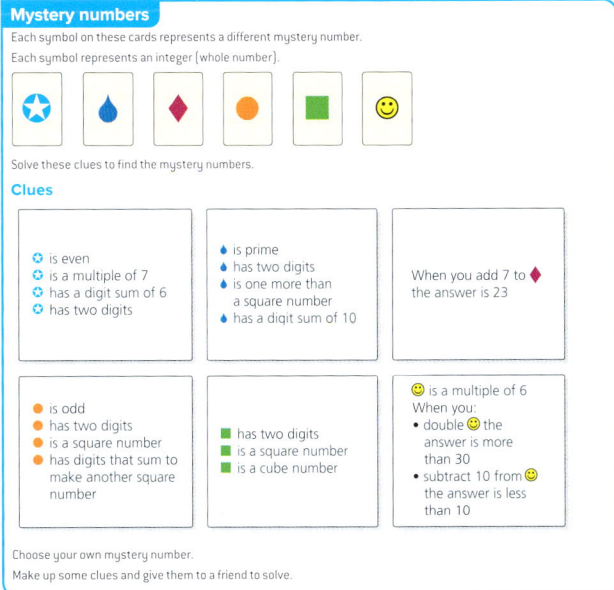

- activities, investigations and whole-class discussion points
- a *Review exercise* at the end of the chapter; this encompasses all of the units covered in the chapter.

Each of the **individual units** within the chapter includes:

- a *Skill checker* – simple diagnostic questions to test basic understanding in preparation for the unit

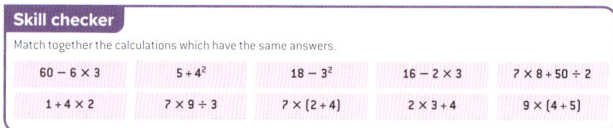

- a clear and detailed explanation of the topic
- plenty of worked examples with solutions
- a focus on fluency, with a carefully structured approach that takes into account cognitive load theory
- helpful hints and guidance on misconceptions and pitfalls to watch out for
- *Now try these* exercise questions, which:
  - develop conceptual understanding and communication using symbols
  - are split up into three bands of increasing demand: Band 1 questions are for those students who are working towards age-related expectations, Band 2 are for those at age-related expectations, and Band 3 are for those working beyond age-related expectations. Most students will engage with Band 2 questions and either Band 1 or Band 3, depending on which is most appropriate.
  - are carefully calibrated to enable the whole class to understand each question and answer before moving on
  - give the opportunity to apply skills, including working systematically, modelling, breaking problems down into stages, visualising, working backwards, and trial and improvement.

  Non-calculator questions are indicated.
- a list of key words (highlighted in the text). These are fully explained in a glossary at the back of the book.

There are six sets of consolidation questions throughout the book, each of which appears after a sequence of two or three chapters. These are designed to cover approximately half a term's work.

The book encourages learners to use physical equipment (manipulatives) and representations as well as visual and abstract representations, for example using cards, bar model diagrams and physical number lines to aid the development of understanding.

Opportunities to link with Science and Technology, Humanities, Expressive Arts, Health and Well-being, Languages, Literacy and Communication teaching and learning are included in the *Cross-curricular activities* panels. Scattered throughout the books are examples that we hope will encourage exploration of historical Welsh mathematicians and contexts.

All answers are provided online at **www.hoddereducation.co.uk/MasteringMathematicsWales** and are freely accessible. You can also find an editable course planner here, with lesson suggestions and time built in for consolidation, assessment and application lessons. A suggestion for how content in these resources can be mapped to the Wales curriculum's What Matters statements and Progression Steps has been included in an editable format, to enable schools to create their own structure, as well as a full set of links to other areas of the curriculum across all subjects.

1 Sequences

Coming up...
- Using function machines
- Understanding arithmetic sequences
- Sequences in diagrams

Patterns

The diagram below shows three shapes in a pattern made using matchsticks.

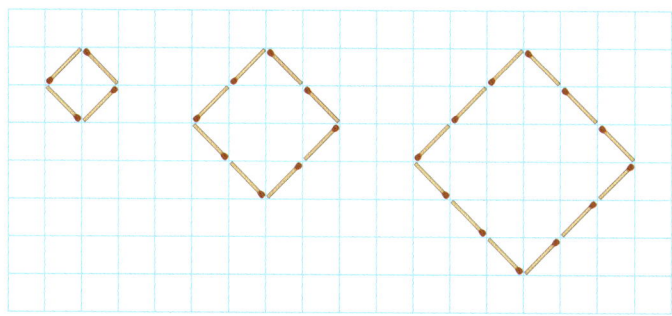

Without drawing the shape, work out how many matchsticks will be needed to make the next shape in the pattern.

Draw the next shape to see whether you were correct or not.

The number of matchsticks in each shape in this pattern forms an arithmetic sequence, which you will learn about in Section 1.1.

1.1 Function machines

Skill checker

① Complete each of the calculations below.
    a   16 + 27   b   96 − 39   c   18 × 6

② Find the unknown values in each of the calculations below.
    a   18 + ☐ = 39   b   ☐ − 13 = 58   c   ☐ × 9 = 63

▶ Function machines

A function machine applies a **rule** to any numbers which are **input** into the machine.

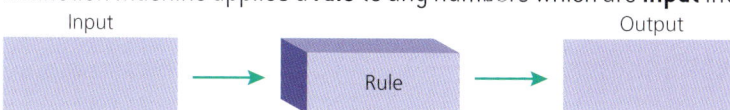

The **output** of the function machine is determined by what operation the machine performs. This is called its **rule**.

This function machine multiplies the **input** number by 3.

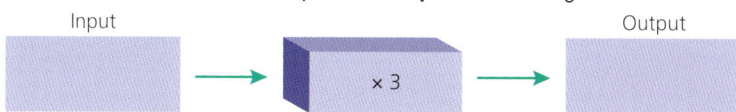

Worked example

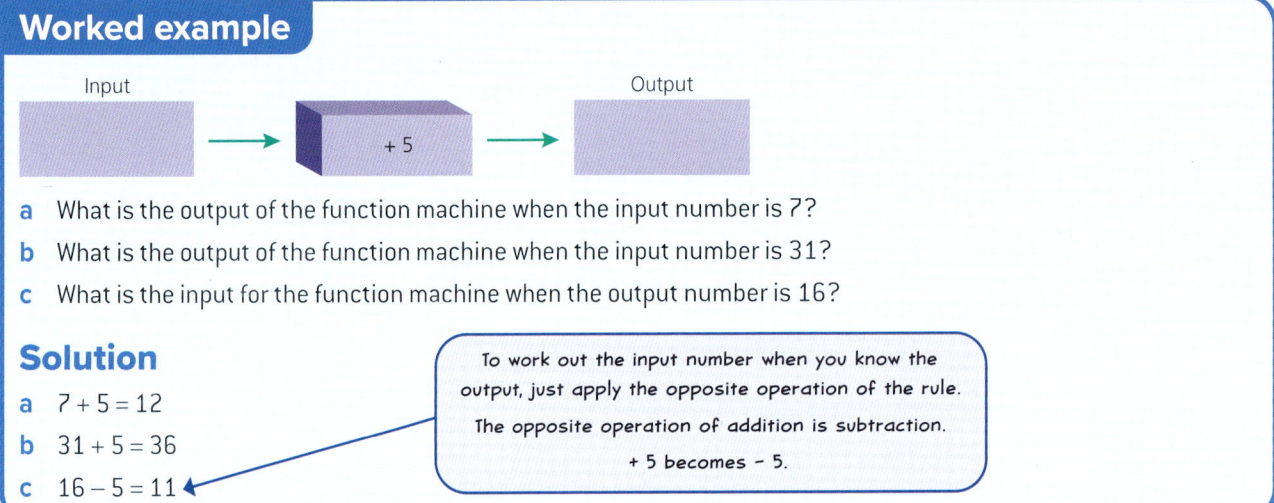

a What is the output of the function machine when the input number is 7?
b What is the output of the function machine when the input number is 31?
c What is the input for the function machine when the output number is 16?

Solution
a $7 + 5 = 12$
b $31 + 5 = 36$
c $16 - 5 = 11$

> To work out the input number when you know the output, just apply the opposite operation of the rule.
> The opposite operation of addition is subtraction.
> + 5 becomes − 5.

Worked example

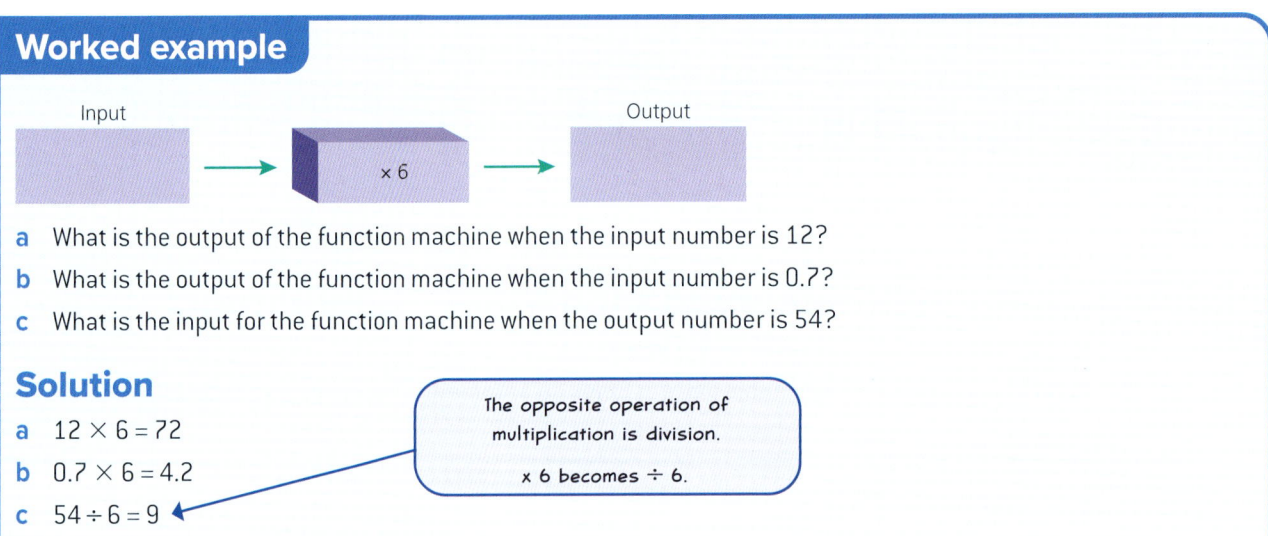

a What is the output of the function machine when the input number is 12?
b What is the output of the function machine when the input number is 0.7?
c What is the input for the function machine when the output number is 54?

Solution
a $12 \times 6 = 72$
b $0.7 \times 6 = 4.2$
c $54 \div 6 = 9$

> The opposite operation of multiplication is division.
> × 6 becomes ÷ 6.

Activity

1. Owain thinks of a number.
   He adds 17 to his number and gets an answer of 25.
   What number did Owain first think of?

2. Elin thinks of a number.
   She multiplies her number by 5 and gets an answer of 60.
   What number did Elin first think of?

3. Tomos thinks of a number.
   He divides his number by 4 and gets an answer of 6.
   What number did Tomos first think of?

1 Sequences

1.1 Now try these

Band 1 questions

1. Work out the output for each of these function machines.

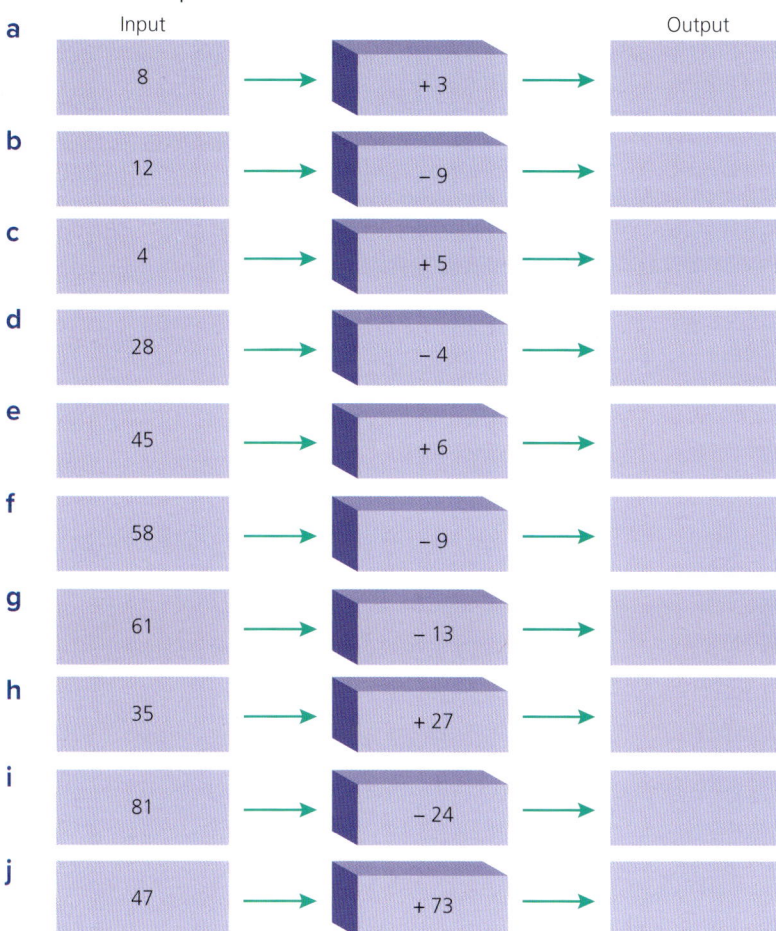

2. Work out the input for each of these function machines.

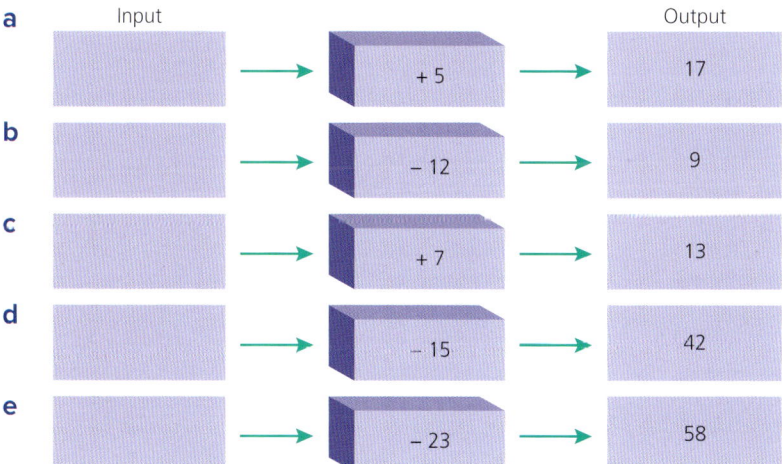

Band 2 questions

3 Work out the output for each of these function machines.

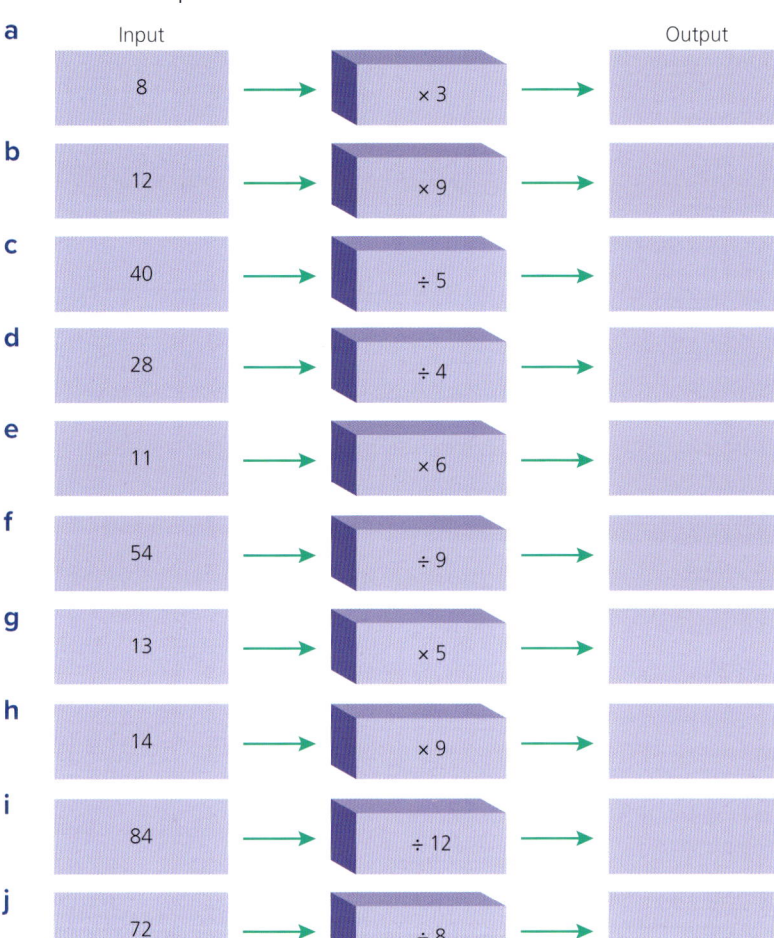

4 Work out the input for each of these function machines.

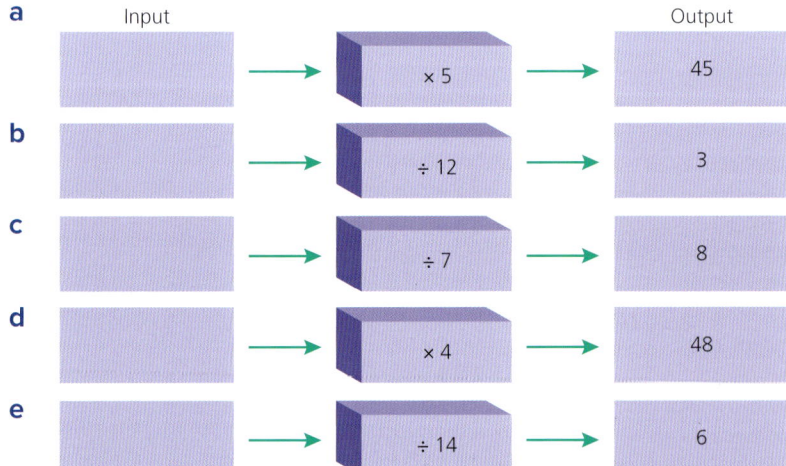

1 Sequences

Band 3 questions

5 Work out the output for each of these function machines.

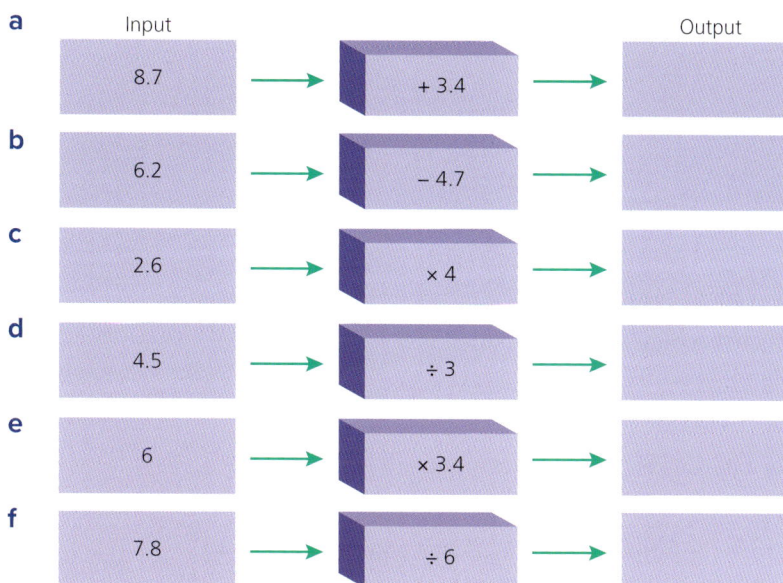

a. 8.7 → + 3.4 →
b. 6.2 → − 4.7 →
c. 2.6 → × 4 →
d. 4.5 → ÷ 3 →
e. 6 → × 3.4 →
f. 7.8 → ÷ 6 →

6 Work out the input for each of these function machines.

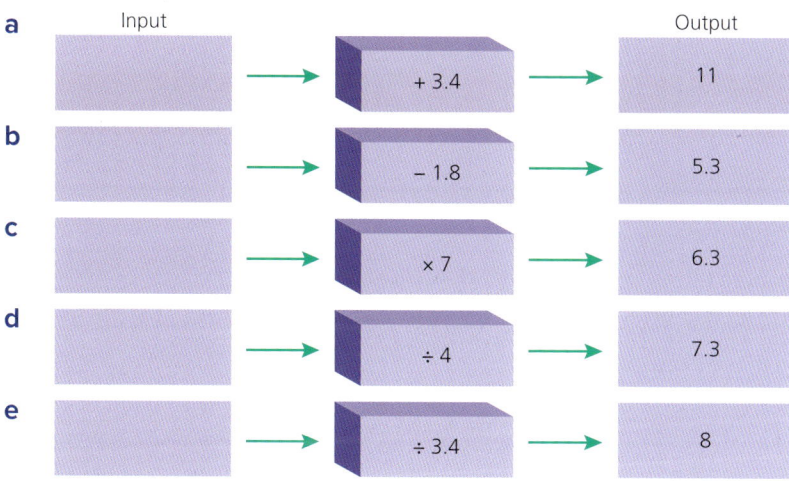

a. → + 3.4 → 11
b. → − 1.8 → 5.3
c. → × 7 → 6.3
d. → ÷ 4 → 7.3
e. → ÷ 3.4 → 8

7 Work out the rule for each of these function machines.

a.
- 3 → → 6
- 8 → → 11

b.
- 3 → → 15
- 7 → → 35

c.
- 10 → → 5
- 8 → → 4

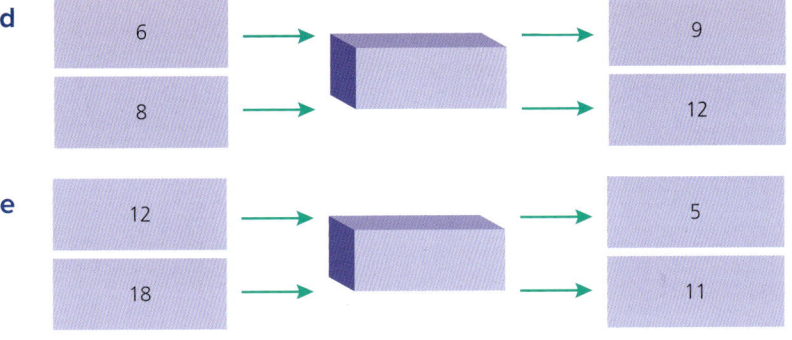

d

e

⑧ Dewi input 3 into a function machine and got an output of 12.
Angharad input 1.5 into the same function machine and got an output of 6.
What is the rule of the function machine?

⑨ Aled input 3 into a function machine.
He got an output number which was five times as big as the input number.
What are the two possible rules of the function machine?

⑩ Bethan input 6 into a function machine and the output was a single-digit square number.
Catrin input 11 into the function machine and the output was also a single-digit square number.
What is the rule of the function machine?

1.2 Arithmetic sequences

Skill checker

① Continue these sequences for three more terms.
   a 3, 6, 9, 12, …   b 45, 50, 55, 60, …   c 13, 23, 33, 43, …   d 11, 22, 33, 44, …

② Substitute $n = 1$, $n = 2$ and $n = 3$ into these expressions.
   a $3n$ b $n + 4$ c $5n - 1$ d $100n - 2$

▶ Arithmetic sequences

Look at this sequence of numbers:

   3, 6, 9, 12, …

This sequence is based on the three times table.

The 1st term is $3 \times 1 = 3$.

The 4th term is $3 \times 4 = 12$.

The 10th term is $3 \times 10 = 30$.

The nth term is $3 \times n = 3n$.

> $3n$ is called the 'position-to-term rule' for the sequence. It is also sometimes called the nth term of the sequence.

You can use the **position-to-term rule** to find any term of the sequence.

For example, to find the 20th term:

   $3 \times 20 = 60$.

A sequence in which you add or subtract a fixed amount to move from one term to the next is called an **arithmetic sequence**.

In this arithmetic sequence, the **term-to-term rule** is 'add 3'.

Worked example

Look at this sequence:

  4, 7, 10, 13, ...

a What is the term-to-term rule?
b Find the position-to-term rule for the sequence.
c What is the 50th term of the sequence?
d Which term of the sequence is 100?
e What type of sequence is this?

Solution

a The term-to-term rule is 'add 3'.

b Each term is 1 more than the three times table, so the position-to-term rule is $3n + 1$.

The sequence goes up in 3s, so is based on the three times table.

Look at this comparison with the previous example.

$3n$	3	6	9	12
+1	+1	+1	+1	+1
$3n + 1$	4	7	10	13

You can **check** this position-to-term rule works. Use $n = 1, 2, 3$ and 4 in the rule.

n	1	2	3	4
$3n + 1$	$3 \times 1 + 1 = $ **4**	$3 \times 2 + 1 = $ **7**	$3 \times 3 + 1 = $ **10**	$3 \times 4 + 1 = $ **13**

The table shows that the position-to-term rule gives the first four terms of the sequence.

c You can find the 50th term using the rule.

nth term = $3n + 1$
50th term = $3 \times 50 + 1$
         = 151

d To find which term of the sequence is 100, use the rule again.
Solve the equation to find n.

nth term = $3n + 1$
$100 = 3n + 1$
$100 - 1 = 3n$
$3n = 99$
$n = 33$

The 33rd term of the sequence is 100.

e To move from one term to the next, you add 3. This is an arithmetic sequence.

> **Remember**
>
> The difference tells you which times table the sequence is based on. For example, if there is a difference of 3 between each term, then $3n$ will appear in the position-to-term rule.

Communication using symbols

1 Sequences

Curriculum for Wales Mastering Mathematics: Book 2

Worked example

a Find the position-to-term rules for these sequences:
  i $-2, -4, -6, -8, \ldots$
  ii $-1, -3, -5, -7, \ldots$
b What type of sequence are these?

Solution

a i This is the -2 times table, so the position-to-term rule is $-2n$.
  ii Each term is one more than the terms in the -2 times table.
  The position-to-term rule is $-2n + 1$.

Another way to write this is $1 - 2n$.

b In both sequences you subtract 2 to move from one term to the next. They are both arithmetic sequences.

Activity

Read this newspaper report about Steel City.
Somebody has spilt coffee on the page and some of the numbers are unreadable.

STEEL CITY BOOM TOWN!
It's a new year and the residents of Steel City are waking up this morning to the sound of yet more construction work.
Work has begun on the huge SC Tower, which will become the tallest building so far in Steel City.
We have contacted the city planners. They say that there are currently 14 skyscrapers in Steel City and there will be 44 skyscrapers in the city in five years' time.

They have given permission for new skyscrapers EVERY YEAR!

If the city's plans become a reality, by the end of this year there will be skyscrapers.

If construction work in Steel City continues at this rate, there will be 62 skyscrapers in years' time.

In ten years' time the city will have new skyscrapers!

Can the residents of Steel City put up with this much noise? Does Steel City have enough sandwich shops to feed all of these hungry office workers? Let us know what you think.

a Can you replace the coffee marks with the correct numbers?
b What is the term-to-term rule for the sequence of the number of skyscrapers?

1.2 Now try these

Band 1 questions

1 Find the next three terms in these arithmetic sequences.
  a $2, 3, 4, 5, \ldots$
  b $0, 4, 8, 12, \ldots$
  c $-4, -2, 0, 2, \ldots$
  d $-5, -8, -11, -14, \ldots$
  e $4, 0, -4, -8, -12, \ldots$
  f $0, -2, -4, -6, \ldots$
  g $-0.1, -3.1, -6.1, -9.1, \ldots$

1 Sequences

2 Find the missing numbers in the arithmetic sequences below.
  a 2, 4, 6, 8, ☐, 12, ...
  b 5, 9, 13, ☐, 21, ...
  c −3, −6, ☐, −12, −15, ...
  d 21, 25.5, ☐, 34.5, 39, ...
  e 0, −2, −4, ☐, −8, ...
  f −1.5, −3, −4.5, ☐, −7.5, ...
  g 30, 27.5, 25, ☐, 20, ☐, ...
  h 8, ☐, 24, ☐, 40, ...
  i 6, ☐, 18, ☐, 30, ...
  j ☐, −15, ☐, −25, −30, ...

3 Find the *n*th term of these sequences.
In each case the rule is ☐*n*.
  a 2, 4, 6, 8, 10, ...
  b 4, 8, 12, 16, ...
  c 5, 10, 15, 20, ...
  d 1, 2, 3, 4, 5, ...
  e 100, 200, 300, 400, 500, ...
  f 10, 20, 30, 40, 50, ...
  g 8, 16, 24, 32, ...

4 Find the *n*th term of these sequences.
In each case the rule is ☐*n* + 1.
  a 6, 11, 16, 21, 26, ...
  b 11, 21, 31, 41, 51, ...
  c 101, 201, 301, 401, 501, ...
  d 5, 9, 13, 17, ...

5 Find the *n*th term of these sequences.
In each case the rule is ☐*n* − 1.
  a 99, 199, 299, 399, ...
  b 9, 19, 29, 39, 49, ...
  c 4, 9, 14, 19, 24, ...
  d 7, 15, 23, 31, ...
  e 3, 7, 11, 15, ...

6 Copy this spiral pattern onto isometric paper.
Start with the red dot near the centre of the page.

  a Count the number of dots on each line that you draw.
     Write them as a sequence: 2, 3, ...
  b Predict how many dots will be on the next line that you draw.
     Check by drawing and counting.
  c Describe the pattern in the number of dots.

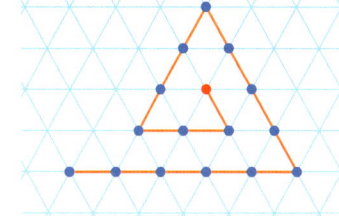

7 Find the first five terms of these sequences, given their position-to-term rules. Use the tables to help you.

  a The sequence with position-to-term rule $n + 4$.

n	1	2	3	4	5
$n + 4$	1 + 4 = 5	2 + 4 = __			

  b The sequence with position-to-term rule $n - 5$.

n	1	2	3	4	5
$n - 5$	1 − 5 = −4	2 − 5 = __			

  c The sequence with position-to-term rule $2n$.

n	1	2	3	4	5
$2n$	2 × 1 = 5	2 × 2 = __			

  d The sequence with position-to-term rule $-3n$.

n	1	2	3	4	5
$-3n$	−3 × 1 = __				

  e The sequence with position-to-term rule $2n + 3$.

n	1	2	3	4	5
$2n + 3$	2 × 1 + 3 = 5	2 × 2 + 3 = __			

9

f The sequence with position-to-term rule $2n - 4$.

n	1	2	3	4	5
$2n - 4$					

8 Ahmed has £16 in his money box at the start of the year.

He gets £5 pocket money each week and puts it into his money box without spending any of it.

   a Beginning with £16, write down the amount of money in Ahmed's money box as a sequence for 5 weeks.

   b What is the term-to-term rule for this sequence?

   c What is the position-to-term rule?

9 Siôn is making patterns from matchsticks.

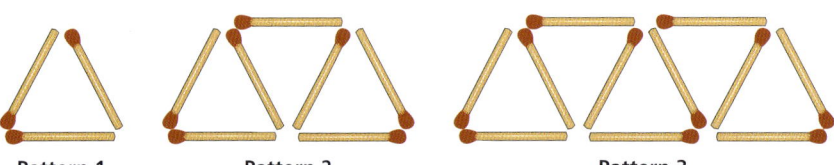

Pattern 1   Pattern 2   Pattern 3

   a Copy and complete this table for the number of matchsticks used in each pattern.

Include the number of matchsticks Siôn would use if he continued to Pattern 4.

Pattern number, n	1	2	3	4
Number of matchsticks	3			

   b If the pattern is continued, which pattern will use 23 matchsticks?

   c Find the term-to-term rule for this sequence.

   d Find the position-to-term rule for the number of matchsticks.

> **Hint**
> You add four matchsticks to move from one pattern to the next. This means that $4n$ will be in your position-to-term rule.

Band 2 questions

10 Find the nth term of these sequences.

   a 4, 8, 12, 16, …   ← This is the four times table.

   b 5, 9, 13, 17, …   ← These numbers are all one more than the numbers in the four times table.

   c 3, 7, 11, 15, …

   d 6, 10, 14, 18, …   ← These numbers are all one less than the numbers in the four times table.

   e 9, 13, 17, 21, …

11 Find the nth term of these sequences.

   a 2, 4, 6, 8, 10, …   ← This is the two times table.

   b 4, 6, 8, 10, 12, …   ← These numbers are all two more than the numbers in the two times table.

   c 1, 3, 5, 7, 9, …

12 Find the nth term of these sequences.

   a 3, 6, 9, 12, 15, …   **b** 6, 9, 12, 15, 18, …   **c** $-2, 1, 4, 7, 10, …$

13 Find the nth term of these sequences.

   a 3, 4, 5, 6, 7, …   **b** $-1, 3, 7, 11, 15, …$ **c** $-1, -2, -3, -4, -5, …$ **d** $-3, -6, -9, -12, -15, …$

1 Sequences

14 The Olympic Games take place every four years. The first modern Olympic Games took place in 1896.

A position-to-term rule to find the year of the nth Olympic Games is:

$4n + 1892$

The table shows you how to use the position-to-term rule to find the year of the 2nd Olympic Games.

> Some Olympic Games did not actually take place. For example, the 13th Olympic Games in 1944 was cancelled because of the Second World War, but the 1948 Games was still called the 14th Games. The 2020 Games were postponed until 2021 because of the Covid-19 pandemic.

a Copy and complete the table to find the years of the 3rd, 5th and 10th Olympic Games.

n	2	3	5	10
Year	$4 \times 2 + 1892 = 1900$	$4 \times 3 + 1892 = $ __		

b Use the rule to find what number Olympic Games took place in London in 2012.

15 Christophe is making patterns of houses from matchsticks.

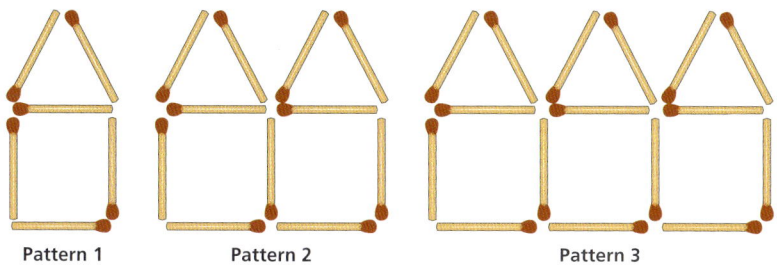

Pattern 1   Pattern 2   Pattern 3

a Copy and complete this table for the number of matchsticks used in each pattern.
Include the number of matchsticks Christophe would use if he continued to Pattern 4.

Pattern number, n	1	2	3	4
Number of matchsticks	6			

b If the pattern is continued, which pattern will use:

  i 26 matchsticks   ii 201 matchsticks?

c What is the term-to-term rule for this sequence?

16 Look at the patterns made up from square tiles.
Each pattern is in the shape of a cross.
Each tile measures 1 cm by 1 cm. The area of one tile is 1 cm².
The total area of Pattern 1 is 5 cm².
The total area of Pattern 2 is 9 cm².

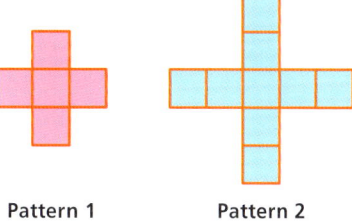

Pattern 1   Pattern 2

a If the pattern is continued, what will be the area of Pattern 4?

b Which pattern will have an area of 101 cm²?

The perimeter of Pattern 1 is 12 cm.

The perimeter of Pattern 2 is 20 cm.

c If the pattern is continued, what will be the perimeter of Pattern 4?

d Which pattern will have a perimeter of 100 cm?

17 Pedr is making patterns with square tiles. Some of the tiles are green, some are shaded.

In Pattern 1, there is one green tile and there are eight shaded tiles.
In Pattern 2, there are two green tiles and ten shaded tiles.
In Pattern 3, there are three green tiles and twelve shaded tiles.

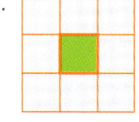 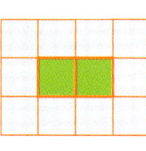 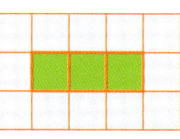

Pattern 1   Pattern 2   Pattern 3

a How many green tiles will there be in Pattern 6?

b How many shaded tiles will there be in Pattern 6?

c How many tiles will there be in Pattern 6 in total?

d If Pedr continued these patterns, which pattern would contain 80 shaded tiles?
How many green tiles would this pattern have?

Curriculum for Wales Mastering Mathematics: Book 2

Band 3 questions

18 Find the nth term of these sequences.

    a $-1, -3, -5, -7, -9, \ldots$ **b** $3, 2, 1, 0, -1, \ldots$

    c $-6, -8, -10, -12, -14, \ldots$ **d** $-8, -13, -18, -23, -28, \ldots$

    e $-5, -8, -11, -14, -17, \ldots$ **f** $-4, -9, -14, -19, -24, \ldots$

    g $0, -1, -2, -3, -4, \ldots$

19 For each of the sequences in question 18:

    a   Enter the terms of the sequence into a table, like the one below.

           Part **a** has been done for you.

           The terms are $-1, -3, -5, -7, -9$.

n	1	2	3	4	5
Term	-1	-3	-5	-7	-9

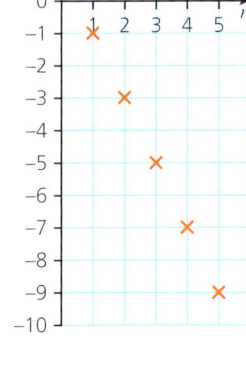

    b   Plot the first five terms in the sequence against its position number, n, on a graph.

           For part **a**, the points $(1, -1), (2, -3), (3, -5), (4, -7), (5, -9)$ have been plotted.

    c   What do you notice about all six of the graphs you have drawn?

20 Dayah is making patterns with square tiles. Some of the tiles are orange, some are grey.

      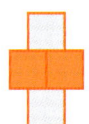   

Pattern 1   Pattern 2   Pattern 3

    a   How many orange tiles will there be in Pattern 6?

    b   How many grey tiles will there be in Pattern 6?

    c   If the patterns are continued, which pattern will have a total of 118 tiles?

    d   Fill in the blank space below with a number.

           The first __ patterns could be made using 30 grey and 21 orange tiles.

    e   Dayah draws a graph to show the total number of tiles used for each pattern.

           Copy and complete Dayah's graph.

    f   What do you notice about the points?

    g   Is Dayah's sequence arithmetic?

    h   Do you think she should join the points? Why/why not?

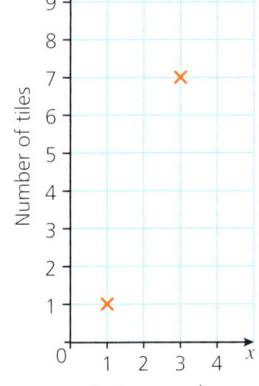

21 Mr Jones is driving from Rhyl to Cardiff

The total distance is 220 miles.

He drives at 55 miles per hour.

    a   Copy and complete this table.

Time taken (hours)	0	1	2		
Distance travelled (miles)	0	55			
Distance left (miles)	220				

    b   How many hours does it take Mr Jones to drive to Cardiff?

    c   **i**   Write down the numbers in the second row of the table as a sequence.

           ii   What is the term-to-term rule for this sequence?

           iii   What is the position-to-term rule?

d **i** Write down the numbers in the third row of the table as a sequence.
  ii What is the term-to-term rule for this sequence?
  iii What is the position-to-term rule?

22 Mr Coulston has 500 plums in his fridge after buying them in a special offer from his local supermarket.
He is now trying to give them away.

a On Day 1, Mr Coulston gives away 40 plums. He has 460 plums left.
On Day 2, Mr Coulston gives away another 40 plums. How many plums does he have left?

b He continues giving away 40 plums each day. Copy and complete this table.

Day, n	1	2	3	4	5	6	7
Number of plums left in fridge	460						

c The number of plums left forms a sequence. Find the term-to-term rule for the sequence.

d Find the position-to-term rule for the number of plums.

23 **a** Make as many arithmetic sequences with five terms as you can from these numbers:

  1   3   5   6   10   15   21   20   25   9   16   8   2   32   12   4

b For each sequence you have made, work out the position-to-term rule.

24 Pria's clock has gone crazy!
At 12:01 p.m. it said 5:32 p.m.
One minute later it said 5:34 p.m.
One minute after that it said 5:36 p.m.
It seems to be moving forward 2 minutes every minute!

a Copy and complete this table.

Real time	12:01 p.m.	12:02 p.m.	12:03 p.m.			
Time shown on Pria's clock	5:32 p.m.	5:34 p.m.				

b Look at just the minutes part of the times shown on the clock.
They form this sequence:

  32, 34, 36, ...

What is the term-to-term rule for this sequence?

c What is the position-to-term rule?

d What time is shown on Pria's clock when the real time is 12:14 p.m.?

e What is the real time when Pria's clock shows 5:50 p.m.?

Key words

Here is a list of the key words you met in this chapter.

Arithmetic sequence	Function machine	Input	Output
Position-to-term rule	Square numbers	Term	Term-to-term rule

Use the glossary at the back of this book to check any you are unsure about.

Review exercise: sequences

Band 1 questions

1 Work out the output for each of these function machines.

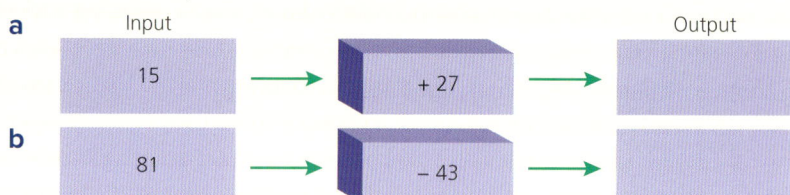

2 Work out the input for each of these function machines.

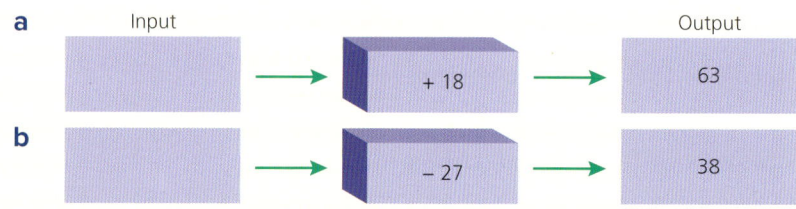

3 Tomos is stacking tins of paint in a DIY shop.

He begins by making a triangle with the tins as shown in Pattern 1.

He then makes a second stack of tins as shown in Pattern 2.

Tomos notices there are two tins in the bottom row of his first pattern and three tins in the bottom row of his second pattern.

The table below shows these results.

Pattern number	1	2
Number of tins in bottom row	2	3

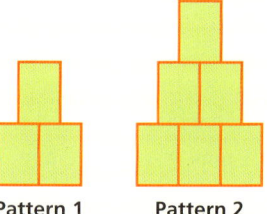

Pattern 1   Pattern 2

a If Tomos keeps making triangle patterns, which pattern number will give him six tins in the bottom row?

b Which pattern number will have a total of 15 tins?

4 Find the next term and the term-to-term rule for each of these sequences.

a 2, 5, 8, 11, 14, …   b 50, 47, 44, 41, …   c 9, $10\frac{1}{2}$, 12, $13\frac{1}{2}$, …

5 a A sequence has the position-to-term rule $2n - 6$.

Find the first five terms of this sequence. Use the table below to help you.

n	1	2	3	4	5
$2n - 6$	$2 \times 1 - 6 = -4$	$2 \times 2 - 6 = \_\_$			

b Match the position-to-term rules on the left with the correct sequences on the right.

You can use tables like the one above to help you.

$2n - 6$	$-2, -4, -6, -8, -10, …$
$3n + 2$	$5, 8, 11, 14, 17, …$
$-2n$	$-4, -2, 0, 2, 4, …$
$4n$	$2, 3, 4, 5, 6, …$
$n + 1$	$4, 8, 12, 16, 20, …$

1 Sequences

6 A car is travelling at 60 miles per hour along a straight road.

The driver sees a red traffic light ahead and applies the brakes.

The car slows down, losing 5 miles per hour of its speed every second.

   a Copy and complete this table to show how fast the car was travelling throughout the following 7 seconds.

Number of seconds after braking	1	2	3	4	5	6	7
Speed (miles per hour)	55						

   b How fast is the car going after 10 seconds?

Band 2 questions

7 Work out the output for each of these function machines.

   a
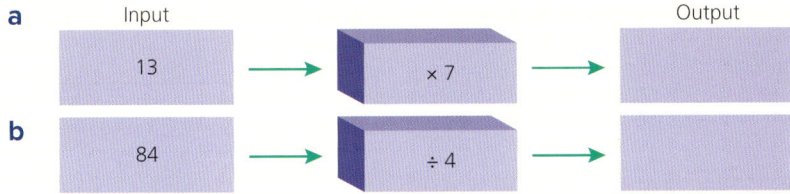

   b

8 Work out the input for each of these function machines.

   a
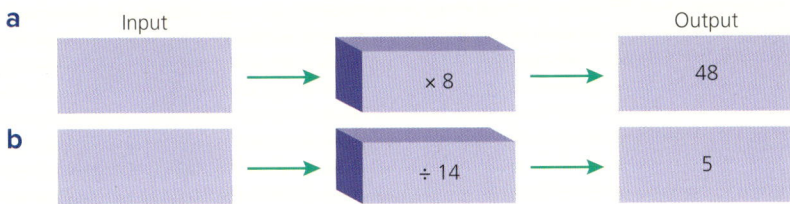

   b

9 Jake is climbing up the stairs at home. There are 21 stairs.

He starts at ground level, up zero stairs.

Jake's sister Eva claps. Every time Eva claps, Jake climbs five steps higher, but then comes two steps back down.

   a Starting with 0, write down the first six terms in the number sequence that describes Jake's height just before each clap.

   b Eventually Jake reaches the top of the stairs. How many times did Eva clap?

10 The first two terms of a sequence are 1 and 4.

How many different term-to-term rules can you think of that would give this sequence?

11 Look at the two term-to-term rules below.

   Add $\frac{1}{2}$ Add $1\frac{1}{2}$

Now look at these number sequences.

Match each of the sequences to one of the term-to-term rules above.

Two of the sequences do not match a rule. What are the rules for these ones?

   a $1\frac{1}{2}, 2, 2\frac{1}{2}, ...$ Rule: _____

   b $1\frac{1}{2}, 2\frac{1}{2}, 4\frac{1}{2}, ...$ Rule: _____

   c $2\frac{1}{2}, 5\frac{1}{2}, 11\frac{1}{2}, ...$ Rule: _____

   d $2\frac{1}{2}, 4, 5\frac{1}{2}, ...$ Rule: _____

Band 3 questions

12 Bronwen input 4 into a function machine.

She got an output number which was three times as big as the input number.

What are the two possible rules of the function machine?

13 Dyfrig input 9 into a function machine and the output was a single-digit odd number.

Kate input 27 into the function machine and the output was also a single-digit odd number.

What is the rule of the function machine?

14 Complete this cross-number using the clues below.

Across:

1. The first four terms in the arithmetic sequence with the term-to-term rule 'Add 2' and first term 2
3. The first three terms in the arithmetic sequence with the term-to-term rule 'Add 0.5' and first term 1
5. The eighth number in the sequence with position-to-term rule $11n$
7. The first three odd numbers
8. The fifth number in the geometric sequence in 1 Down
9. The seventh square number

Down:

1. The first three terms in the geometric sequence with the term-to-term rule 'Multiply by 2' and first term 2
2. The ninth square number
4. The first four terms in the arithmetic sequence with the term-to-term rule 'Add 1.5' and first term 2
6. The first three terms in the sequence with position-to-term rule $-3n + 11$
7. The first three square numbers
8. The sixth square number

15 A bus leaves the city centre bus station with 80 people on board.

At each stop on its route, the bus drops off five people and picks up one person.

a. Write down the number of people on the bus after it leaves the first five stops.

Begin with 80 and write the numbers as a sequence.

b. How many people are left on the bus after nine stops, including the city centre bus station?

c. Find the number of people left on the bus after n stops.

d. After how many stops are there 28 people on board?

16 Look at this sequence of hexagonal matchstick patterns.

a. Draw the next pattern.

b. Copy and complete this table for the first five patterns.

Number of hexagons	1	2	3	4	5
Number of matchsticks	6				

c. Predict the number of matchsticks in six hexagons using the patterns in the table.

d. What is the term-to-term rule for the number of matchsticks?

e. What is the position-to-term rule?

f. Zoe has 43 matchsticks.

Can she use all of them to make one pattern?

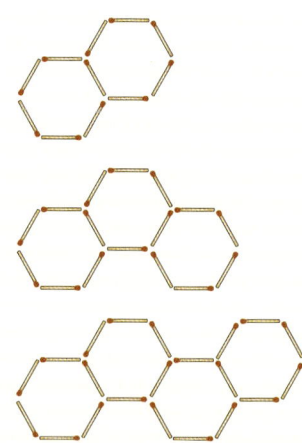

2 Graphs

Coming up...
- Plotting straight line graphs
- Using real life graphs

Four in a row

Draw a grid like the one below.

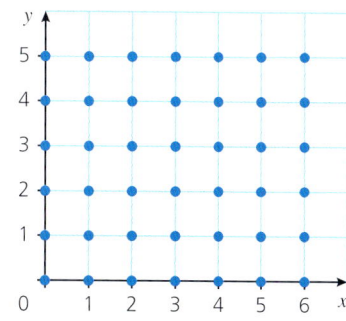

One person has 21 yellow counters; one person has 21 red counters.

Take it in turns to lay a counter. The winner is the first to get a row of four.

Remember to place your counters on the blue dots, not inside the squares!

After the game, see if you can work out the equation of the winning line. For example, in this game, yellow wins.

There is a line of four yellow counters in these positions:

(2, 2), (3, 3), (4, 4), (5, 5)

The line that goes through these points is $y = x$.

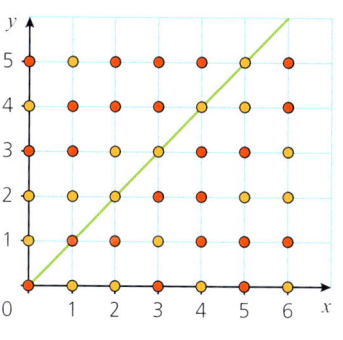

2.1 Straight line graphs

Skill checker

Crack the code below to answer this question:

> Why should you watch out when you see your Maths teacher holding pens and graph paper?

(2, 0), (0, 2), (2, 3), (2, −1) / (0, 1), (1, 2), (−1, 2), (0, 2), (2, 0) / (−1, 3), (2, 3) / (−2, 0), (−1, 1), (2, 1), (2, 0), (2, 0), (1, 2), (1, 1), (−1, 2) / (1, 0), (2, 1), (0, 1), (2, 3), (2, 0), (0, 2), (1, 2), (1, 1), (−1, 2) !

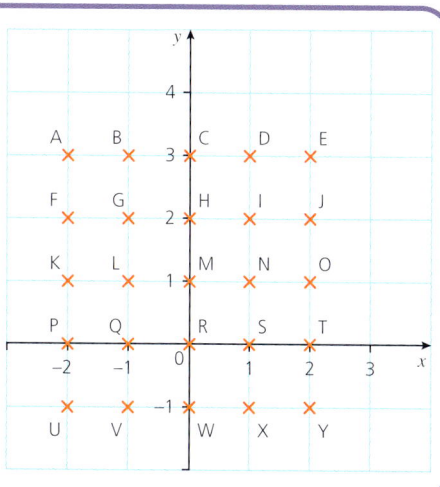

17

2.1 Plotting straight line graphs

> **Worked example**
>
> Look at this equation.
>
> $y = 2x + 3$
>
> For every value of x, there is a single value of y.
>
> a What is the value of y when $x = 1$?
>
> b What is the value of y when $x = 3$?
>
> **Solution**
>
> a When $x = 1$
> $y = 2 \times 1 + 3$
> $y = 5$
>
> b When $x = 3$
> $y = 2 \times 3 + 3$
> $y = 9$

There are several ways you can display this information.

It can be shown in a **table of values**.

x	1	3
y	5	9

You can also plot a **graph**. The next example demonstrates this.

> **Worked example**
>
> Look at this equation linking x and y.
>
> $y = x + 1$
>
> a Complete this table of values.
>
> | x | 0 | 1 | 2 | 3 | 4 |
> |---|---|---|---|---|---|
> | y | 1 | | | | |
>
> b Plot the points and join them with a straight line.
> c If you continued the line, would the point (10, 12) lie on the line?
>
> **Solution**
>
> a
> | x | 0 | 1 | 2 | 3 | 4 |
> |---|---|---|---|---|---|
> | y | 1 | 2 | 3 | 4 | 5 |
>
> b Since this graph is a straight line, y is a **linear** function of x.
> c Include an extra pair of values in the table.
>
> $y = x + 1$.
> When $x = 10$, $y = 11$.
>
> | x | 10 |
> |---|---|
> | y | 11 |
>
> The point (10, 11) would lie on the line; the point (10, 12) would not.

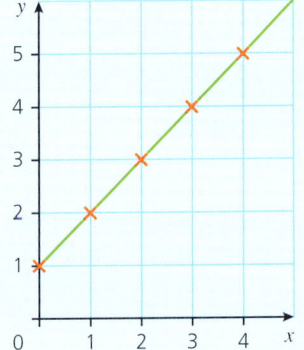

When plotting a straight line, always use three or more points.
Using only two points is dangerous. If one point is wrong, you could draw an incorrect straight line through the two points.
Using a third point works as a check. If one point is wrong, the three points will probably form a triangle rather than a line. Then you will know something has gone wrong!

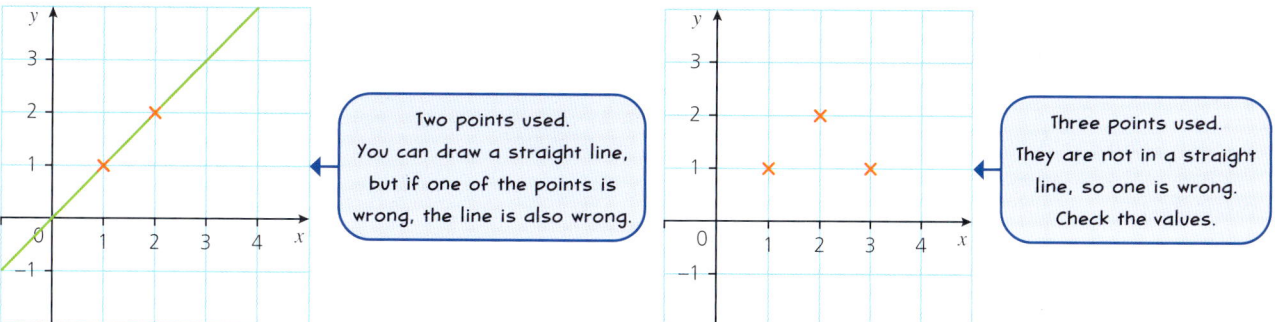

Remember

▶ When you draw a table of values, x always goes on the top row and y on the bottom row.
▶ On the graph, use ✗ to mark each point.
▶ When you join the points with a straight line, it continues through the first and last points. The line doesn't stop at the first and last ✗ you have drawn.
▶ A straight line goes on forever in both directions! If it doesn't, it is called a **line segment**.
It's a common mistake to use the word **line** when you really mean **line segment**.

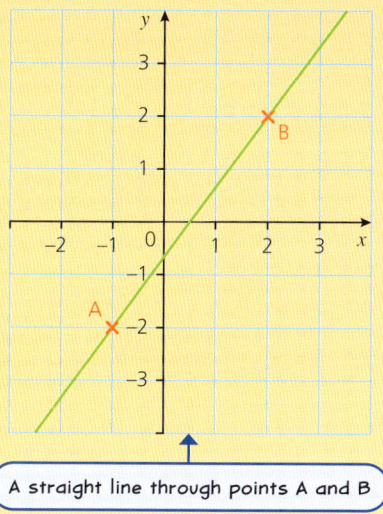

A straight line through points A and B

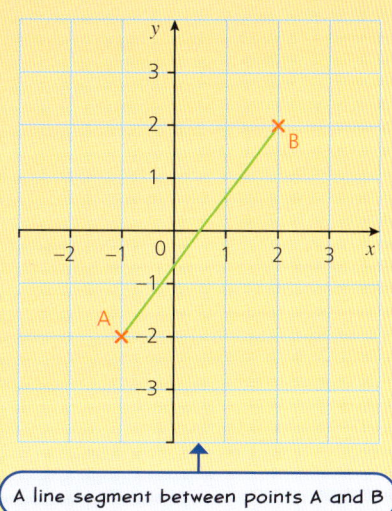

A line segment between points A and B

Maths in context

Euclid was a famous Greek mathematician known for his work on geometry.
He is sometimes called 'the founder of geometry'.
He made some unproven assumptions in his work. Two of these are:
• Given two points, there is only one straight line that joins them.
• A straight line segment can be extended indefinitely.

Worked example

In a science experiment, the length of a piece of elastic is recorded when different masses are hung on it.
Here are the results.

Mass, x (g)	0	10	20	30	40
Length of elastic, y (cm)	30	40	50	60	70

a Plot the graph.

b Using your graph, find the mass that stretches the elastic to 56 cm.

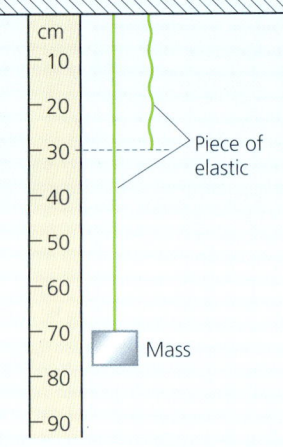

Solution

a

[Graph showing Length of elastic (cm) vs Mass (g), with points plotted and joined by a line. Dashed lines show reading 56 cm across to the line and down to approximately 26 g.]

Join the points up with a line.

Remember to label the axes.

b From 56 cm on the length axis, draw across horizontally until you reach the line.
Then draw a line vertically downwards to read off the value on the mass axis, as shown in the diagram.
The mass that stretches the elastic to 56 cm is 26 g.

Worked example

Plot these straight lines.

a $y = 4$
b $x = 3$

Solution

a A line with the equation $y = c$ is a horizontal line.
For $y = 4$, the y-coordinate is always 4.

b A line with the equation $x = c$ is a vertical line.

For $x = 3$, the x-coordinate is always 3.

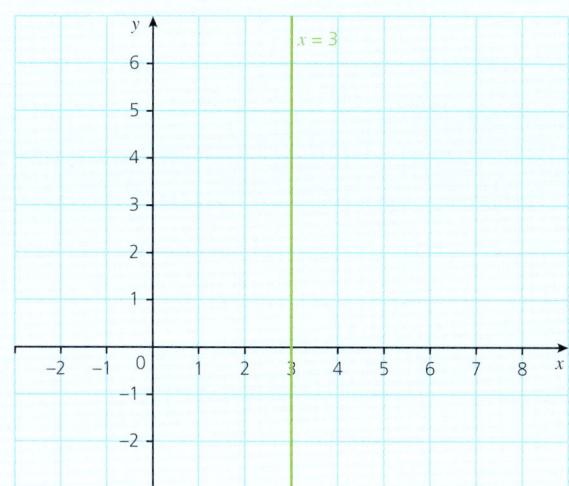

2.1 Now try these

Band 1 questions

1 a Draw a graph from this table of values.

x	0	1	2	3	4	5
y	1	2.5	4	5.5	7	8.5

b How can you tell from the graph that y is a linear function of x?

2 For each equation:

  i Copy and complete the table of values.

  ii Plot the points on a graph and join the points with a straight line.

a $y = 2x$

x	0	1	2	3
y		2		

b $y = 3x + 1$

x	0	1	2	3	4
y					

c $y = x - 2$

x	0	1	2	3	4
y			0		

d $y = 2x - 1$

x	0	1	2	3
y			3	

e $y = \frac{1}{2}x + 4$

x	0	2	4	6
y		5		

f $y = x - 1$

x	0	1	2		
y			1		

3 Plot these straight lines **without** a table of values.

    a $y = 2$ **b** $x = 5$ **c** $y = -3$ **d** $x = -9$

4 **a** Copy and complete this table using the equation $y = 3x$.

x	−3	−2	−1	0	1	2	3
y		−6				6	

    b Copy the graph below and plot the points from the table above.

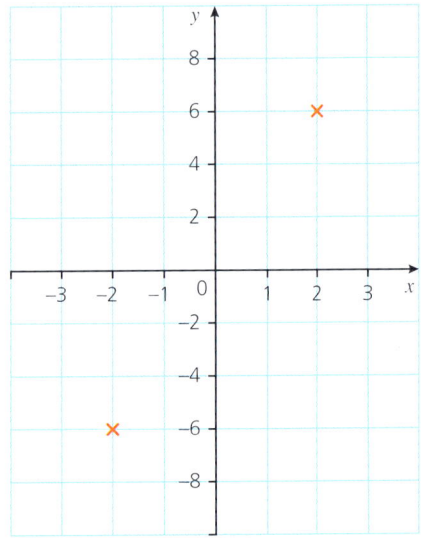

    c Using your graph, find the value of y when $x = 1\frac{1}{2}$.

    d Find the value of x when $y = 0$.

    e If you continued the line, would the point (5, 20) lie on it?

Band 2 questions

5 **a** Copy and complete these tables.

    i $y = x + 1$

x	0	1	2	3	4
y			3		

    ii $y = x + 2$

x	0	1	2	3	4
y					

iii $y = x + 3$

x	0	1	2	3	4
y					

iv $y = x + 4$

x	0	1	2	3	4
y					

b Draw a blank graph using x-axis values of 0 to 4 and y-axis values of 0 to 8.

c On your graph, draw a straight line for each of the equations in parts **a i** to **iv**.

d What do you notice about your four lines?

6 Look at this sign.

Use w for the weight of strawberries picked and C for the cost.

a Complete this formula for the cost in £.

   $C =$ _____

b Copy this table and use your formula to complete it.

w (kg)	0	1	2	3	4	5	6
C (£)	0				8		

c Copy the graph.

Use your table to complete it.

Join the points with a straight line.

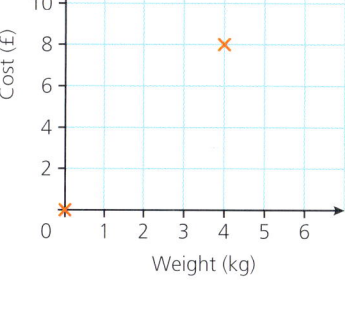

d How much does it cost for 3.5 kg of strawberries?

e John spends £9 on strawberries. How many kilograms does he buy?

7 Here is the graph of $y = 2x + 2$.

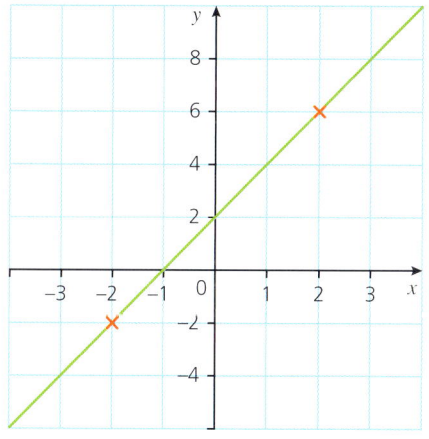

a Find the value of y when:

   i $x = 1$ **ii** $x = 0$ **iii** $x = -1$ **iv** $x = 1.5$

b Find the value of x when:

   i $y = 8$ **ii** $y = 7$ **iii** $y = -4$

8 Look at the equation $y = 4 - x$.

a Make a table of values using x values from 0 to 4.

b When are the values of x and y the same?

c Draw the graph of y against x.

9. Fencing is made up of pieces of timber.

   The width of this piece of fencing with one section is 80 cm.

   The width of this piece of fencing with two sections is 150 cm.

   a Copy and complete this table.

Number of sections, n	1	2	3	4	5
Width, w (cm)	80	150			

   b What is the width of a vertical fence post?

   c What is the length of the horizontal pieces of timber?

   d What is the formula for the width of a piece of fencing?

   e Plot a graph of w against n.

10. a Draw these two straight lines on the same graph.

    i $y = 3x - 1$ ii $y = 3 - x$

    b Where do the two lines cross?

11. A car is travelling at 60 miles per hour along a straight road.

    The driver sees a red traffic light ahead and applies the brakes.

    The car slows down, losing 5 miles per hour of its speed every second.

    a Copy and complete this table to show how fast the car was travelling throughout the following 7 seconds.

Number of seconds after braking	0	1	2	3	4	5	6	7
Speed (miles per hour)		55						

    b Plot a graph of the car's speed against time, in seconds.

       Use the numbers 0 to 7 on the time axis and 0 to 60 on the speed axis.

    c Using your graph, work out how fast the car is travelling after 10 seconds.

    d After how many seconds does the car come to a stop?

12. Using an x-axis from -3 to 3 and a y-axis from -3 to 9:

    a Plot the line $y = 2x + 3$.

    b Plot the line $y = -2x + 3$.

    c Where do the two lines pass through the y-axis?

2 Graphs

Band 3 questions

13 A boy sits on a clifftop. He throws a stone downwards with a speed of 5 metres per second.

Every second, the speed of the stone increases by 10 metres per second.

The stone takes 3 seconds to reach the sea below.

a   Copy and complete the table below.

Time after stone is dropped (seconds)	0	1	2	3
Speed (metres per second)	5			

b   Plot a graph of the stone's speed against time in seconds.

c   Use your graph to work out how fast the stone is moving after 2.5 seconds.

14 a   Draw a blank graph using an x-axis from -4 to 4 and a y-axis from -4 to 4.

b   Complete the table of values for these four straight lines.

$y = x - 6$ $y = x + 6$ $y = 6 - x$ $y = -6 - x$

x	-4	-3	-2	-1	0	1	2	3	4
$y = x - 6$									
$y = x + 6$									
$y = 6 - x$									
$y = -6 - x$									

c   Plot these four lines on your graph. (It is not possible to plot all the points.)

d   Then plot these four straight lines.

$y = 4$ $y = -4$ $x = 4$ $x = -4$

e   What is the name of the shape enclosed by the eight lines on your graph?

15 Look at the charges for using the Cwmtir Leisure Centre pool and Fitness Club.

a   Copy and complete this table of values for non-members using the swimming pool.

Number of visits to the pool	0	2	4	6	8	10
Total cost (£)		11				

b   Plot the information in your table from part **a** on a graph.

c   Copy and complete this table of costs for members of the Fitness Club using the swimming pool.

Number of visits to the pool	0	2	4	6	8	10
Total cost (£)	45	50				

d   Plot the information in your table from part **c** on the same graph as part **b**.

e   Dev joins the Fitness Club. He uses the pool twice a week.

Bronwen is a non-member. She also visits the pool twice a week.

After how many weeks has Dev paid less than Bronwen?

2.2 Real life graphs

Skill checker

A pot of porridge is heated on a cooker.

A graph of the porridge's temperature against time is shown.

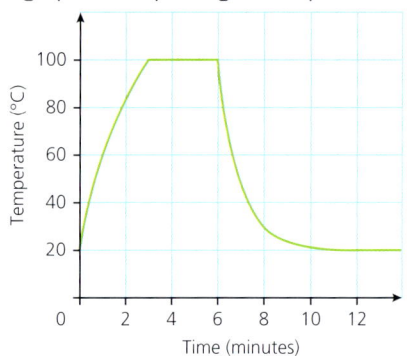

a   Before the cooker is switched on, the porridge is at room temperature.
    What is the temperature of the room?
b   How hot is the porridge after 1 minute?
c   How long does it take for the porridge to start boiling?
d   For how long does the porridge stay at boiling point?
e   At what time has the porridge cooled to a temperature of 50 °C?
f   What is the temperature of the porridge after 8 minutes?

Worked example

The diagram shows the shape of Iwan's bath.

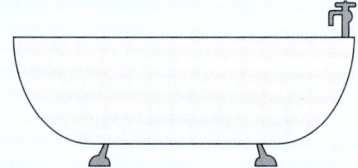

Water runs in at a constant rate.

a   Which sketch graph best shows the relationship between the depth of the water and time?

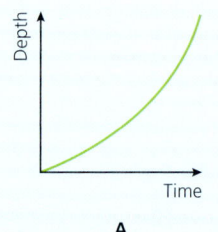 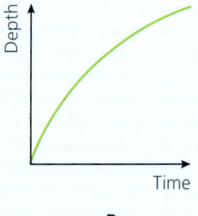 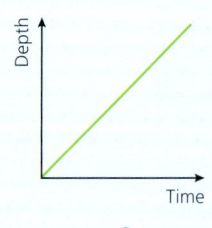

    A   B   C

b   Explain your answer.

Solution

a   Graph B is correct.
b   Initially, the cross-section of the bath is small and the depth of water increases quickly. As time passes, the cross-sectional area increases and the bath takes longer to fill. The water level increases more slowly.

Worked example

A company declares its profits on 1st June each year.

The profit for the years 2018 to 2022 is shown in the table.

Year	2018	2019	2020	2021	2022
Profit (£)	30 000	27 000	32 000	25 000	18 000

a Plot a graph of this data with time on the horizontal axis.

b Is it sensible to join the points up? Explain your answer.

Solution

a
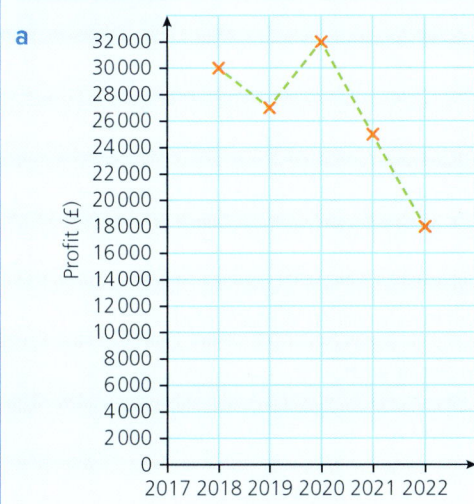

b You should not join the points with a solid line. You would only do that if you could use the line to find the profit at points between the yearly figures given.

Instead the points can be joined with a dashed line, as shown.

Worked example

Jamila plants a shrub.

She measures its height on 1st January every year.

She draws this graph.

a What is the height of the shrub 3 years after planting?

b In which year does the plant grow the most?
How do you know?

c In which year does the shrub grow the least?

d Do you think the shrub will grow much more?
Explain your answer.

e Jamila's mother says, 'You should join the points with a solid line!'
Jamila replies, 'No, for some of the year the shrub doesn't grow at all.'
Comment on this conversation.

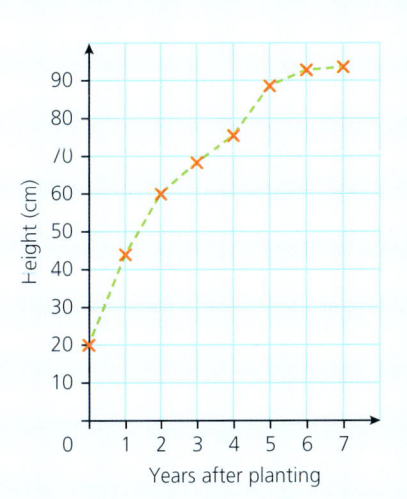

Solution

a Roughly 68 cm

b During the first year the shrub grows the most.

Its height increases by about 24 cm from 20 cm to 44 cm.

c In the 7th year the shrub only grows roughly 1 cm from 92 cm to 93 cm.

d The shrub probably won't grow much more. The increases from one year to the next have become very small.

e Jamila is right. Most of the growing takes place in spring and summer.

Joining the points with a solid line would suggest steady growth throughout the year. In this case, the line could be used to read off the shrub's height at any time.

Activity

Look at this newspaper article. Can you fill in the blanks using the information in the graph?

The Hendre Star
What A Scorcher
Hendre shot into the record books yesterday as temperatures reached a record high.
Yesterday was the hottest day since records began, as temperatures soared to a maximum of ____°C.

By 10 a.m. the temperature in Hendre was already ____ °C.
Drenewydd, however, which is only 20 miles down the road, saw temperatures of only ____°C at the same time.
Temperatures peaked in both towns at ____ o'clock. By that time there was a temperature difference of about ____ °C between the two towns.
In Hendre temperatures remained above 30°C for about ____ hours.
By 10 o'clock last night, temperatures in Hendre were still about ____ °C and only fell to 18°C during the night.
Enjoy the sunshine while it lasts! Next week temperatures are predicted to plummet, with Hendre's afternoon high being only ____°C, a full 20 degrees cooler than yesterday.

2.2 Now try these

Band 1 questions

1 Dilip plants a plum tree. The graph shows its growth.

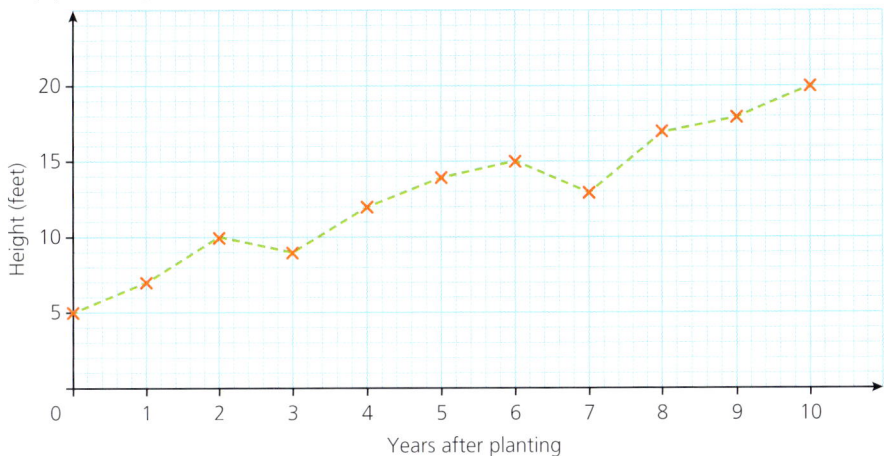

- a How high is the tree when it is planted?
- b How high is it after 5 years?
- c When is the tree 17 feet high?
- d In which years does Dilip prune it?
- e In which year does the tree grow the most?

2 These graphs show the temperature for four different days. Match each description **a–d** with one of the graphs **i – iv**.

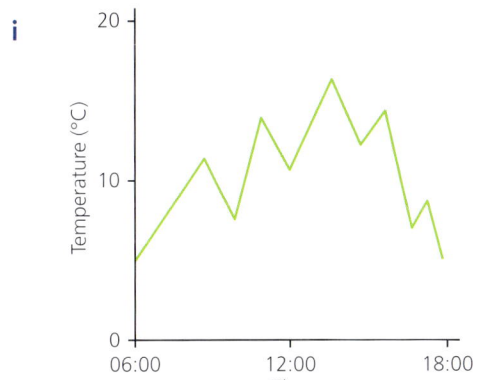

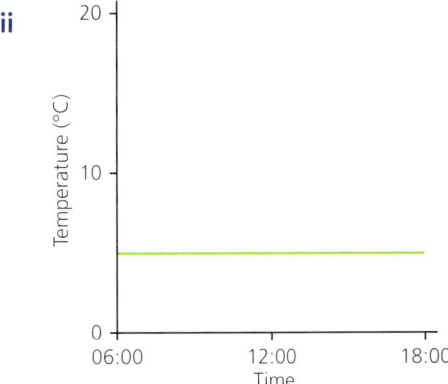

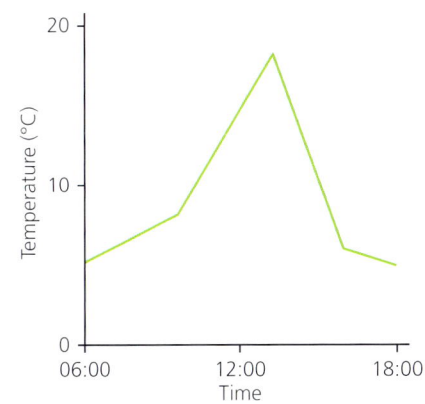

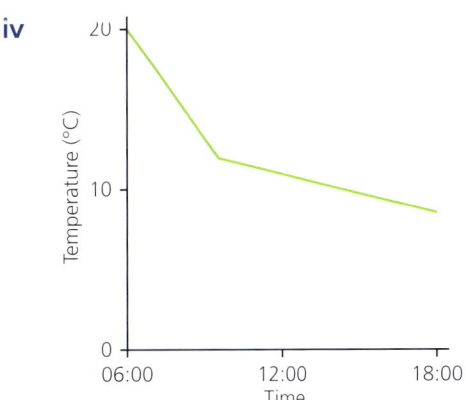

a The temperature stays cool all day.
b The temperature starts cool, warms up and then cools down.
c The temperature starts warm and then cools down.
d The temperature goes up and down a lot all day.

3 The table gives the number of guests at a small bed and breakfast in the months of one year.

Month	Jan	Feb	Mar	Apr	May	Jun	Jul	Aug	Sep	Oct	Nov	Dec
Number of guests	12	20	26	60	40	45	54	75	20	15	4	42

a Show this data on a graph.
Plot the months on the horizontal axis and the number of guests on the vertical axis.
Think carefully about whether you will use a solid line or a dashed line to join the points.
b In which month were there the fewest guests?
c In which month were there the most guests?
d Explain the shape of the graph.

4 The table shows the hourly temperature at the holiday resort Palma on a particular day.

Time	9 a.m.	10 a.m.	11 a.m.	12 noon	1 p.m.	2 p.m.	3 p.m.	4 p.m.
Temperature °C	10	15	29	30	35	33	30	28

a Show this information on a graph.
b Between which times does the temperature rise most quickly?
c Between which times does the temperature fall most quickly?
d Estimate for how long the temperature remains above 29 °C.

5 Angharad visited a music festival in her local park.
The graph shows the number of visitors at various times throughout the day.

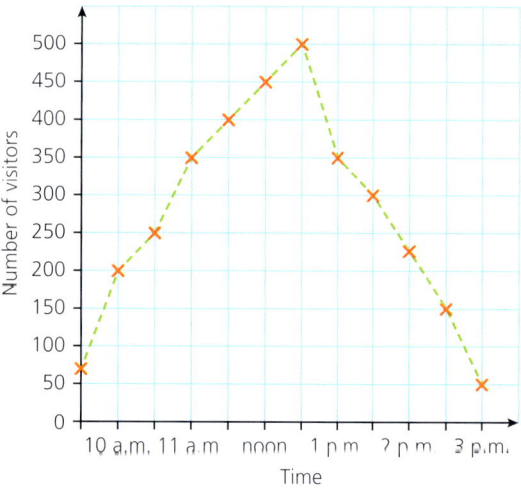

a What is the largest number of people at the festival?
b At what time are most people there?
c How many people are there at 2 p.m.?
d Angharad leaves at 12:45 p.m. Estimate how many people were still at the festival then.

Band 2 questions

6 The graph shows the income tax paid on different salaries.

Raj's salary is £30 000. He pays £6000 in tax.

a   Siôn's salary is £20 000. How much tax does he pay?

b   Anette's salary is £9000. How much tax does she pay?

c   Mari's salary is £22 000. Estimate how much tax she pays.

7 The table below shows the level of oil in an oil tank on the first of each month throughout one year.

Date	1 Jan	1 Feb	1 Mar	1 Apr	1 May	1 Jun	1 Jul	1 Aug	1 Sep	1 Oct	1 Nov	1 Dec
Depth of oil (mm)	510	370	695	585	495	360	685	660	570	400	275	625

a   Plot the data points on a graph and join them with a dashed line.

b   When is the level of oil at its lowest?

c   The oil level on 1st March is high because the tank was refilled during February.
    When do you think the oil tank was next refilled?

d   How many times during the year was the tank refilled?

8 Each of the situations **a**, **b** and **c** below can be represented using a graph.

Match each of the situations to one of the graphs **i**, **ii** and **iii**.

i 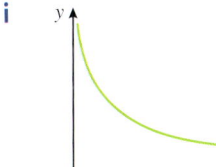   ii 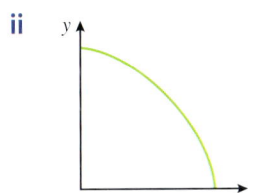   iii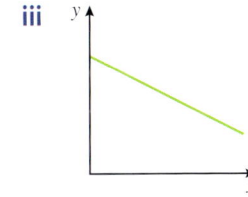

a   The amount of fuel left in the tank of a car travelling at constant speed on a motorway.

b   The temperature of a cup of tea left standing on a table.

c   The number of bacteria left in a human body after treatment has begun. The treatment works slowly at first, but then speeds up.

9 Sheena is in hospital with a fever.

On Tuesday a nurse takes Sheena's temperature every hour. The table below shows the readings.

Time	07:00	08:00	09:00	10:00	11:00	12:00	13:00	14:00
Temperature (°C)	37.5	37.7	37.9	38.0	38.1	38.0	37.8	37.8

Time	15:00	16:00	17:00	18:00	19:00	20:00	21:00
Temperature (°C)	37.6	37.5	37.5	37.4	37.2	37.1	37.0

a Draw a graph to show Sheena's temperature readings.
b What is her highest temperature?
c For how long is Sheena's temperature 38 °C or higher?
d At what time does Sheena's recovery begin?
e Using your graph, estimate when Sheena's temperature is at 37.3 °C.
f Normal body temperature is 37 °C. At what time does Sheena's temperature return to normal?

10 Water is poured at a constant rate into each of the three containers below.

For each container, a graph has been drawn of the water depth against time. Match each container with one of the depth–time graphs below.

a 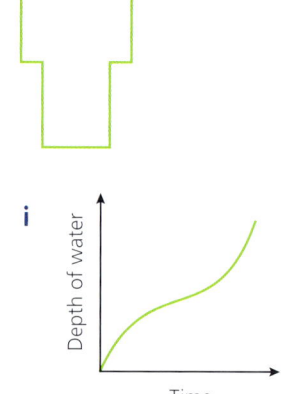   **b** 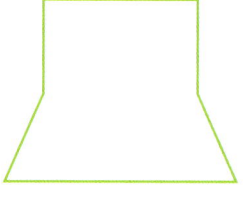   **c**

i 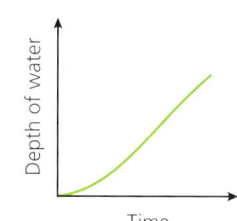   **ii** 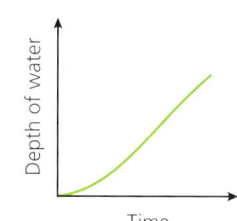   **iii**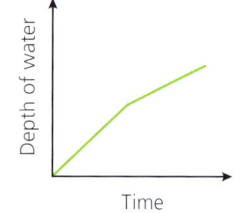

Band 3 questions

11 Water is poured at a constant rate into each of the three containers below.

For each container, sketch a graph of the water depth against time.

a 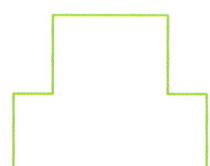   **b** 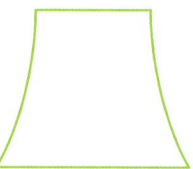   **c**

d What shape of container would give a graph this shape?

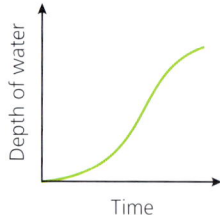

12 In their Chemistry lesson, Osian and Rhian are studying changes of state.

They heat some solid paraffin wax in a test tube until it melts.

Then they measure its temperature every 2 minutes as it cools and solidifies.

The table shows their results.

Time (minutes)	Temperature (°C)
0	86
2	70
4	62
6	57
8	55
10	55
12	55
14	55
16	55
18	55
20	55
22	53
24	50
26	45
28	41
30	37
32	33
34	30
36	28
38	26
40	24
42	23
44	22
46	22
48	22
50	22

a Plot these points on a graph, with time in minutes on the x-axis and temperature in °C from 0 to 90 on the y-axis.

b Join the points with a smooth curve.

This graph is the cooling curve for paraffin wax.

c What is the temperature of the paraffin wax after:

    i 5 minutes   **ii** 27 minutes?

d How long does it take the wax to cool from its initial temperature to 65 °C?

e When a liquid cools to its melting point, the temperature stays the same until it has all become solid.

What is the melting point of paraffin wax?

f What is room temperature

g Osian and Rhian repeat the experiment with a different wax.

This wax has a melting point of 64 °C.

Starting from the same initial temperature as the paraffin wax, sketch a cooling curve for this wax.

13 Ajay and Siân are carrying out a traffic survey.

They record the number of people travelling in each car on a main road into the city.

Unfortunately, after doing the survey, they lost their data. Before they lost it, however, they wrote down some important information.

a Use the clues below to find the frequencies: how many cars had one person, how many had two people, and so on.

b Draw a graph to illustrate the data.

We surveyed 100 cars altogether

 The number of cars with 2 people was greater than the number with 3 people

 The number of cars with 2 people was a square number

 There were 17 cars with 3 people

 No cars had more than 5 people

 The number of cars with an even number of people was 31

 Three quarters of all the cars had only 1 or 2 people

14 The faster you drive, the more space you need to stop.
The Highway Code gives this information about stopping distances.

Typical stopping distances

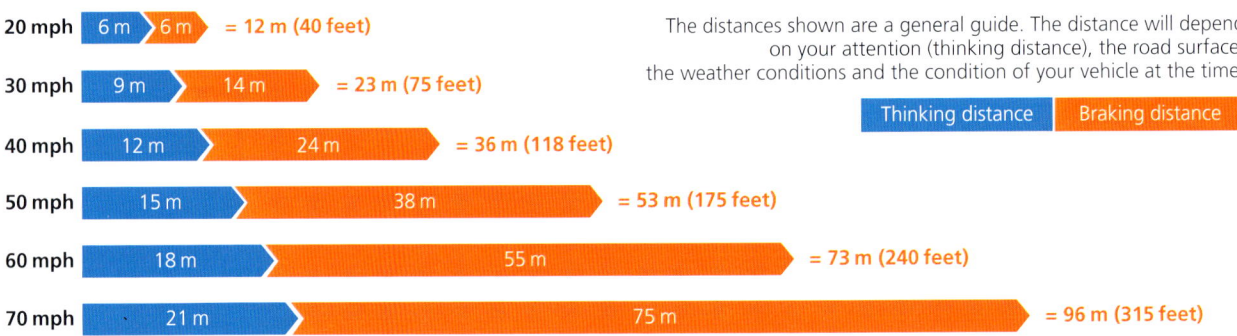

The distances shown are a general guide. The distance will depend on your attention (thinking distance), the road surface, the weather conditions and the condition of your vehicle at the time.

Thinking distance   Braking distance

a   Bleddyn is driving at 45 miles an hour (mph) and Bethan is driving at 65 mph.
    What are their approximate stopping distances?

b   Draw a graph of stopping distances.
    Use the horizontal axis for the speed, going up to 80 mph (scale: 1 cm to 10 mph).
    Use the vertical axis for the distance, going up to 500 feet (scale: 1 cm to 50 feet).
    Plot the six points from the Highway Code and join them with a smooth curve.
    Use your graph to check your answers to part **a**.

c   What is the stopping distance for 52 mph?

d   What is the stopping distance for 15 mph?
    (You will need to extend your curve.)

e   You are driving 200 feet behind Bleddyn.
    What is your maximum safe speed?

f   Do you think the curve should go through (0, 0)?
    Explain your answer.

Key words

Here is a list of the key words you met in this chapter.

Curve   Graph   Line   Line segment   Linear   Real life graph   Straight line graph

Use the glossary at the back of this book to check any you are unsure about.

Review exercise: graphs

Band 1 questions

1 **a** Copy and complete these tables.

  i $y = 2x - 1$

x	0	1	2	3	4
y			3		

  ii $y = 2x$

x	0	1	2	3	4
y					

  iii $y = 2x + 1$

x	0	1	2	3	4
y					

  iv $y = 2x + 2$

x	0	1	2	3	4
y					

  b Draw a blank grid using x values from 0 to 4 and y values from -1 to 10.

  c On your grid draw the straight line graph for each of the equations in parts **i** to **iv**.

  d What do you notice about your four lines?

2 **a** Copy and complete this table for $y = 2x - 4$.

x	−2	−1	0	1	2	3
y			−4			

  b Use your table to plot the graph of $y = 2x - 4$.

  c Find the value of x when $y = 1$.

3 Draw axes with x values from -2 to 2 and y values from -5 to 7.

  a Plot the graphs of $y = 3x + 1$ and $y = -3x + 1$ on your axes.

  b How are your graphs related?

4 **a** Construct a table of values for $y = 4x + 3$.

  b Use your table to plot the graph of the line $y = 4x + 3$.

  c Find the value of y when $x = 2\frac{1}{4}$.

  d Find the value of x when $y = -3$.

5 Look at the equation $y = 7 - x$.

  a Make a table of values using x values from 0 to 7.

  b Draw the graph of y against x.

  c Do x and y ever take the same value?

Hint
You may have to think about fractions!

Band 2 questions

6 Draw an x-axis from -3 to 3 and a y-axis from -9 to 9.

  a Plot the line $y = 3x$.

  b Plot the line $y = -3x$.

  c Where do the two lines pass through the y-axis?

7 The time, t minutes, taken to cook a cake depends on its weight, m kilograms. The formula used to calculate time is:

$t = 25m + 20$

a Copy and complete this table to find the cooking time for cakes of various weights.

m(kg)	1	1.5	2	2.5	3
t(mins)	25 × 1 + 20 = 45				

b Draw the graph of t against m.

c Use your graph to find the cooking time for a cake of weight 1.3 kg.

d A cake takes 90 minutes to cook.
How much does it weigh?

8 a Draw the line $y = 2x - 5$ with x values from 0 to 4 and y values from -5 to 3.

b Do these points lie on the line?
  i (1, −3)   ii (3, 2)

c If you extended the line, would these points lie on it?
  i (10, −25)   ii (−3, −11)

Band 3 questions

9 Draw axes with x values from -1 to 3 and y values from -1 to 5.

a Plot the graphs for $y = x + 2$ and $y = x + 3$ on the same axes.

b Where does each graph cross the y-axis?

c Where do you think the graph for $y = x + 1$ would cross the y-axis?

d Draw the graph $y = x + 1$ on the axis. Were you correct?

10 Do the points (2, 6), (4, 12) and (6, 18) lie on a straight line?
Explain how you know.

11 Abdul has a leaky bath and needs to call a plumber.

He gets quotes from three local plumbers.

Their charges are summarised in the table.

Plumber	Fixed charge	Rate per hour
Tristan	£20	£10
Dylan	£10	£30
Seren	£5	£15

a Draw a graph of cost against time taken for all three plumbers, on the same set of axes.
On the time axis go up to 4 hours.

b How much would Seren charge in total if she worked for 3 hours?

c If the job takes less than 3 hours, which plumber is the cheapest?

d If the job takes more than 3 hours, which plumber is the cheapest?

e Which plumber would charge £70 in total for working 2 hours?

12 The formula for the perimeter of this rectangle is:

Perimeter = 2 × length + 6 cm

3 cm

a Copy and complete this table:

Length (cm)	1	2	3	5	10
Perimeter (cm)	2 × 1 + 6 = 8				

b Draw axes similar to the ones shown here.

Plot the perimeter of the rectangle against length.

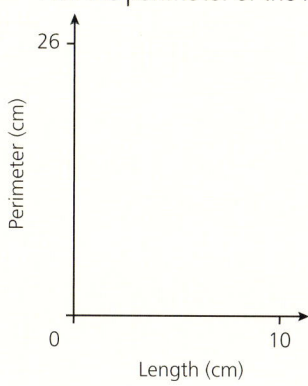

13 Ffion puts £200 into a savings account for 8 years.

She checks the balance every year.

Time (years)	0	1	2	3	4	5	6	7	8
Balance (£)	200	220	242	266	293	322	354	389	428

a Plot the balance against time on a graph, joining the points with a smooth curve.

Use your graph to answer the following questions.

b How much is Ffion's investment worth after $4\frac{1}{2}$ years?

c How long does it take for Ffion's investment to double in value?

Consolidation 1: Chapters 1–2

Band 1 questions

1. Find the next three terms in these arithmetic sequences.
   - a 3, 8, 11, 14, …
   - b 30, 50, 70, 90, …
   - c 20, 18, 16, 14, …
   - d −11, −7, −3, 1, …
   - e −15, −21, −27, −33, …

2. Find the missing term in each of the arithmetic sequences below.
   - a 7, 9, 11, 13, ☐, 17, …
   - b 3, 7, 11, ☐, 19, …
   - c −4, −7, ☐, −13, −16, …
   - d 10.5, 14, ☐, 21, 24.5, …
   - e −10, −25, −40, ☐, −70, …

3. Find the nth term of these sequences.
   In each case the nth term is in the form ☐n.
   - a 6, 12, 18, 24, …
   - b 30, 60, 90, 120, …
   - c −5, −10, −15, −20, …
   - d $1\frac{1}{2}$, 3, $4\frac{1}{2}$, 6, …

4. Find the nth term of these sequences.
   In each case the nth term is in the form ☐$n + 1$ or ☐$n − 1$.
   - a 6, 13, 20, 27, …
   - b 10, 19, 28, 37, …
   - c 101, 201, 301, 401, …
   - d −6, −11, −16, −21, …
   - e 24, 49, 74, 99, …

5.
   - a Work out if the number 67 appears in the sequence with nth term $11n + 1$.
   - b Work out if the number −27 appears in the sequence with nth term $−7n + 1$.

6. Find the first five terms of these sequences, given their position-to-term rules. Use the tables to help you.
   - a The sequence with position-to-term rule $n + 7$.

n	1	2	3	4	5
$n + 7$	1 + 7 = 8	2 + 7 = ___			

   - b The sequence with position-to-term rule $n − 9$.

n	1	2	3	4	5
$n − 9$	1 − 9 = −8	2 − 9 = ___			

7. For each equation:
   - i Copy and complete the tables below using the equations given.
   - ii Draw a graph of y against x. Draw the x-axis from 0 to 5 and the y-axis from 0 to 20.
   - a $y = 4x$

x	1	2	3	4	5
y	4 × 1 = 4	4 × 2 = ___			

   - b $y = 7x − 2$

x	1	2	3	4	5
y	7 × 1 − 2 = 5	7 × 2 − 2 = ___			

Consolidation 1

8 a Using values of x from 0 to 6, construct a table of values using the equation $y = 3x - 4$.
b Plot the graph of y against x.

9 Work out the output for the function machine below.

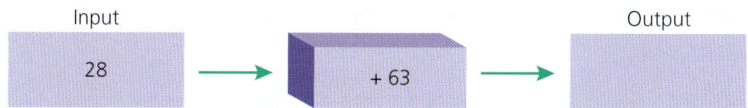

10 Work out the input for the function machine below.

Band 2 questions

11 Are these sequences arithmetic or geometric? In each case, what is the term-to-term rule?
a $-6, -3, 0, 3, 6, 9, \ldots$
b $60, 30, 15, 7.5, \ldots$
c $50, 46, 42, 38, 34, 30, \ldots$
d $-9, 18, -36, 72, \ldots$
e $-13, -15, -17, -19, -21, -23, \ldots$

12 Match the position-to-term rules on the right with the correct sequences on the left.

$3, 5, 7, 9, 11, \ldots$	$n - 4$
$\frac{1}{2}, 1, 1\frac{1}{2}, 2, 2\frac{1}{2} \ldots$	$\frac{1}{2}n$
$-3, -6, -9, -12, -15, \ldots$	$-3n$
$-3, -2, -1, 0, 1, \ldots$	$2n + 1$

13 a Construct a table of values for $y = 2x + 2$ with x from -2 to 5 and y from -2 to 12.
b Draw the graph of $y = 2x + 2$.
c $A(0, 2)$ and $B(3, 8)$ are two points on this line.
C is the point $(3, 0)$ and O is the origin $(0, 0)$.
Draw the quadrilateral OABC.
What type of quadrilateral is OABC?

14 The cost, £y, of x invitation cards to Bethan's wedding is given in the table.

Number of invitation cards, x	20	40	60	80	100
Cost, y (£)	16	26	36	46	56

a Plot a graph of this information.
b Use your graph to estimate:
   i the cost of 55 invitation cards
   ii how many cards can be bought for £42.

15 Work out the output for the function machine below.

Input: 48 → ÷ 12 → Output

16 Work out the input for the function machine below.

Input → × 9 → Output: 108

Band 3 questions

17 The thickness, T millimetres, of some children's books is given by:

$$T = 4 + \frac{n}{5}$$

where n is the number of pages.

a Copy and complete the table below.

n	10	15	20	25	30	35	40
T (mm)	$4 + \frac{10}{5} = 6$	$4 + \frac{15}{5} =$ ___					

Each book is made up of pages, a front cover and a back cover.

b What is the thickness of one page?

c What is the thickness of each cover?

18 Draw axes with x values from -1 to 7 and y values from -4 to 7.

a Plot the line $y = 2x - 3$.

b Mark the point A where this line crosses the y-axis.

c Plot the line $y = 6 - x$.

d Mark the point B where this line crosses the y-axis.

e Mark the point C where the two lines cross.

f Find the area of triangle ABC.

19 Little Explorers is a holiday club for primary school children.

Llinos is the leader. She wants to hire a bus to take the children on a day trip.

The cost £C of hiring the bus for n hours is given in the table below.

The cost forms an arithmetic sequence.

Number of hours, n	1	2	3	4	5
Cost, C (£)	100	140			

a Copy and complete the table.

b How much would it cost to hire the bus for 7 hours?

c Llinos pays £420 for the hire of the bus. For how long did the club hire it?

d Find the position-to-term rule for the cost of hiring the bus.

e Plot a graph of cost against the number of hours. Use the horizontal axis for n, with values from 0 to 10.

f Use your graph to find the cost of hiring the bus for $3\frac{1}{2}$ hours.

20 Bethan is making patterns from matchsticks.

Look at the first three patterns.

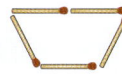
Pattern 1

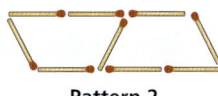
Pattern 2

Pattern 3

a Draw Pattern 4.

b Copy and complete this table for the number of matchsticks used in each pattern.

Include the number of matchsticks Bethan would use if she continued to Pattern 4.

Pattern number, n	1	2	3	4
Number of matchsticks, m	5			

c Write the number of matchsticks as a sequence of numbers.
What is the name for this type of sequence? Explain your answer.

d If the pattern is continued, which pattern will use 29 matchsticks?

e Find the position-to-term rule for the number of matchsticks.

f Bethan has 45 matchsticks.
Can she use all of them to make one pattern?

Hint
You add four matchsticks to move from one pattern to the next. This means that $4n$ will be in your position-to-term rule.

g Plot a graph of m against n using values of n from 1 to 6.

21 The load that a steel beam can support depends on the thickness of the beam.

Ashok is an architect. He uses this table to find out the load supported by beams of different thicknesses.

Thickness of beam (cm)	5	10	15	20
Load (tonnes)	2.25	3	4.25	6

a Plot a graph of the load against the thickness of the beam.
Join the points with a smooth curve.

b Use your graph to find the load supported by a beam with a thickness of 18 cm.

c What thickness of beam would be needed to support a load of 4 tonnes?

22 Carwyn input 20 into a function machine.
The output of the function machine was 5.
What are the two possible rules of the function machine?

23 Delyth input a single-digit even number into a function machine and the output was 12.
Erin input a single-digit odd number into the same function machine and the output was 18.
What is the rule of the function machine?

3 Angles

Coming up...

▶ Understanding and using the relationship between parallel lines and alternate, corresponding and allied angles

▶ Understanding and using the properties of regular polygons and their angles

Around in circles

① **a** Using a pair of compasses, draw a circle of radius 5 cm.

> A pair of compasses is frequently, and incorrectly, referred to as a 'compass'.

Leaving the compasses opened to 5 cm, place the point on the edge of the circle.

Draw an arc on the circle.

Place the point of the compasses where the arc meets the circle, and draw another arc on the circle.

Repeat the last step four more times.

The last arc drawn should coincide with the start point.

Join the points to create a regular hexagon.

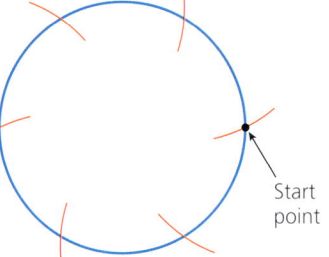

Start point

b Find a pair of parallel sides and label each of them with >.

c Find another pair of parallel sides and label each of them with >>.

d The remaining sides should also be parallel – label each of these with >>>.

3.1 Parallel lines

Skill checker

① What is the sum of the angles in a full turn?
② What is the sum of the angles on a straight line (half turn)?
③ Draw a pair of vertically opposite angles.

3 Angles

Parallel lines never meet.

Use arrows to indicate that lines are parallel.

Use more arrows if a diagram includes more than one set of parallel lines.

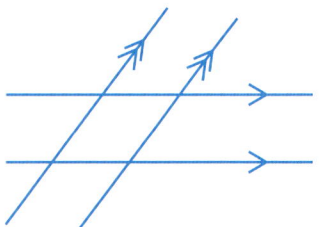

Activity

1. How many angles are created when two lines meet?

2. a Draw three parallel lines and another set of three parallel lines crossing the first three.

   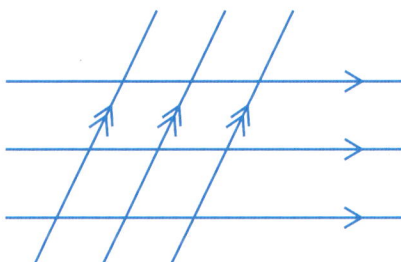

   b How many angles have been created?

   c Draw an arc at each angle created.

   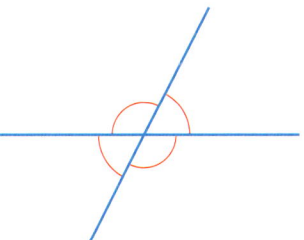

   d Colour each acute angle red and each obtuse angle blue.

   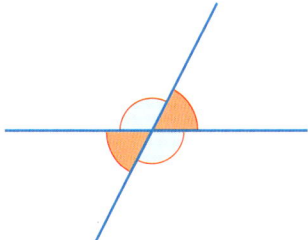

43

e   Measure the size of each angle.
   i   What do you notice about all the red angles?
   ii What do you notice about all the blue angles?
f   Find the total of one red angle and one blue angle.
   Is this the same total for all pairs of red and blue angles?

Alternate angles are equal. The shaded angles in this diagram are a pair of alternate angles.

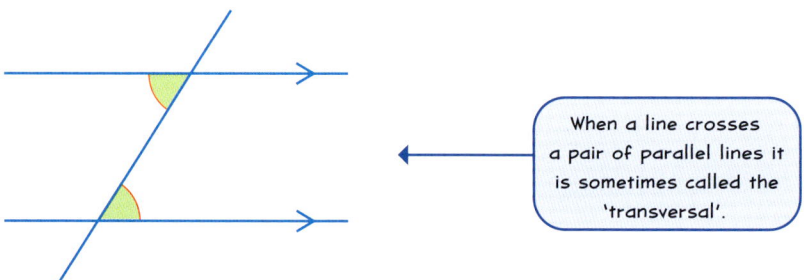

When a line crosses a pair of parallel lines it is sometimes called the 'transversal'.

Corresponding angles are equal. The shaded angles in this diagram are a pair of corresponding angles.

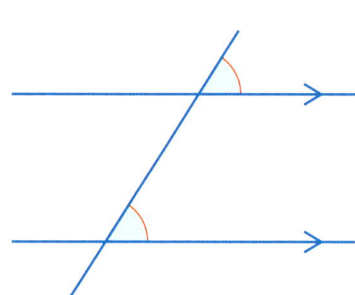

Note

If you refer to alternate, corresponding or allied angles by their shapes (F, Z, C, etc.) you will not get the reason marks

Allied angles add up to 180°. The shaded angles in this diagram are a pair of allied angles.

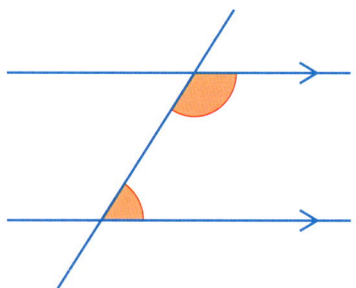

Allied angles are sometimes referred to as 'co-interior angles'.

Remember, **vertically opposite angles** are equal.

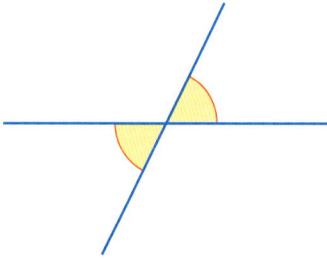

Remember

Angles in a full turn add up to 360°.
Angles in a half turn add up to 180°.
Angles in a triangle add up to 180°.
Angles in a quadrilateral add up to 360°.

3 Angles

Worked example

The diagram shows a pair of parallel lines and an intersecting line.

Work out the size of the lettered angles.

Give a reason for each of your answers.

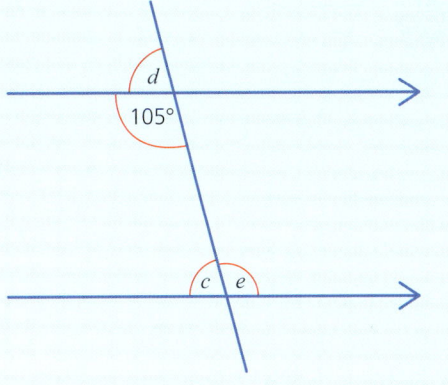

Solution

$e = 105°$ (alternate angles)

$d + 105° = 180°$ (angle sum on a straight line is $180°$)

$\quad d = 75°$

$c = 75°$ (c and d are corresponding angles)

> To find angle c you could have used $c + e = 180$ instead, because the angle sum on a straight line is $180°$.

3.1 Now try these

Band 1 questions

1 Find the size of angles a, b and c in these diagrams.

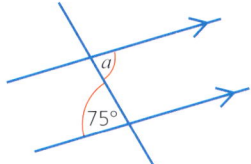 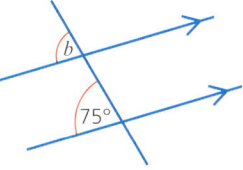 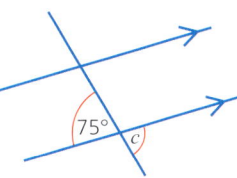

2 Find the size of each lettered angle in these diagrams.

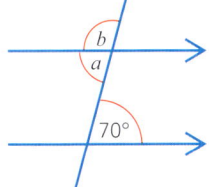 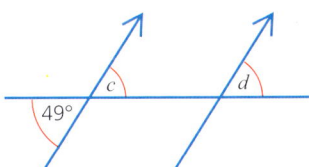 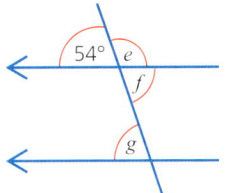

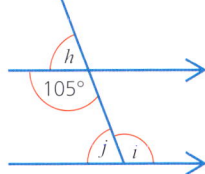 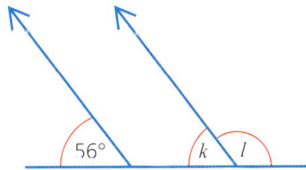 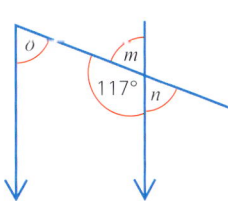

3 Find the size of each lettered angle in these diagrams.

a

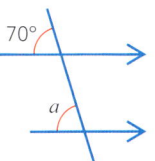 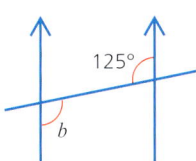 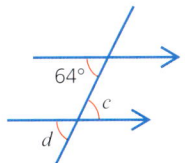

45

b

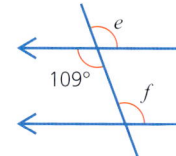
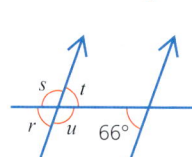
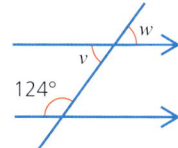

c

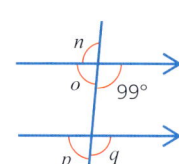

④ Calculate the size of each lettered angle and write down your reason.

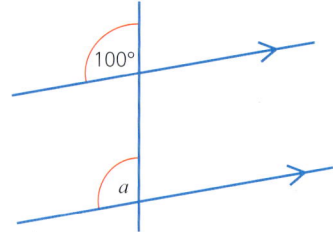
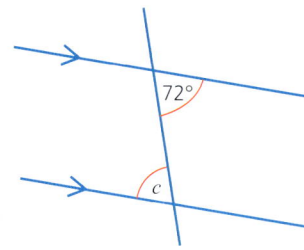

⑤ Copy the diagram. Shade all the angles equal in size to the shaded angle.

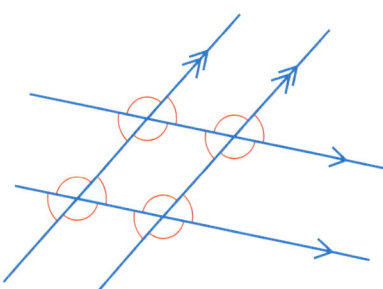

⑥ Bryn shows that angles a and b add up to $180°$.

> When a pair of angles add up to $180°$, they are said to be 'supplementary angles'.

Copy Bryn's working. Fill in the missing reason.

$b = c$ (_____)

$a + c = 180°$ (angles on a straight line add up to $180°$)

so $a + b = 180°$

3 Angles

Band 2 questions

7 a Find the size of each lettered angle.

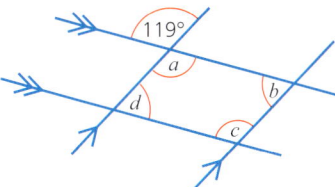

  b What is $a + b + c + d$?

  c What type of quadrilateral has been formed by the parallel lines?

8 Find the size of each lettered angle in these diagrams.

For each one write down the angle fact(s) that you use.

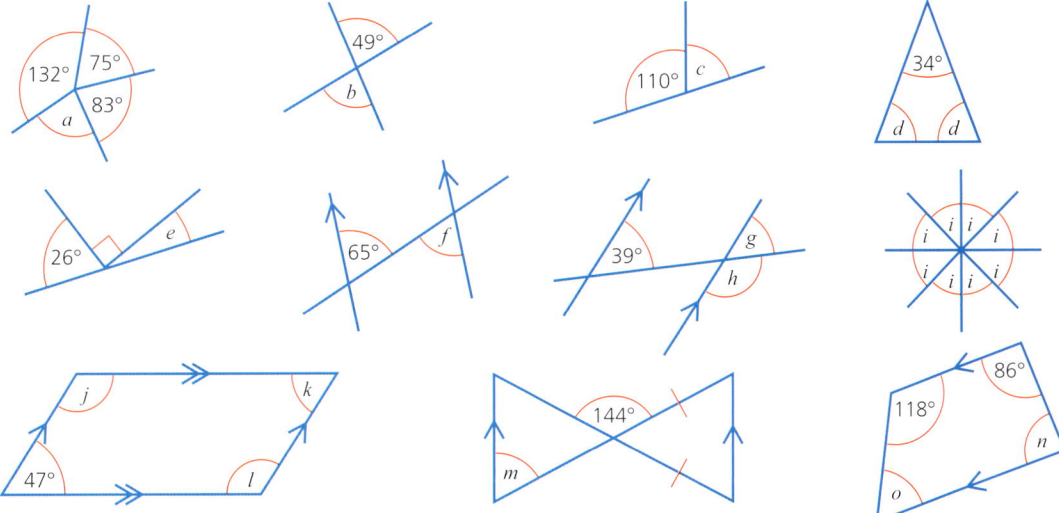

9 a Write down the size of each lettered angle. Give a reason for each one.

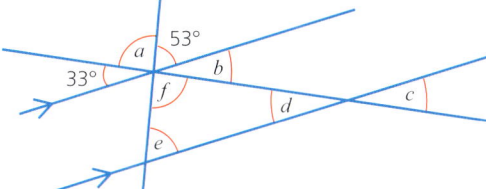

  b What is $f + d + e$?

10 Find the size of each lettered angle in these diagrams.

For each one write down the angle fact(s) that you use.

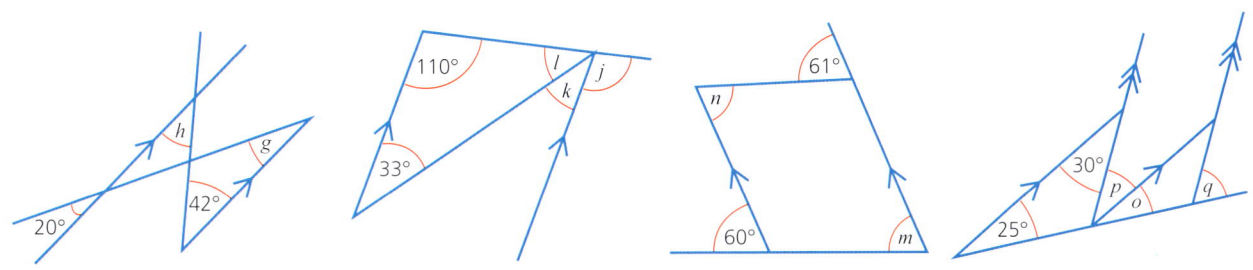

⑪ Calculate angles *a* and *b* shown in the diagram representing an electricity pylon.

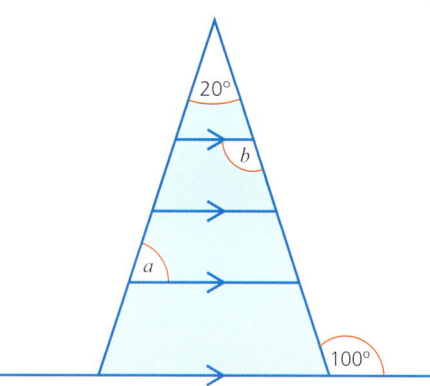

⑫ Ceri shows that triangle ABC has the same angles as triangle ADE.
Copy Ceri's working. Fill in the missing reasons.

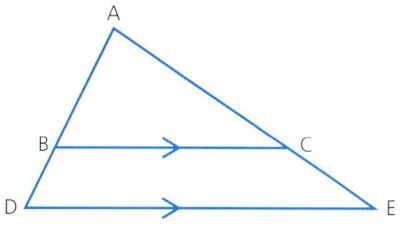

BÂC = DÂE (common to both triangles)
AB̂C = AD̂E (   )
AĈB = AÊD (   )
So triangle ABC has the same angles as triangle ADE.

Band 3 questions

⑬ Tomos says that triangle PQR has the same angles as triangle PTS.
Show that Tomos is right, giving a reason for each pair of equal angles.

See question 12 for a way to name angles.

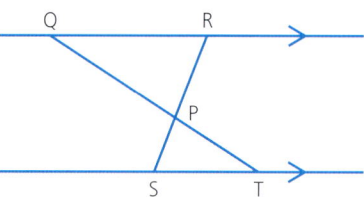

⑭ Beca is making a rabbit hutch.
This is a scale drawing of one end. It is made from three pieces of wood cut from 3 m lengths that are 40 cm wide.
Make a scale drawing of a piece of wood.
Show where Beca should cut it to make the end from the least amount of wood.

⑮ Find the size of angle *x*.

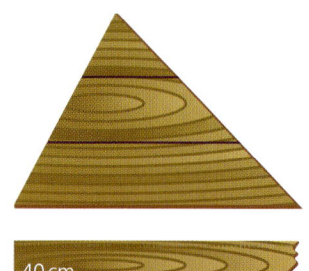

40 cm
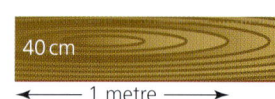
1 metre

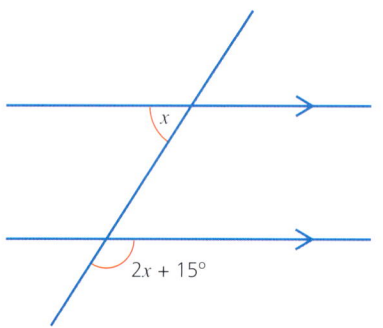

$2x + 15°$

48

16 Find the value of y.

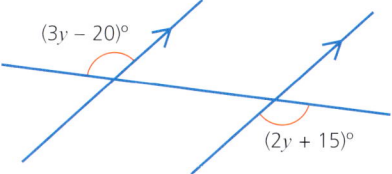

17 Find the size of angle w.

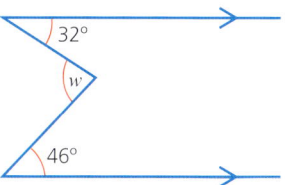

> **Hint**
> Draw another parallel line through angle w.

18 Find the size of angle m.

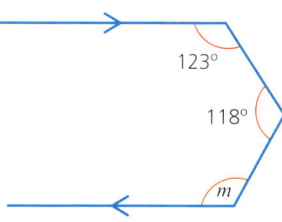

3.2 Polygons

Skill checker

① What is a polygon?
② What is the sum of the three angles inside any triangle?
③ What is the sum of the four angles inside any quadrilateral?
④ Write down the first ten multiples of 180.

The words 'poly' and 'gon' come from the Greek language. They mean many and angle.

An **interior angle** is the angle inside the vertex (or corner) of a shape.
An **exterior angle** is the angle that has to be turned through to move from one side to the next.
At each vertex:

 interior angle + exterior angle = 180°

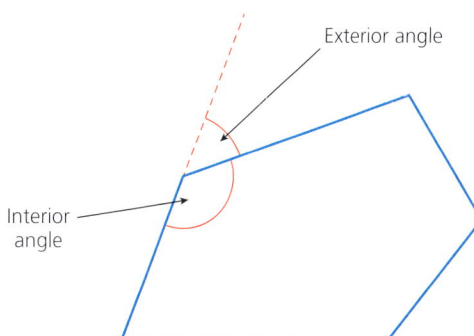

The exterior angles of a polygon always make one complete turn and so **add up to 360°**.

Exterior angles can only be calculated when a polygon is convex. If an interior angle of a polygon is reflex, then the polygon is concave. The exterior angle would then need to be negative for these two rules to work! This pentagon is an example of a concave shape. The indicated angle is reflex.

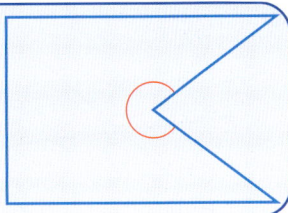

Conceptual understanding

Activity

a Draw any polygon.
b Using a protractor, measure all the interior angles.
c Calculate the sum of the interior angles.
d Choose one vertex of your polygon.
   Draw straight lines from that vertex to all other vertices.
e How many triangles have been formed?
f Write down the angle sum of any triangle.
g Calculate 180 × the number of triangles.
h Compare your answer to part **g** with your answer to **c**.
i By joining vertices in the same way as in part **d**, how many triangles can be formed in a 15-sided polygon?
j Calculate the interior angle sum for a 15-sided polygon.
k Calculate the interior angle sum for a 42-sided polygon.

The 'angle sum' of a triangle or quadrilateral usually refers to the interior angles only.

The interior angles of any polygon always add up to:
   180 × (number of sides − 2)
This formula can also be written as:
   180 × number of sides − 360

Worked example

a Write down the sum of the exterior angles of a hexagon.
b Calculate the sum of the interior angles of any hexagon.

Solution

a The sum of the exterior angles of any convex shape is 360°.
b The sum of the interior angles of any hexagon is 180 × 6 − 360 = 720°.

Remember

A hexagon has six sides.

▶ Regular polygons

For a regular polygon all interior angles are equal, all exterior angles are equal, and all sides are the same length.

If a shape is not regular, then it is irregular.

The number of lines of symmetry of any regular polygon is the same as its number of sides.
The order of rotational symmetry of any regular polygon is the same as its number of sides.

3 Angles

Worked example

For a regular pentagon, calculate:

a the size of each exterior angle, a
b the size of each interior angle, b
c the number of lines of symmetry
d the order of rotational symmetry.

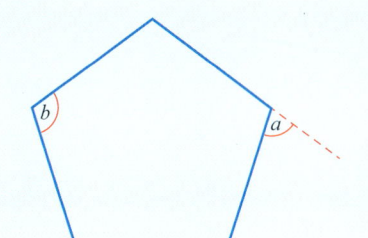

Solution

a Exterior angle = 360 ÷ 5 = 72°
b Interior angle = 180 − 72 = 108°
c Five lines of symmetry

Alternatively, the interior angle sum (3 × 180° = 540°) could be calculated and then divided by 5 to find the size of each interior angle.

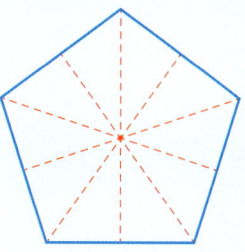

d Rotational symmetry order 5

Remember
A pentagon has five sides.

Remember
A seven-sided shape is a heptagon, and a nine-sided shape is a nonagon.

Polygon name	Number of sides
Triangle (Triongl)	3
Quadrilateral (Pedrochr)	4
Pentagon (Pentagon)	5
Hexagon (Hecsagon)	6
Octagon (Octagon)	8
Decagon (Decagon)	10

3.2 Now try these

Band 1 questions

1 a Calculate the exterior angle a of a regular hexagon.
    b Calculate the interior angle b of a regular hexagon.

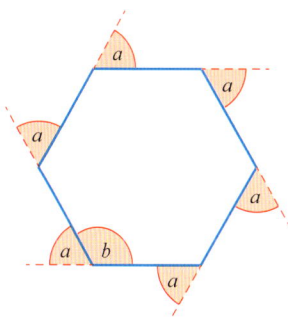

2 a Calculate the exterior angle a of a regular octagon.
    b Calculate the interior angle b of a regular octagon.

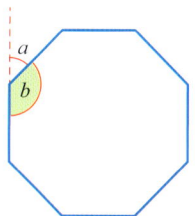

3 Calculate the size of each lettered angle.

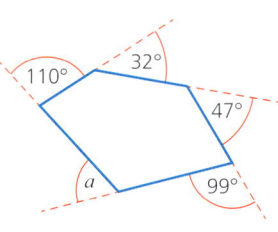 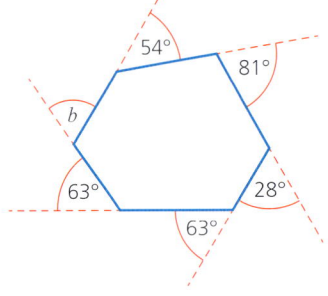 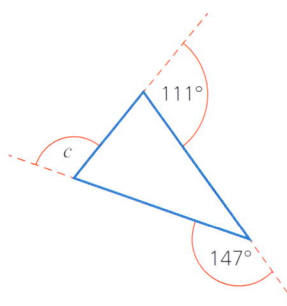

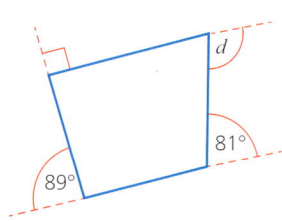 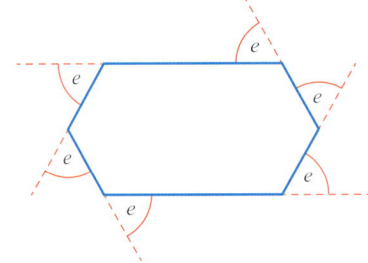

4 The diagram shows a pentagon split into five triangles.
What is the sum of:
  a all the angles in all the triangles
  b the angles at the centre
  c the interior angles of the pentagon?

5 **a** Calculate the size of each exterior angle of a regular decagon.
  b Calculate the size of each interior angle of a regular decagon.
  c Find the number of lines of symmetry of a regular decagon.
  d Find the order of rotational symmetry of a regular decagon.

6 **a** Write down the name of this shape.
  b Calculate the size of angle m.

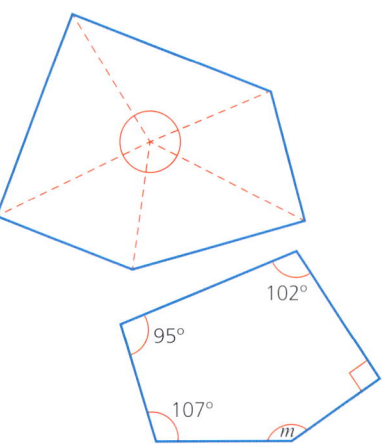

Band 2 questions

7 The exterior angles of a pentagon are $78°, 84°, 45°, x$ and $2x$.
  a Find the size of angle x.
  b Is it a regular pentagon?
  How do you know?

8 A 12-sided shape is called a dodecagon.
  a Find the size of angle y.
  b Find the size of angle z.
  c If all the exterior angles of a polygon are equal, can we assume it is regular?
  Explain your answer.

9 The interior angles of a hexagon are $104°, 129°, 135°, 97°, (3y + 20)°$ and $(2y - 10)°$.
Find the value of y.

10 Calculate the sum of the interior angles of:
  a a hexagon
  b a nonagon (nine sides)
  c a polygon with 23 sides
  d a quadrilateral
  e a triangle
  f a polygon with 501 sides.

11 A regular polygon has an exterior angle of 15°.
How many sides does the polygon have?

12 A regular polygon has an interior angle of 160°.
How many sides does the polygon have?

Band 3 questions

13 The diagram shows an irregular hexagon.
Two of its interior angles are right angles.
  a Draw irregular hexagons with:
    i one right angle
    ii three right angles
    iii four right angles
    iv five right angles.
  b Prove that the interior angles of a hexagon cannot be six right angles.

14 Three adjacent vertices of a regular polygon form a triangle.
One of the angles of the triangle is 5°.
Find the number of sides of the regular polygon.

15 The diagram shows a regular heptagon (seven sides) with a diagonal joining two of its vertices.
Find the size of angle n to two decimal places.

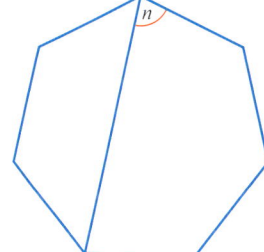

16 ABCDE is a regular pentagon.
Calculate the size of CÂD.

17 Explain why the exterior angle of a polygon cannot be 17°.

18 The interior angles of a hexagon form an arithmetic sequence.
The smallest interior angle is 80°.
Find the biggest interior angle.

Key words

Here is a list of the key words you met in this chapter.

Allied	Alternate	Co-interior	Concave	Convex	Corresponding	Decagon
Heptagon	Hexagon	Interior	Irregular	Nonagon	Octagon	Pair of compasses
Parallel	Pentagon	Polygon	Regular			

Use the glossary at the back of this book to check any you are unsure about.

Review exercise: angles

Band 1 questions

1 Find the size of each lettered angle.

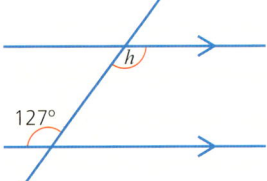

2 A heptagon has seven sides.
  a How many lines of symmetry does a regular heptagon have?
  b What is the order of rotational symmetry of a regular heptagon?

3 a Find the size of angle h.
  b Give a reason for your answer.

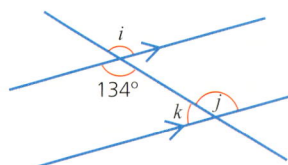

4 A nonagon has nine sides.
  a Find the exterior angle of a regular nonagon.
  b Find the interior angle of a regular nonagon.

5 a Find the size of angles i, j and k.
  b Give one reason for each of your answers.

6 Find the size of angle l.

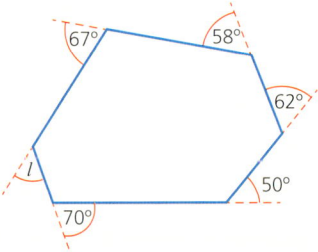

Band 2 questions

7 Four of the interior angles of a hexagon are 100°, 115°, 132° and 121°.
The other two interior angles are equal. Find the size of these equal angles.

8 Find the size of angle n.

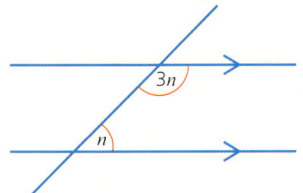

9 A regular pentagon shares two of its sides with two other congruent regular pentagons. Find angle m.

10. Lowri shows that angle *s* is 124°.

    Copy her working and include the reasons.

    $r = 180 - 56$ (....)

    $r = 124°$

    $s = 124°$ (....)

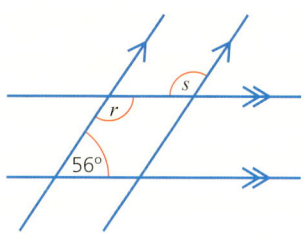

11. Copy and complete this table.

    Write angles correct to one decimal place where necessary.

Polygon	Number of sides	Exterior angle	Interior angle	Number of lines of symmetry	Order of rotational symmetry
Equilateral triangle	3	120°	60°	3	3
Square	4				
Regular pentagon	5				
Regular hexagon	6				
Regular heptagon	7				
Regular octagon	8				
Regular nonagon	9				
Regular decagon	10				

12. A regular polygon has an exterior angle of 9°.

    How many sides does the polygon have?

Band 3 questions

13. Three congruent regular polygons are drawn sharing sides.

    The gap formed is an equilateral triangle.

    How many sides does the large regular polygon have?

14. The interior angle of a regular polygon is 11 times greater than the exterior angle.

    How many sides does the polygon have?

15. Find the sizes of angles *v* and *w* in the diagram on the right.

16. Calculate the value of *x*.

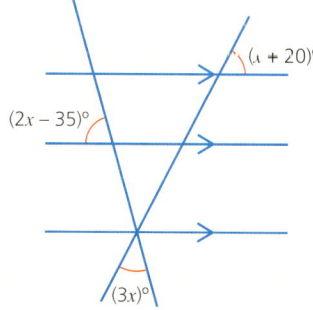

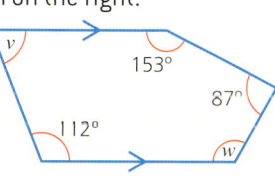

17. Explain why the interior angle of a regular polygon cannot be 169°.

18. Points A, B, C, D, E, F and G form a regular pentagon ABCDE and a square DEFG.

    Find the size of EÂF.

4 Constructions

Coming up...

▶ Using scale factors, scale diagrams and maps
▶ Drawing and measuring line segments and angles in geometric figures
▶ Drawing and interpreting scale drawings
▶ Constructing accurate diagrams using a ruler and a protractor

Have a guess

Without measuring, draw a random straight line.

Each person then estimates the length of the line, hiding their estimates from the others in the group until all have finished.

Then measure the length with a ruler.

Award points to each person, depending on how close their estimate is.

For example, in a group of five, award 5 points to the closest, 4 points to the next closest, and so on.

Take turns to draw and measure the line.

Try estimating angles also, including obtuse and reflex angles.

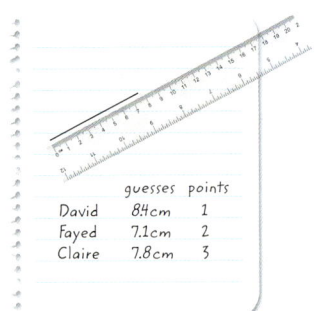

4.1 Bearings

Skill checker

① Draw, separately, angles of 40°, 155° and 307°.
② Calculate the angles labelled with a letter

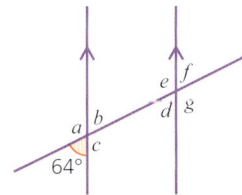

▶ Bearings

A **bearing** is an angle measured **clockwise** from North.

Bearings are always given using **three figures** to reduce the risk of confusion.

The directions North, East, South-West, and so on, can be replaced with numbers.

This allows us to describe other directions between each of those shown here.

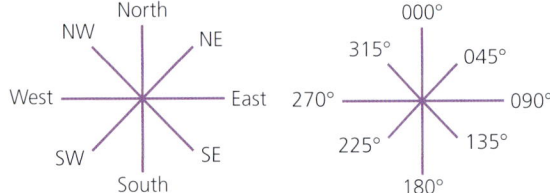

4 Constructions

Worked example

B is 3 cm from A on a bearing of 120°.
Draw the bearing of B from A.

The start point usually follows the word 'from'.

Solution

First, draw a North line at A.

Then measure an angle of 120° clockwise from North at A.

Finally, draw a line at this angle from A. Use a ruler to mark point B 3 cm along this line.

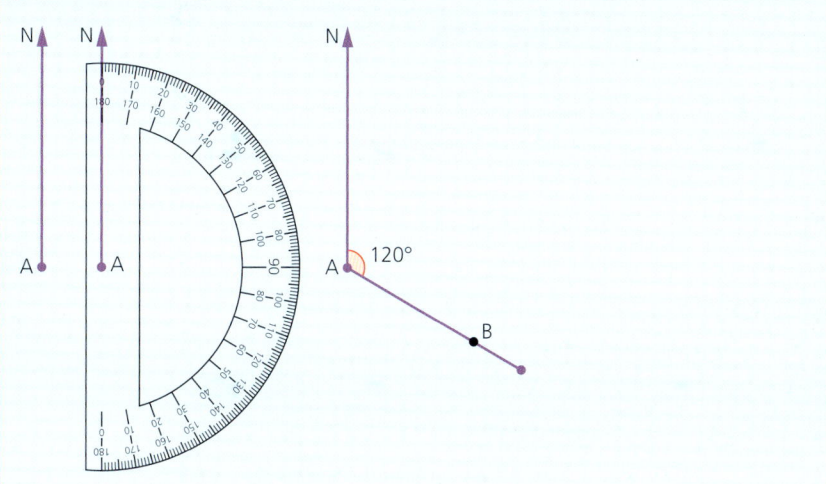

Remember

When drawing or measuring a bearing, the centre of the protractor is placed on the start point.

For example, when measuring **from P to Q**, the centre of the protractor must be placed on P.

Worked example

Here is a map of Great Britain.
Find the bearing of:

a London from Penzance
b Birmingham from London.

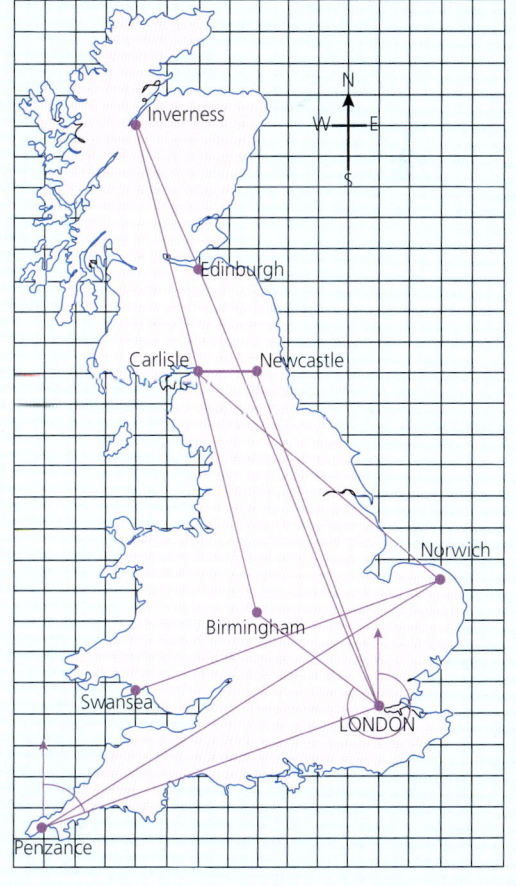

Solution

a   The bearing is from Penzance so place your protractor on the map with the centre at Penzance and the zero line vertically up.

Use the grid lines to help you.

Measure the angle clockwise from North.

As 70° only has two digits, write your answer with a zero in front.

070°

b   The bearing from London to Birmingham is a reflex angle.

Place your protractor on the map with the centre at London and the zero line vertically up.

If you have a 360° protractor you can measure the angle directly.

If you have a 180° protractor, measure the angle anticlockwise from North.

Subtract this angle from 360°.

360° − 53° = **307°**

4.1 Now try these

Band 1 questions

1 Write down the bearing for each of these compass directions.
  a South
  b East
  c North-West
  d South-East

Remember
A bearing is three figures.

2 Write down the compass directions with these bearings.
  a 270°
  b 135°
  c 225°
  d 000°

3 Measure the bearing from A to B in each of these diagrams.

a 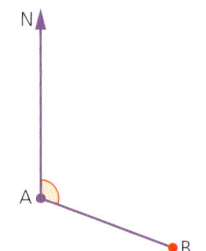   b 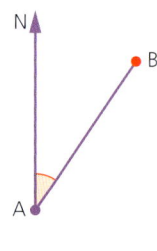   c    d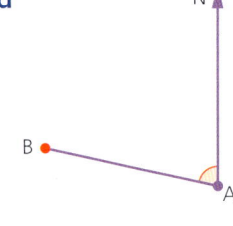

4 Calculate the bearing of each point from P.

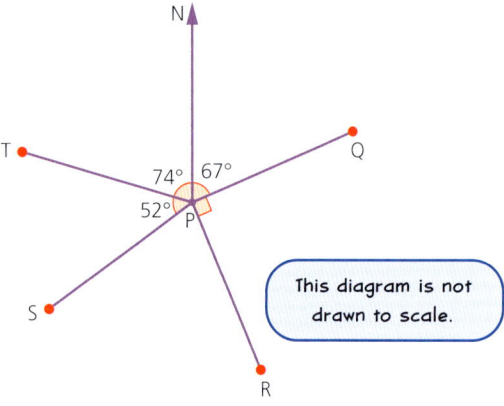

This diagram is not drawn to scale.

5 Draw a point A with a North line starting at A.
Point B is on a bearing of 030° from A.
Point C is on a bearing of 145° from A.
Point D is on a bearing of 202° from A.
  a Using a protractor and ruler draw points B, C and D, 5 cm from A.
  b Measure the lengths BC, CD and DB.

6 Measure these bearings:
  a Newcastle from Carlisle
  b Norwich from Swansea
  c Edinburgh from Inverness
  d Carlisle from Birmingham
  e Penzance from Norwich.

7 a Which city is on a bearing of 130° from Carlisle?
  b Which city is on a bearing of 340° from London?

Band 2 questions

8
  a Write down the bearing from A to B.
  b Calculate the size of angle x.
  c Find the bearing from B to A.

9
  a Find the bearing from D to C.
  b Find the bearing from C to D.

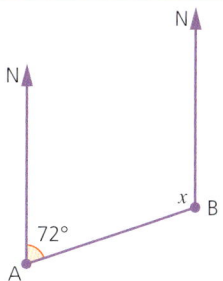

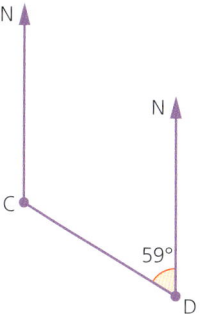

These diagrams are not drawn to scale.

10
  a The bearing from E to F is 124°. Find the bearing from F to E.
  b The bearing from G to H is 207°. Find the bearing from H to G.
  c If the bearing from I to J is known, explain how the bearing from J to I can be calculated.

A bearing in the opposite direction is sometimes called a reverse bearing or a back bearing.

11 Point K is on a bearing of 145° from L.
Point L is on a bearing of 325° from M.
Which point is the most northerly? Explain your answer using a diagram.

12 Points P, Q and R are all 2 km from each other.
Q is due North of P. R is the most easterly of the three points.
  a What type of triangle is PQR?
  b Find the bearing from P to R.
  c Find the bearing from Q to R.

Band 3 questions

13 Find the following bearings:
  a from Alun to Bryn
  b from Bryn to Ceri
  c from Ceri to Alun.

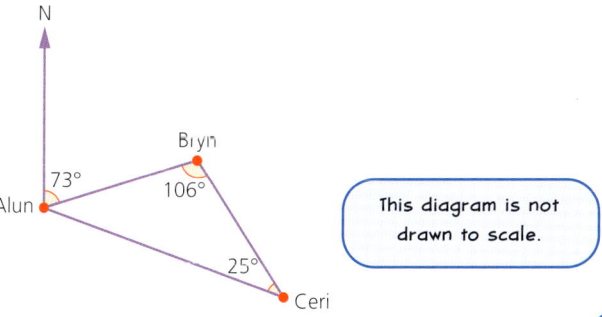

This diagram is not drawn to scale.

14 Point X is due North of point Y.
X and Y are both 15 km from point Z.
X is on a bearing of 070° from Z.
  a What type of triangle is XYZ?
  b Find the bearing of Y from Z.

Due North means 'directly North', rather than just being in a more northerly position.
For example, Pontypridd is North of Cardiff, but it is not due North. Look at a map of Wales to compare them.

15 Weston is 3 km from the centre of Otley on a bearing of 300°.
Farnley is 3 km from the centre of Otley on a bearing of 030°.
Find the bearing of Farnley from Weston.

16 A yacht sails 10 km on a bearing of 060°.
It then sails 10 km due South.
 a How far is the yacht from its starting point?
 b Find the bearing on which it must travel to return to its starting point.

17 Sioned walks 5 km on a bearing of 040°.
She then walks 5 km on a bearing of 060°.
Find the bearing on which she must travel to return to her starting point.

18 ABCDE is a regular pentagon with sides of length 8 km.
C is due East of D, and A is the most northerly of the five points.
 a Find the bearing of B from E. b Find the bearing of C from B. c Find the bearing of A from D.

4.2 Scale drawings

Skill checker

1. How many millimetres are in 1 centimetre?
2. How many centimetres are in 1 metre?
3. How many metres are in 1 kilometre?
4. How many millimetres are in 1 kilometre?

A scale drawing is the **same shape** as the original but a **different size**.
All the lengths are in the same **ratio**.
On a scale drawing where 1 cm on the drawing represents 2 m on the actual object, the scale can be written as:

$\frac{1}{200}$ or 1 cm : 2 m or 1 : 200

> If two shapes are the same except for one being bigger than the other, they are said to be 'similar'. You will study similar shapes in Book 3.

Worked example

Megan has a toy car on a scale of $\frac{1}{50}$.
The toy car is 6 cm long.
The real car is 1.7 m wide.

a How long is the real car? b How wide is the toy car?

Solution

a Length of real car = 6 cm × 50
  = 300 cm

> To find the length of the real car, multiply by 50.

The real car is 300 cm or 3 m long.

b 1.7 m = 170 cm

> It is easier to convert to centimetres first.

Width of toy car = 170 cm ÷ 50
  = 3.4 cm

> To find the width of the toy car, divide by 50.

The toy car is 3.4 cm wide.

4 Constructions

Worked example

Adebola is working out distances between places on a map.

The map has a scale of 1 : 50 000.

On the map, the bowling alley is 4.7 cm from the aqueduct.

a How far is the bowling alley from the aqueduct in kilometres?

b Adebola's house is 3.8 km from her school.
How far is her house from her school on the map?

Solution

a 4.7 × 50 000 = 235 000 cm

235 000 ÷ 100 000 = 2.35 km

b 3.8 × 100 000 = 380 000 cm

380 000 ÷ 50 000 = 7.6 cm

Notes

- To find a distance on the ground, multiply by the scale factor of 50 000.
- There are 100 000 cm in 1 kilometre.
- Convert to centimetres first.
- To find a distance on the map, divide by the scale factor of 50 000.

4.2 Now try these

Band 1 questions

1 A model aeroplane is built using a scale of 1 : 20.

The wingspan of the model is 1 metre.

The length of the real plane is 30 metres.

   a Find the wingspan of the real plane.
   b Find the length of the model.

2 Two model trains look the same, but one is bigger than the other.

The length of the small train is 10 cm, and the length of the big train is 20 cm.

   a Find the scale ratio between the two trains.

The height of the small train is 4 cm.

   b Find the height of the big train.

Each wheel on the big train is 12 mm wide.

   c Find the width of the wheel on the small train.

3 The distance between the rails on a train track is 1435 mm.

A model train set is built using a scale of 1 : 35.

   a Find the distance between the tracks on the model.
   b The length of one of the model carriages is 40 cm. Find the length of the real carriage.

4 The plan of the house shown below is not drawn to scale.

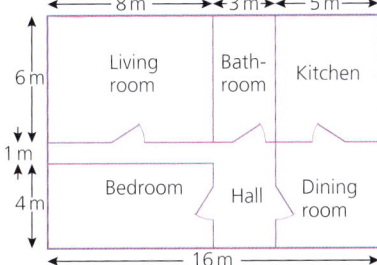

Using the measurements given on the plan, draw an accurate scale drawing.

Use the scale 1 cm : 2 m.

5 A model of a building is built to a scale of $\frac{1}{100}$.

Copy and complete this table showing the lengths of various parts of the building.

Building part	Model measurement	Real measurement
Roof length	20 cm	
Door height		2 m
Window width	32 mm	
Chimney height		15 m
Fire escape	11.2 cm	

Band 2 questions

6 A square patio is 4 metres wide.
  a Find the scale of the plan.
  b Find the actual width of the circular pond.

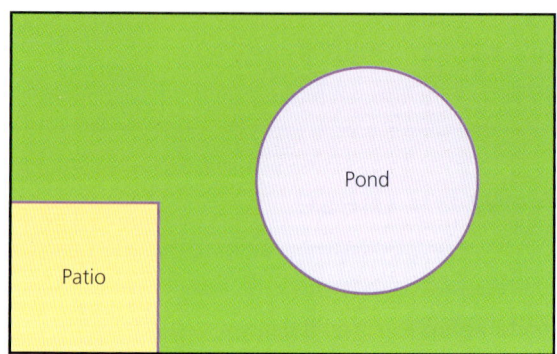

7 A map has a scale of 1 : 500.

The actual length of a footpath is 0.2 km.
  a How many metres is 0.2 km?
  b How long is the footpath on the map?

8 The scale of a map is 1 : 25 000.
  a The perimeter of a field on the map is 3 cm. Find its actual perimeter, using appropriate units.
  b A road is 5 km long. Find its length on the map, using appropriate units.

9 A footpath of length 2 km is drawn on a map. Its length on the map is 5 cm.

Carwyn's homework question asks him to work out the scale of the map.

> 5 cm : 2 km
> = 5 cm : 2000 m
> = 5 cm : 2 000 000 m
> = 1 : 400 000

  a Find the error in Carwyn's working.
  b Find the correct map scale.

Band 3 questions

10 The scale on a globe is 1 : 20 000 000.
  a The width of a country on the globe is 10 cm. What is the actual width of the country?
  b The actual length of a railway track is 2300 km. Find its length on the globe.

11 Lowri sets out from High Trees.

She walks 5 km South-East then 7 km North-East to reach the lake at point P.

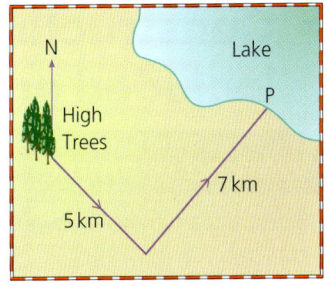

   a Make a scale drawing to find:
      i how far point P is from High Trees
      ii the bearing from High Trees of point P.
   b What is the bearing of High Trees from point P?

12 The distance between two points on a map is 3 cm.

The actual distance between the two points is 3 km.

Find the scale of the map in the form $1:n$.

13 Tegan is drawing a plan of the school canteen which measures 43 m by 31 m.

She wants to make her plan as big as possible, but it must fit on one side of a sheet of A4 paper.

She also wants to use a scale of the form $1:m$, where m is a whole number.

Find the scale she should use.

14 Geraint is drawing a plan of his bedroom.

He measures the lengths of two adjacent walls as 255 cm and 360 cm.

Using a scale of $1:50$, he then draws this plan on graph paper.

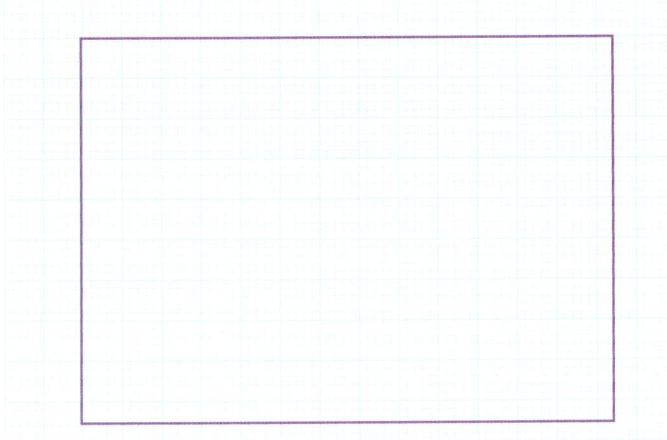

Cross-curricular activity

Draw a scale drawing of your bedroom, using an appropriate scale of your choice.

What could you use this scale drawing for?

   a What assumption has Geraint made?
   b What other measurements should he have taken?

4.3 Constructions

Skill checker

1 Draw an angle of 123°.

2 What is the sum of the angles in a triangle?

3 **a** How many sides does a pentagon have?
    b Find the interior angle of a regular pentagon.

When constructing a triangle, first draw one of the sides.

Measure the angles and other lines from this side.

Curriculum for Wales Mastering Mathematics: Book 2

▶ Constructing a triangle when two sides and the included angle are given

Worked example

Construct a full-size accurate drawing of this triangle.
Measure the third side length and the other two angles.

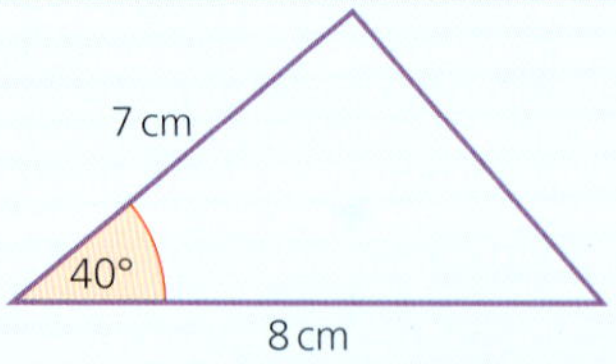

Solution

▶ Use a ruler to draw one side of the correct length and label it.
8 cm

You may find it easier to start with the longest side. Then your diagram is less likely to disappear off the page!

▶ Use a protractor to draw the correct angle at one end of the side.

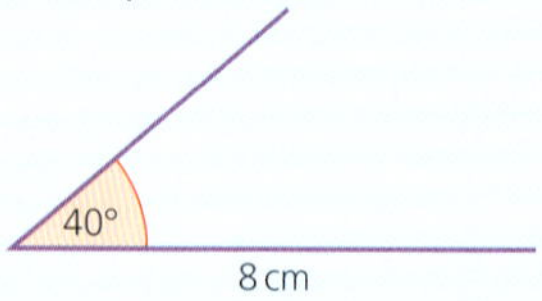

40°
8 cm

▶ Use a ruler to draw the second side to the correct length.

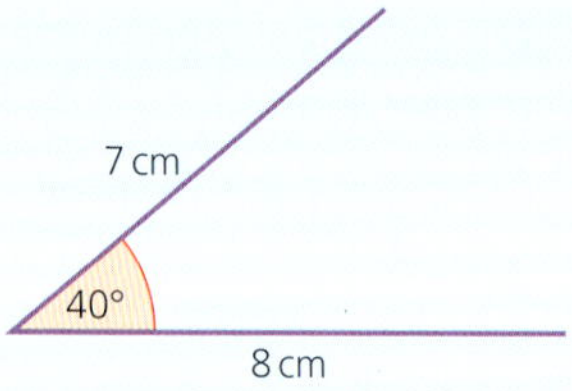

7 cm
40°
8 cm

▶ Draw the third side to form the triangle.

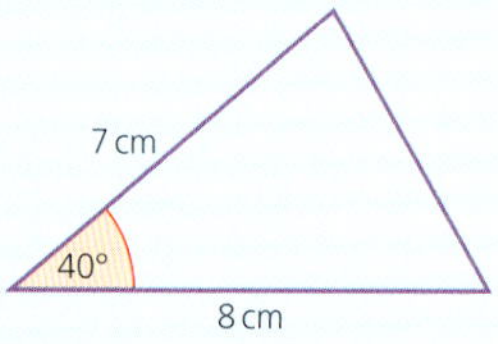

7 cm
40°
8 cm

The length of the third side is 5.2 cm.

The other angles are 80° and 60°.

4.3 Now try these

Band 1 questions

1 Use a ruler and protractor to make an accurate full-size drawing of this triangle.
Measure and label the remaining side and angles.

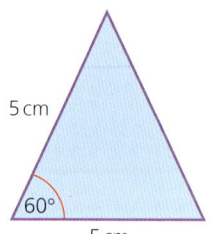

64

4 Constructions

2 Use a ruler and a protractor to make accurate full-size drawings of these triangles.

In each case, measure the remaining sides and angles.

a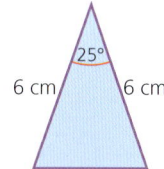

b 3 cm 120° 3 cm

Band 2 questions

3 Hywel's dad is designing a slide for the garden.
This is his sketch of the slide.

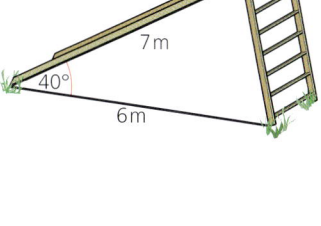

a Using a scale of 1 cm to represent 1 m, make an accurate scale drawing of the side of the slide.

b How long must Hywel's dad make the ladder?

4 Barton is 6 km due North of Afonffordd.

Carville is 8 km due East of Afonffordd.

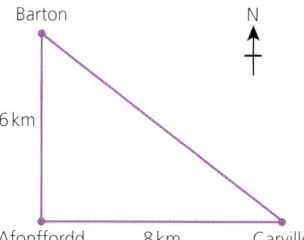

a Using a scale of 1 cm to 1 km, make a scale drawing.

b How far is it from Barton to Carville?

5 a Ten people are told to draw this triangle.
Will they all draw it exactly the same?

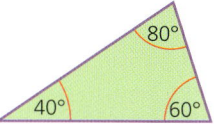

b Is the answer the same for this triangle?

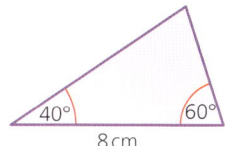

6 Construct these triangles.

In each case, say what is special about the triangle.

a AB = 8 cm, BC = 8 cm, AB̂C = 60°

b LM = 8 cm, MN = 4 cm, LM̂N = 60°

c XY = 9.9 cm, YZ = 14 cm, XŶZ = 45°

7 a Draw these two triangles accurately.

b Measure the other two angles in each triangle.
What do you notice about the angles?

c Measure the third side of each triangle.
What do you notice about the sides?

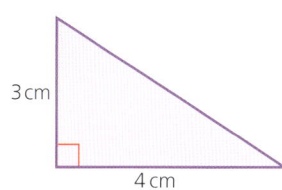
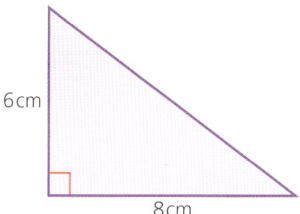

65

Band 3 questions

8 **a** Draw this triangle at its full size. Measure the length of AB.

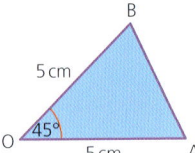

  b Now add seven more triangles to make a regular octagon.
    Measure the angles:
     i CÔF   **ii** CÔE   **iii** DÔF.

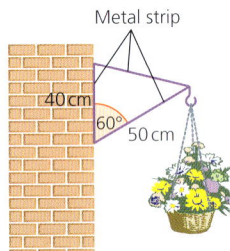

  c What is the perimeter of the octagon?

9 Wes is making a hanging basket support for his Design and Technology project.
  He has made this sketch.
  He is going to use a ready-formed hook.
  a Make an accurate scale drawing of the frame.
  b Use the diagram to find the total length of metal strip that Wes needs.

10 **a** Using a ruler and a protractor, construct a regular pentagon with sides 5 cm.
   b Why would a regular heptagon be more difficult to draw accurately?

> A heptagon has seven sides.

Key words

Here is a list of the key words you met in this chapter.

Bearing   Clockwise   Construct   Included side/angle   Ratio   Scale

Use the glossary at the back of this book to check any you are unsure about.

4 Constructions

Review exercise: constructions

Band 1 questions

1. Write down the compass directions for each of these bearings.
   a 180°   b 090°   c 315°   d 045°

2. A model of a bridge is built using a scale of 1 : 40.
   The height of the model is 1 metre.
   The length of the real bridge is 800 metres.
   a Find the height of the real bridge.   b Find the length of the model.

3. Write down the bearings of each of these compass directions.
   a West   b South-West   c North-East   d North

4. Using a ruler and a protractor, construct an accurate full-size drawing of this triangle.
   Measure, and label, the remaining sides and angle.

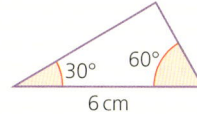

5. Measure the bearing from P to Q in each of these diagrams.
   a 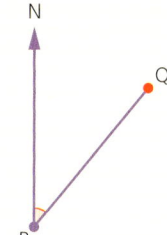   b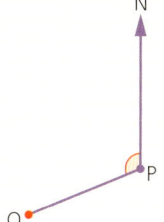

6. Max says that the bearing from C to D is 130°.

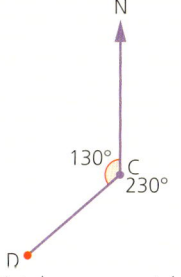

   Not drawn accurately

   a What mistake has Max made?
   b What is the bearing from C to D?

Band 2 questions

7. The floor of a summer house is built in the shape of a square with a triangle removed.
   The floorplan has one line of symmetry.
   Using the scale 1 : 50, draw an accurate plan of the summer house floor.

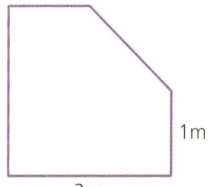

8 a Find the bearing from E to F.
b Find the bearing from F to E.

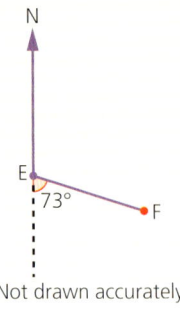

Not drawn accurately

9 The bearing of Apperley from Grasston is 164°.

Find the bearing of Grasston from Apperley.

10 Two points are marked on this map of an island.

Treasure is buried on a bearing of 160° from the lighthouse and a bearing of 078° from the hut.

Copy the island and mark the position of the treasure.

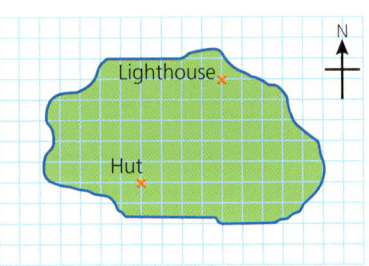

11 Point U is 5 km from W on a bearing of 140°.

Point T is 5 km from U on a bearing of 220°.

a Draw an accurate diagram showing the three points.
b Find the distance between T and W.
c What is the bearing of T from W?
d What is the bearing of W from T?

Band 3 questions

12 ABCDEF is a regular hexagon.

A is due North of B, and due West of E.

a Find the bearing of C from F.
b Martine says that D is South-East of A.
   Is she right? Explain your answer.
c Find the bearing of C from E.

13 Barnswick is 10.3 km from Ashworth on a bearing of 040°.

Cabtree is on a bearing of 100° from Barnswick.

Ashworth is on a bearing of 255° from Cabtree.

a Draw a scale diagram to find the direct distances between Ashworth and Cabtree, and Barnswick and Cabtree.
b Ioan says he can calculate the distance between Barnswick and Cabtree without drawing a scale diagram and without using a calculator.

   Explain Ioan's method.

4 Constructions

14 Josh flies his helicopter around England, Scotland and Wales.

Starting in Penzance, he visits seven other places marked on the map, finishing in Cambridge.

Copy and complete the table showing each leg of his journey.

Leg	Start	End	Distance (km)	Bearing
1	Penzance	Swansea		034°
2	Swansea		175	058°
3		Carlisle		
4	Carlisle		260	
5		Grimsby		160°
6	Grimsby		140	
7		Cambridge		245°

15 Sailing ships cannot sail directly into the wind.

However, they can sail towards the wind, but only at an angle to its direction.

Captain Joe's square rigger cannot sail closer than 67° into the wind.

A 'square rigger' is a sailing ship.

It can sail in all other directions (shown by the green arc).

In the diagram, the wind is from the North – it is a 'North wind'.

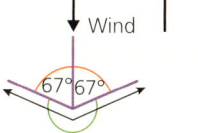

Between which bearings can Captain Joe not sail if the wind is:

a a North wind (as in the diagram)

b a South wind

c an East wind

d a South-West wind?

5 Calculations

Coming up...

▶ Using the four operations, including formal written methods, on integers
▶ Using the four operations on decimals

Number puzzles

① If you add up the numbers in any row, column or diagonal in this magic square, you should always get the same total. Unfortunately, one of the numbers is incorrect. Can you find it?

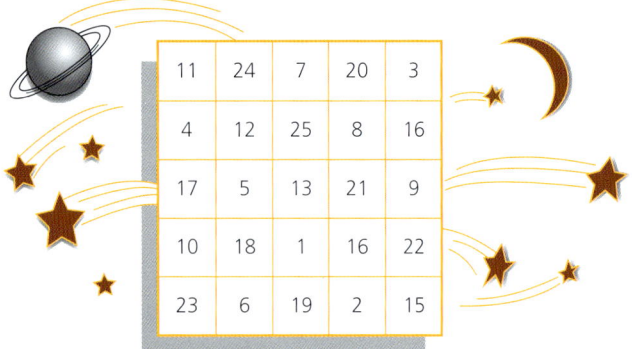

A score must be a non-negative whole number.

② In one very strange cricket match, the scores of all 11 batsmen are consecutive numbers.
   a   What is the smallest total they could have scored?
   b   What are the scores of the batsmen if the total is:
       i   121 runs   ii   374 runs   iii   616 runs?
   c   Could the total be:
       i   84 runs   ii   95 runs?

③ Copy the triangle diagram shown on the right.
Arrange the numbers 1, 2, 3, 4, 5 and 6 in the circles so that each line of three numbers has the same total.

④ Copy the diagrams below. Arrange the numbers 1, 2, 3, 4, 5, 6 and 7 in the circles so that each line of three numbers has the same total.

   a

   b

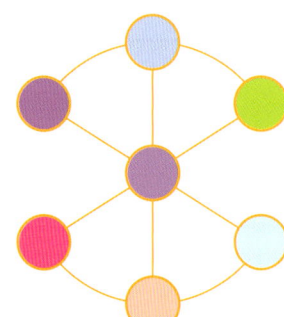

5 Calculations

5.1 Calculations review

Skill checker

① Copy and complete this multiplication table.

×	1	2	3	4	5	6	7	8	9	10
1										
2										
3										
4										
5										
6										
7										
8										
9										
10										

② Calculate $1 + 2 + 3 + 4 + 5 + 6 + 7 + 8 + 9 + 10 =$ ___

There are different methods you can choose from when **adding** numbers.

Worked example

$326 + 79$

Solution

Column method

Line up the units digits and then add the columns from right to left, carrying digits as necessary.

$$\begin{array}{r} 3\ 2\ 6 \\ +_1\ _17\ 9 \\ \hline 4\ 0\ 5 \end{array}$$

$6 + 9 = 15$
$2 + 7 + 1 = 10$

Partitioning

Break up one of the numbers into units, tens, hundreds, and so on. Then add them one at a time.

$326 + 70 = 396$

Add 70 and then add 9.

$396 + 9 = 396 + 4 + 5$
$ = 405$
$326 + 79 = 405$

9 has been split into 4 and 5 because 4 is needed to increase 396 to 400.

Note

If the number is a decimal, then it should be split into tenths, hundredths, etc.

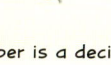

You can use the same methods to **subtract** numbers.

Worked example

73.6 − 4.82

Solution

Column method

Line up the units digits with the smaller number below the higher number.

For each column where the least digit appears on the top then add 10, subtracting 1 from the digit to its left.

Then subtract each column.

If both numbers include a decimal point, then it's easier to line the decimal points up.

Add zeros to the end of the higher number if necessary.

```
      12 15
   6⁷ ⁷2. ⁶5 ¹0
 −    4. 8  2
   ───────────
   6  8. 7  8
```

10 − 2 = 8, 15 − 8 = 7, 12 − 4 = 8, 6 − 0 = 6

Partitioning

Break up the lower number into units, tens, hundreds, and so on. Then subtract them one at a time.

$$73.6 - 4 = 73.6 - 3 - 1$$
$$= 69.6$$
$$69.6 - 0.8 = 69.6 - 0.6 - 0.2$$
$$= 68.8$$
$$68.8 - 0.02 = 68.78$$
$$73.6 - 4.82 = 68.78$$

There are different methods you can choose from when **multiplying** numbers.

Worked example

A school buys 17 computers at £258 each.

Calculate the total cost.

Solution

Long multiplication

Multiply one number by the units column of the other.

Then multiply it by the tens column.

Then the hundreds column, and so on.

Add these values together to get the answer.

```
        2  5  8
   ×   ⁴1 ⁵1  7
   ─────────────
     1  8  0  6    ← 7 × 258
    ₁2  5  8  0    ← 10 × 258
   ─────────────
     4  3  8  6
```

72

Grid method

Break up both numbers into units, tens, hundreds, and so on.

Form a grid with one number at the top, and the other at the side.

For each cell of the grid, find the product of each number.

Finally, add the products.

×	200	50	8
10	2000	500	80
7	1400	350	56

2000 + 1400 + 500 + 350 + 80 + 56 = 4386

Lattice method

Draw a grid with the number of columns and rows matching the number of digits in each number.

Write one number along the top, and the other on the right.

Draw diagonals in each cell from bottom left to top right.

Fill each cell with the product of the corresponding digits, with each digit on either side of the diagonal.

Add along the diagonals, starting with the bottom right.

The answer can be found by reading the digits anticlockwise from top left to bottom right.

> When using the lattice method, it is important to draw the diagonals carefully.

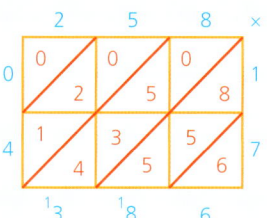

The cost is £4386.

You can **divide** numbers using short or long division.

Both methods involve the same calculations.

Worked example

Share £351 equally amongst 13 people.
How much should each receive?

> The divisor is the number you are dividing by.

Solution

Divide each digit by the divisor, carrying as necessary.

```
        0 2 7
  1 3 ) 3 ³5 ⁹1
```

13 does not divide into 3.
13 divides into 35 two times, remainder 9.
13 divides into 91 seven times.

Each person receives £27.

Chunking method

You may prefer to repeatedly subtract the divisor, counting the number of subtractions.
To speed up the process, multiples of the divisor can be subtracted instead.

```
    3 5 1
  - 1 3 0    1 0
    2 2 1
  - 1 3 0    1 0
      9 1
  -   5 2      4
      3 9
  -   3 9      3
        0
```

10 + 10 + 4 + 3 = 27

5.1 Now try these

Band 1 questions

Fluency

1. a 23 + 61   b 523 + 461   c 429 + 38   d 578 + 386
2. a 85 − 23   b 867 − 431   c 938 − 51   d 625 − 287
3. a 47 × 8   b 36 × 89   c 248 × 72   d 589 × 274
4. a 91 ÷ 7   b 828 ÷ 6   c 1947 ÷ 11   d 299 ÷ 13
5. a 345 + 78 + 1789 b 4378 − 946   c 8236 − 587   d 2346 × 17
    e 3567 ÷ 29
6. a 45.2 + 3.7   b 18.37 + 2.652   c 17.56 − 2.3   d 25.8 − 7.26
    e 5.27 + 34.6 + 0.378

Band 2 questions

Strategic competence

7. a Find the remainder when 473 is divided by 31.
    b Tins of beans are sold at 31p each. How many tins can be bought with 473p?
    c 473 children are going on a trip to Conwy Castle. Each coach can take 31 children.
       How many coaches will be needed to take all 473 children?
    d £473 is shared equally between 31 people.
       How much, to the nearest pound, does each person receive?
       How much is left over?

Logical reasoning

8. Eleri adds £578 and £312 but makes a mistake.
    Her answer is £8810.
    a Here is Eleri's working. What mistake has she made?
    b What is the correct answer?

```
    5 7 8
  + 3 1 2
    8 8 10
```

Strategic competence

9. a £23.89 + 76p
    b 3.7 metres + 186 centimetres
    c 2.13 kg − 567 g
10. Copy this addition. Find the missing digits.

```
    6 2 □
  + 2 □ 6
    9 2 3
```

5 Calculations

11 Copy these calculations. Find the missing digits.
   a ☐ ☐ × 3 = 5 4
   b ☐ ☐ ☐ ☐ ÷ 1 7 = 8 9
   c ☐ 3 × 7 = 9 ☐
   d 2 5 ☐ ÷ ☐ 2 = 2 1

12 Copy this subtraction. Find the missing digits.

      8 ☐ 5
   − 2 7 ☐
      ☐ 4 9

Band 3 questions

13 A school buys 12 computers for £1608.
   The price for each computer is the same.
   How much would 13 computers cost?

14 In these number grids, two numbers are added to give the number above.
   Copy the grids and find the missing numbers.

   a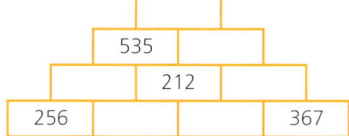

   b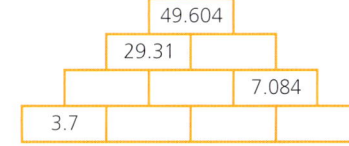

15 Dewi tells Sanjeet that he likes to solve missing-digit problems, but only when they are additions.
   He says the subtraction problems are too difficult.
   Sanjeet tells him a method for changing the questions he doesn't like into addition problems.
   a What is Sanjeet's method?
   b Demonstrate Sanjeet's method on this missing-digit problem.

      1 ☐ 4 6
   −   8 6 ☐
        4 ☐ 9

16 Geraint takes four tests.
   His total mark is used to work out his grade.
   Geraint's marks on the first two tests are 129 and 137.
   He can't remember his other two marks, but he knows they were both the same.
   Geraint was awarded a grade B overall.

Total	Grade
From 500 to 600	A
From 400 to 499	B
From 300 to 399	C
From 200 to 299	D
From 100 to 199	F
99 or less	U

   a Find the lowest possible mark Geraint could have in each of the 3rd and 4th tests.
   b Find the highest possible mark Geraint could have in each of the 3rd and 4th tests.

17 When Abdul's house number is multiplied by 19, the answer is a three-digit number beginning with 4 and ending with 7.
   Find Abdul's house number.

5.2 Multiplying decimals

Skill checker

① a 1.4 × 2 b 2.3 × 3 c 4.2 × 2
② a 4.23 × 2 b 1.21 × 3 c 3.21 × 2
③ a 1.243 × 2 b 3 × 2.312 c 2 × 3.123

There are different methods you can choose from when multiplying decimals.

Worked example

Find the area of this rectangle.

2.3 m
1.45 m

Solution

Long multiplication

Ignore the decimal points until the final digits have been calculated.

Count the number of digits to the right of the decimal point to find the position of the decimal point in the final answer.

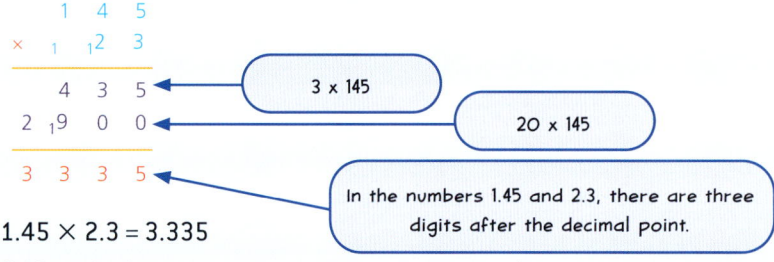

3 × 145
20 × 145
In the numbers 1.45 and 2.3, there are three digits after the decimal point.

1.45 × 2.3 = 3.335

Grid method

Break up both numbers into units, tenths, hundredths, and so on.
Form a grid with one number at the top, and the other at the side.
For each cell of the grid find the product of each number.
Finally, add the products.

×	1	0.4	0.05
2	2	0.8	0.1
0.3	0.3	0.12	0.015

2 + 0.3 + 0.8 + 0.12 + 0.1 + 0.015 = 3.335

The area is 3.335 m².

Alternatively, you could find the position of the decimal point by estimating the answer.

For example, the answer to 1.45 × 2.3 will be bigger than 1 × 2 = 2 and smaller than 2 × 3 = 6.

1.45 is between 1 and 2.
2.3 is between 2 and 3.

5 Calculations

5.2 Now try these

Band 1 questions

1. a 0.3×7 b 3×1.4 c 1.82×9 d 7×0.368
2. a 13×2.4 b 6.2×35 c 4.9×2.7 d 1.8×4.5
3. a 1.43×2.1 b 7.1×0.286 c 3.92×25.4
4. a 0.5×0.7 b 0.03×0.8 c 0.04×0.009
5. Work out the cost of each of these.
   a 6 rolls of wallpaper at £4.35 per roll
   b 4 toothbrushes at £1.59 each
   c 5 packets of toffees at 79p each
   d 9 rolls of parcel tape at £1.62 each
6. a 0.8^2 b $0.7 \times 0.3 \times 0.8$ c 23.67×0.94

Band 2 questions

7. Use the fact that $28 \times 59 = 1652$ to write down the correct answer to each of these.
   a 2.8×59 b 2.8×5.9 c 0.28×59
   d 0.28×0.59 e 0.028×5.9 f 0.028×0.59
   g 28×0.059 h 2.8×0.59 i 0.59×28

8. Without doing any multiplying, match each multiplication in column A with the one in column B that would give the same answer.

Column A	Column B
a 5.7×98	0.057×9.8
b 5.7×0.98	57×0.98
c 0.57×98	0.57×9.8
d 0.57×0.98	57×9.8

9. On one day the exchange rates are £1 = 1.31 US dollars and £1 = 1.19 Swiss francs.
   Convert these amounts into both US dollars and Swiss francs.
   a £40
   b £73
   c £148

10. a Copy and complete: $0.7^2 = 0.7 \times 0.7 = 0.49$ so $\sqrt{0.49} = \boxed{}$.
    b Find the square root of each of these numbers.
       i 0.64   ii 0.36   iii 1.21   iv 0.81
       v 0.04   vi 1.44   vii 0.0144   viii 0.0025

11 Find the total cost of each of these shopping bills. (Each price shown is for one item.)

a
Discount Shopper

1 × Cereal   @ £ 1.75
2 × Soap   @ £ 0.62
2 × Plant   @ £ 1.99
5 × Cat food   @ £ 0.32
1 × Biscuits   @ £ 0.99
3 × Pen   @ £ 0.17
2 × Milk   @ £ 0.60

Thank you for shopping with us

b
Pets U Like

8 × Cat food   @ £ 0.33
1 × Cat litter @ £ 4.99
2 × Dog chew   @ £ 0.25
2 × Dog bowl   @ £ 2.75
1 × Dog food   @ £ 0.37
3 × Nibbles   @ £ 1.16
2 × Collars   @ £ 1.99

You need it, we have it!

12 A double sheet of newspaper measures 72.5 cm × 60 cm.

a Rewrite these lengths in metres.

b Calculate the area of this double sheet in m^2.

c The newspaper contains six double sheets.
What is the area of paper needed to make one copy of this newspaper?

Band 3 questions

13 Iwan buys 8 lots of fish and chips. The total cost is £52.56.
What is the cost of 11 lots of fish and chips?

14 The corner shop sells cartons of orange juice in packs of 6 for £4.40 per pack.
The supermarket sells cartons of orange juice in packs of 8 for £5.95 per pack.
Rhiannon wants to buy 24 cartons of orange juice.

a Where should she buy them from if she wants to spend as little as possible?

b How much money would she save?

15 The full-sized bricks in this wall are 21.4 cm long and 10.3 cm high.
The mortar (brown line) between each brick is 0.9 cm thick.
There is mortar below the bottom row, but there is no mortar on top of the wall or on the sides.

a How long is the top row of the wall?   b How high is the wall?

Give your answers to the nearest centimetre.

16 Length can be measured in feet and inches.
1 inch is the same length as 2.54 cm.
1 foot is the same length as 12 inches.
Lloyd's height is 5 feet plus 3 inches.
How tall is Lloyd in centimetres?

17 A marathon is 26 miles plus 385 yards long.
1 mile is the same as 1.61 kilometres.
1 yard is the same as 0.91 metres.
How long is a marathon in kilometres?

5 Calculations

5.3 Dividing decimals

Skill checker

1. a $2.4 \div 2$ b $6.3 \div 3$ c $8.6 \div 2$
2. a $2.86 \div 2$ b $9.36 \div 3$ c $4.28 \div 2$
3. a $7.777 \div 7$ b $3.693 \div 3$ c $8.642 \div 2$

When dividing decimals, first write the division in the same form as a fraction and then find an equivalent fraction with whole numbers on the top and the bottom. If possible, simplify the fraction to make the final division as easy as possible.

When dividing, add zeros after the decimal point if necessary.

Worked example

$1.42 \div 0.08$

Solution

$1.42 \div 0.08 = \dfrac{1.42}{0.08} = \dfrac{142}{8} = \dfrac{71}{4} = 71 \div 4$

$$\begin{array}{r} 17.75 \\ 4\overline{)7\,^31.\,^30\,^20} \end{array}$$

$1.42 \div 0.08 = 17.75$

Worked example

Change $\dfrac{5}{8}$ into a decimal.

Solution

$$\begin{array}{r} 0.625 \\ 8\overline{)5.\,^50\,^20\,^40} \end{array}$$

No more zeros are needed because 8 divides into 40 exactly.

$\dfrac{5}{8} = 5 \div 8 = 0.625$

5.3 Now try these

Band 1 questions

1. a $1.6 \div 4$ b $3 \div 0.5$ c $0.8 \div 0.2$ d $3.6 \div 0.3$

Curriculum for Wales Mastering Mathematics: Book 2

Strategic competence

2
a	1.5 ÷ 0.5	b	36 ÷ 3.6	c	2.5 ÷ 0.5	d	1.6 ÷ 0.4
e	0.9 ÷ 0.9	f	3.2 ÷ 0.4	g	2.4 ÷ 1.2	h	3.6 ÷ 0.6
i	2.8 ÷ 0.4	j	48 ÷ 1.2	k	2.7 ÷ 0.3	l	1.3 ÷ 0.1
m	3.8 ÷ 0.2	n	3.6 ÷ 0.3	o	2.4 ÷ 0.1	p	6 ÷ 0.3

Match your answers to the letters to solve this riddle.

I	'	M		A		D		O		T		I		N		P		L		A		C		E	!
19	50	2	10	3	40	5	13	24	60	8	12	70	9	7	6	1	4	20

Fluency

3
a	6.3 ÷ 0.04	b	2.35 ÷ 0.02	c	83.4 ÷ 0.05.	d	0.39 ÷ 0.012

4
a	5.4 ÷ 0.003	b	0.61 ÷ 0.002	c	0.039 ÷ 0.0004	d	0.009 ÷ 0.00015

5
a	567.84 ÷ 0.008	b	235.8 ÷ 2.4	c	5.688 ÷ 3.6

6 The area of a rectangle is 5.04 cm².
Its width is 7.2 cm.
Find its length.

Band 2 questions

Logical reasoning

7 Elin is asked to calculate 17 ÷ 0.4.
Without doing the calculation, Elin says, 'I know the answer is bigger than 17'.
a How does she know?
b Calculate 17 ÷ 0.4.

Strategic competence

8 How many textbooks 1.2 cm thick will fit onto a shelf 132 cm long?

9 Is it possible to fit 26 magazines each 0.7 cm thick into a magazine rack 18 cm wide?
Show how you would work this out.

Logical reasoning

10 Instead of dividing a number by 0.13, Llewelyn divides by 13. His answer is 6.5.
What should his answer be?

11 What problem occurs when you divide 0.262 by 0.12?

12 a i   Divide 4 by 11, giving the first six decimal places.
       iii Add your answers to **i** and **ii** together.
       ii   Do the same for 7 divided by 11.
       iv   Comment on part **iii**.
    b i   Divide 2 by 11, giving the first six decimal places.
       ii   Divide 9 by 11, giving the first six decimal places.
       iii Add your answers to **i** and **ii** together.
    c Find another pair of divisions by 11 that give the same pattern.

Band 3 questions

13 Ffion has attempted to change $\frac{3}{8}$ into a decimal.

Here is her working.

```
     2 . 6  6  6
  3 ⟌ 8 . ²0 ²0 ²0
```

Explain the mistake she has made.

14 The answer to the division 9.84 ÷ 1.2 is 8.2.

Write a decimal division with an answer of 7.2.

15 Angharad's scales only give weights in pounds (lb).

1 kilogram is the same as 2.2 pounds.

1 stone is the same as 14 pounds.

Change 9 stones into kilograms.

Give your answer to one decimal place.

16 Megan correctly says that $\frac{11}{20}$ is 0.55.

How can this information be used to write down the decimal equivalent of $\frac{13}{20}$ without doing a division?

17 **a** **i** Write $\frac{1}{7}$ as a decimal, giving the first six decimal places.

  ii Predict the next six decimal places.

  b **i** Write $\frac{2}{7}, \frac{3}{7}, \frac{4}{7}$ and $\frac{5}{7}$ as decimals, giving the first six decimal places for each fraction.

  ii Compare the six decimal places for each of the five fractions you have found.

  c **i** Predict the first six decimal places of $\frac{6}{7}$.

  ii Divide 6 by 7 to check your prediction.

18 On one day the exchange rates are £1 = $1.30 (US dollars) and £1 = €1.15 (euros).

  a Convert these amounts into pounds (£).

  i $3.90 **ii** €23 **iii** $23 **iv** €37

  b Find the exchange rate between the US dollar ($) and the euro (€) in the form:

  i $1 = € ___

  ii €1 = $ ___ .

Key words

Here is a list of the key words you met in this chapter.

| Column method | Digit | Divisor | Grid method | Long division |
| Long multiplication | Partitioning | Product | Short division | Sum |

Use the glossary at the back of this book to check any you are unsure about.

Review exercise: calculations

Band 1 questions

1. a 347 + 79   b 34.8 + 9.37   c 959.76 + 2670.8
2. a 346 − 179   b 74.2 − 8.39   c 3452.4 − 1907.63
3. a 67 × 7   b 378 × 94   c 3.065 × 2.6
4. a 98 ÷ 7   b 1612 ÷ 13   c 77.7 ÷ 0.84
5. a 45.6 + 239 + 3.89   b 0.23^2 c 0.06 ÷ 0.0003
6. Find the length of this rectangle.

   Area = 7.42 m²   Width = 1.4 m

Band 2 questions

7. a Find the remainder when 654 is divided by 23.
    b Apples are sold at 23p each. How many apples can be bought with £6.54?
    c A pile of 654 bricks is to be moved using a wheelbarrow.
       The wheelbarrow can hold 23 bricks at a time.
       How many times must the wheelbarrow be used to move all 654 bricks?
    d £654 is shared equally between 23 people.
       How much, to the nearest penny, does each person receive?
       How much is left over?

8. Sara multiplies 324 by 26 but makes a mistake.
    Her answer is 2592.
    a Here is Sara's working. What mistake has she made?
    b What is the correct answer?

    ```
           3  2  4
       ×  1 ₂2  6
       1  9  4  4
       1  6 ₁4  8
       2  5  9  2
    ```

9. Siôn pays £249 for a hotel room.
    He decides to pay for an upgrade to a £320 room.
    How much extra does he pay?

10. In a cricket match, India have a score of 218 runs. South Africa are batting. They have 176 runs.
    How many more runs do they need to:
    a draw level
    b win?

11. Jiang has four pieces of wood, each 1 metre long.
    He cuts off the following lengths, one from each piece.
    What lengths remain?
    a 972 mm   b 83.4 cm   c 0.86 m   d 0.732 m

5 Calculations

12 The chart shows the distances (in miles) between some UK airports.

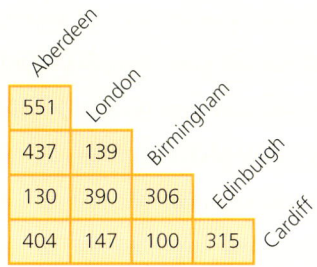

a How far would you travel if you flew from Aberdeen to Edinburgh, and then from Edinburgh to Cardiff?

b How far is it to travel from Aberdeen to Cardiff directly?

c What is the difference between the answers to part **a** and part **b**?

Band 3 questions

13 In these calculations each ☐ is a missing digit.

Can you find them? (One cannot be done.)

a 35 × 5☐ = 1☐☐5

b ☐3 × 59 = 767

c 1☐☐ × 23 = 3013

d 36 × 4☐ = 169☐

e 56 × 7☐ = 4☐32

f ☐9 × ☐7 = 1☐☐3

g 48 × ☐7 = 2☐☐6

h 1☐ × ☐9 = 4☐6

i 36 × 4☐ = 69☐

j ☐3 × ☐7 = 41☐☐

k ☐7 × ☐3 = 1591

14 In each of these calculations, the letters (A, B, etc.) stand for different single digits (but not zero). Work out what they can be. There are two possible sets of answers for parts **a** and **b** but only one for part **c**.

a
```
    A
  × A
  ───
   BA
```

b
```
    C
  × C
  ───
   D6
```

c
```
   2LM
  ×  M
  ────
  LMNL
```

15 Copy this subtraction.

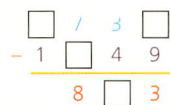

Find the missing digits.

16 You can make different products using the digits 1, 3, 5, 7 and 9 exactly once, for example 317 × 59 or 3 × 1597.

What is the largest product you can make using these digits?

17 In this number grid, two numbers are multiplied to give the number above.
The four numbers on the bottom row are 1, 2, 3 and 4 in some order.
Two numbers have been given.
Copy the grid and find the missing numbers.

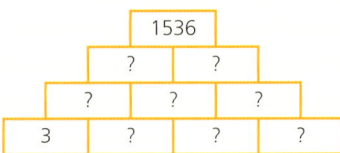

18 On one day the exchange rates are £1 = $1.45 (US dollars) and £1 = €1.20 (euros).

    a Convert these amounts into pounds (£).

        i $2.90   **ii** €24   **iii** $33   **iv** €42

    b Find the exchange rate between the US dollar ($) and the euro (€) in the form:

        i $1 = €__   **ii** €1 = $__ .

Consolidation 2: Chapters 3–5

Band 1 questions

1. Calculate the following.
   a 458 + 23.9 + 1037
   b 288 ÷ 3
   c 45.3 − 9.27
   d 56 × 78

2. Dylan is a plumber.

   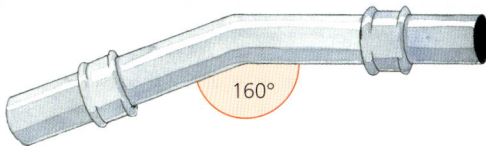

   He uses pipe connectors like this one.
   Why does he call it a 20° connector?

3.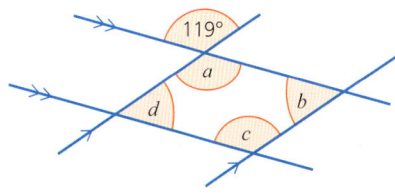

   a Find the size of each lettered angle and write down your reason.
   b What is $a + b + c + d$?

4. Calculate the size of angle x.

   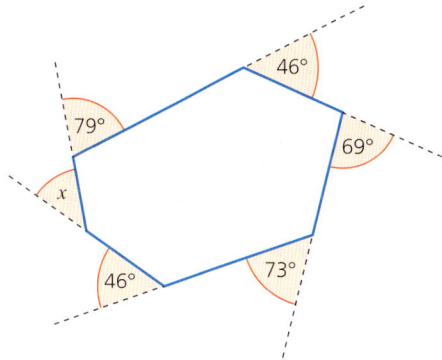

5. Calculate the size of the interior angle of a regular polygon with 120 sides.

6. Work out the following.
   a 9 × £2.78
   b £74.91 ÷ 11
   c €83.21 × 7

7. A map is drawn to a scale of 1 : 1000.
   Find the actual distances of these lengths on the map.
   a 5 cm
   b 8.2 cm
   c 13.4 cm
   d 7 mm

8. Use a ruler and protractor to construct an equilateral triangle with sides 5 cm.

> **Hint**
> The angles in an equilateral triangle are all 60°.

9 The bearing from a ship to a port is shown in the diagram.

a Measure the bearing of the port from the ship.

b What is the back bearing to travel from the port to the ship?

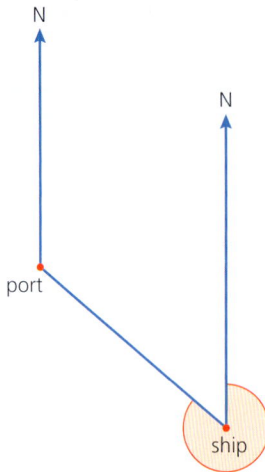

10 Nia and Monique are shopping.

a Nia buys three hair slides.
How much is this altogether?

b Monique buys two perfume sprays.
How much is this altogether?

11 A rectangular carpet measures 4.3 m × 5.7 m.

Calculate the area of the carpet in square metres.

12 This page shows Aled's answers to a Maths test.

```
Maths test
Name: Aled Davies
(1)  3.2 × 0.4 = 12.8
(2)  0.04 × 4.2 = 0.168
(3)  0.8 × 0.6 = 4.8
(4)  5.6 × 0.5 = 2.6
(5)  1.4 × 1.4 = 1.96
(6)  0.05 × 6.8 = 0.34
(7)  4.2 × 0.8 = 3.36
(8)  8.2 × 5 = 4.1
(9)  0.3 × 6.3 = 1.89
(10) 7.5 × 0.2 = 1.5
```

a Copy it out and mark it. What is Aled's mark out of 10?

b Correct any answers that Aled has got wrong.

13 Copy and complete this table:

Number	3.6	6	12.9	17.1	105
÷ 3		2			
÷ 0.3	12				
÷ 0.03					3500

Band 2 questions

14 The digits in these multiplications are correct, but there are no decimal points in the answers.

Copy the calculations and put the decimal points in the correct place.

For example, 7.6 × 5.31 = 40356 can be corrected to 7.6 × 5.31 = 40.356.

- **a** 6.1 × 0.9 = 549
- **b** 15.6 × 0.04 = 624
- **c** 0.79 × 0.03 = 237
- **d** 1.4 × 2.68 = 3752
- **e** 0.816 × 13.7 = 111792
- **f** 0.98 × 3.71 = 36358
- **g** 5.8 × 5 = 29
- **h** 7.92 × 6 = 4752

15 The diagram shows a gutter around a bay window on a house.

Calculate the size of angle x.

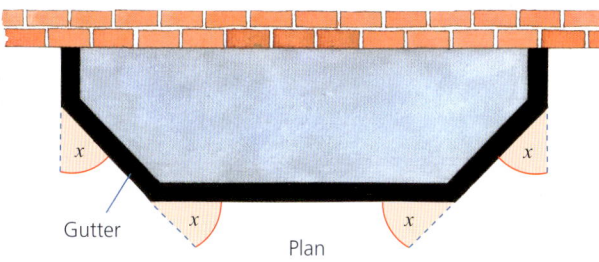

16 a Measure the bearings of these journeys, shown on the map.

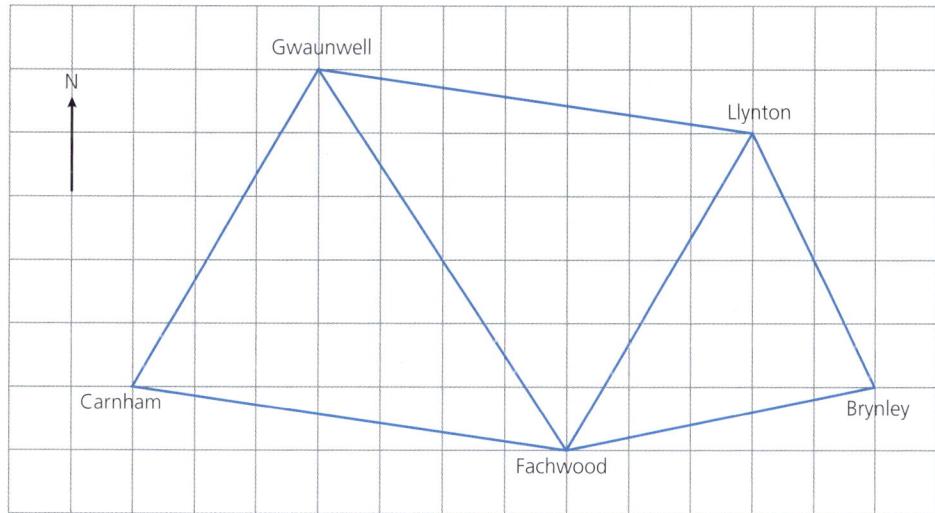

- **i** Fachwood to Llynton
- **ii** Llynton to Brynley
- **iii** Fachwood to Gwaunwell
- **iv** Brynley to Fachwood
- **v** Carnham to Fachwood
- **vi** Brynley to Carnham

b Calculate the back bearing of each journey in part **a**.

c Which journeys drawn on the map have the same bearings?

d Explain how you can tell from the gridlines on the map that they have the same bearings.

17. Five of the rectangular tiles pictured are used to make the pattern below.

    What is the area, in square centimetres, of this pattern?

    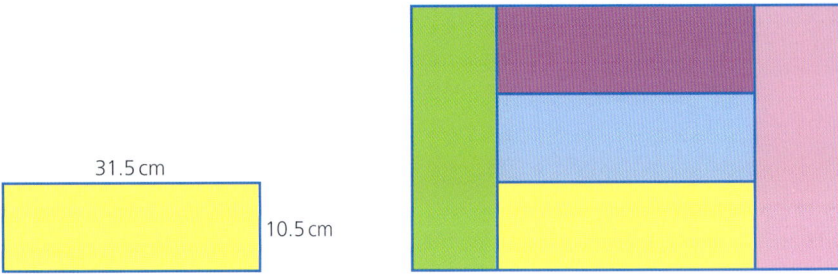

    31.5 cm

    10.5 cm

18. One day an exchange rate is £1 = $1.33 (US dollars).

    Convert these amounts into US dollars.

    a £30   b £78   c £268   d £67.85

19. Copy this subtraction.
    Find the missing digits.

20. a Work out the sum of the interior angles of an icosagon.

    b Work out the size of each of the interior angles of a regular icosagon.

    An icosagon is a 20-sided shape.

21. The bearing of Henporth from Vandiff is 086°.

    Find the bearing of Vandiff from Henporth.

Band 3 questions

22. In this number grid, two numbers are added to give the number above.

    Copy the grid and find the missing numbers.

    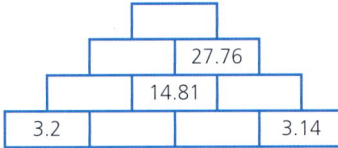

23. The interior angle of a regular polygon is 132° greater than its exterior angle.

    Find the sum of the interior angles of the polygon.

24. Work out the size of angle x.

    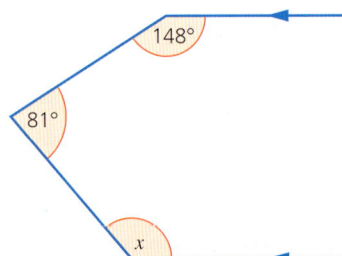

25. A yacht sails 7 km on a bearing of 280°.

    It then sails 7 km on a bearing of 190°.

    Find the bearing on which it must travel to return to its starting point.

26. ABCDEF is a regular hexagon. Work out the size of $A\hat{D}B$.

27. On one day two exchange rates are £1 = 1.40 US dollars and £1 = €1.10 (euros).

    Find the exchange rate between the US dollar ($) and the euro (€) in the form $1 = € __.

6 Negative numbers

Coming up...

- Understanding and using the number line
- Ordering positive and negative numbers, and understanding the inequality signs < and >
- Using the four operations on directed numbers
- Efficiently using a calculator when adding, subtracting, multiplying and dividing negative numbers

Snakes and ladders

Take turns rolling a dice and moving your counter the number of spaces shown on the dice.

If you land at the bottom of a ladder, then move your counter to the top.

If you land at the top of a snake, then move to the bottom.

The winner is the first to land on 99.

a   One of the ladders represents a move of +17. From which space does it start?

b   The longest ladder starts at 53. Describe the climb up the ladder as an addition.

c   How could you describe the movement down a snake as an addition?

d   Describe the movements (as additions) along each of the snakes and each of the ladders.

6.1 Negative numbers

Skill checker

1. Write these numbers in order of size, starting with the lowest.
   17, 36, 0, 25, 9, 42, 6

2. a   Draw a number line from 0 to 20.
   b   Which number is the same distance from 5 as it is from 17?

On a number line, negative numbers are to the left of zero and positive numbers are to the right.

Zero is neither positive nor negative.

Positive numbers don't usually have a sign written in front of them. So +3 is normally written as 3.

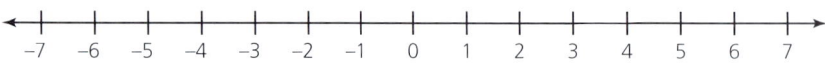

Worked example

Place these numbers in order, from lowest to highest.

$-10, -17, 27, 17, 6, -7, -27$

Solution

Consider the position of each number on the number line.

> Always think about a number line when working with negative numbers.

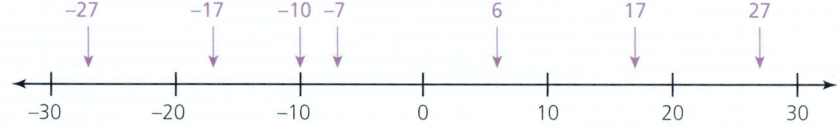

The order is $-27, -17, -10, -7, 6, 17, 27$.

▶ Inequalities

There are two symbols you can use to indicate the relative size of two numbers.

< means '**less than**' and > means '**greater than**'.

Worked example

Insert a suitable sign between these pairs of numbers.

a 2 and 5   b 7 and 3   c -2 and 8   d -5 and -9

Solution

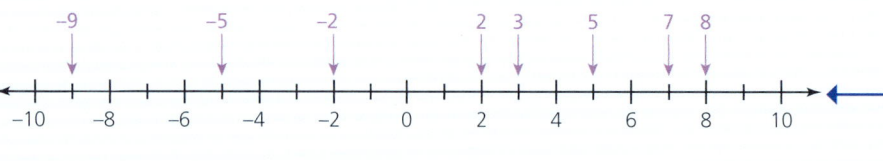

a $2 < 5$ b $7 > 3$ c $-2 < 8$ d $-5 > -9$

> Compare the positions of each number on the number line. The closed side of the inequality sign should point to the lesser of the two numbers. This is the number which appears further towards the left on the number line.

Maths in context

Archimedes and Euclid were both famous ancient Greek mathematicians.

It is reported that Archimedes was born in 287 BC. It is thought he died in 212 BC.

Euclid was said to have been born mid-fourth century BC. It is said that he died mid-third century BC.

Mark 0 BC on a number line.

Show the lives of Achimedes and Euclid on your number line, in a similar way to how you think about negative numbers.

How old was Archimedes when he died?

6 Negative numbers

6.1 Now try these

Band 1 questions

1. Draw a number line from −10 to 10.
   a Place the numbers 8, −4, 0, 5, −1, 4, −7 on the line.
   b Which two numbers are the same distance away from 0 on your line?

2. Temperatures are measured above and below zero.
   Write down each of these temperatures. Use + or −

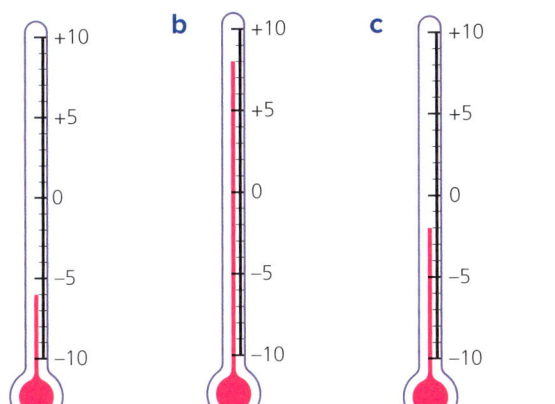

Cross-curricular activity

Degrees Celsius (°C) is a unit of temperature.

In your Science lessons you may have used Kelvin (K) as a unit of temperature.

1. What is 0 K written in degrees Celsius?
2. What is this temperature also known as?

3. For each pair of numbers, identify the lowest.
   a 4 and 9
   b −3 and 8
   c 0 and −2
   d −5 and −7

Band 2 questions

4. Each set of five numbers is written in order of size.
   Copy each list, replacing each ☐ with a possible number.
   a −8, −3, 1, ☐, 7, 18
   b −19, −15, −7, ☐, −2, 8
   c 23, 11, 4, ☐, −3, −12
   d 63, 17, 0, ☐, −18, −29

5. Write each set of numbers in order, starting with the lowest.
   a 6, −3, 0, 18, 90, 2
   b 4, −5, 27, 0, −19, 8
   c −2, 0, 11, −80, 57, −16
   d −18, 7, −6, −1, 0, −23

6. Insert < or > between each pair of numbers.
   a 6 and 3
   b 0 and 7
   c −8 and 2
   d −1 and −5

7. Which of these numbers is furthest from 0?

   −12   8   37   −63   −37   44   62

Band 3 questions

8 Heledd is thinking of two positive numbers.

She says that, when she puts a negative sign in front of each of them, they stay in the same order on the number line.

Is Heledd correct?

Use a picture of a number line to explain how you know.

9 Two numbers are placed on a number line.

They are both the same distance from 0, and are 60 apart from each other.

What are the two numbers?

10 A, B, C, D are four different integers.

Use the clues to decide what each number must be.

- Both D and A are the same distance from 0.
- B is 2 more than D.
- C is 4 away from 0.
- Two numbers are above 0.
- The number that is furthest from 0 is 8.

6.2 Adding and subtracting negative numbers

Skill checker

① Copy and complete these addition grids.

a
+	1	3	4	7
2				
5				
6				
9				

b
+		1		8
1		2		
3				
6				
			12	15

Activity

Chalk a number line on the floor, with numbers increasing from the left to the right.

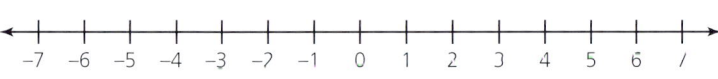

Stand at the starting number.

- The operation 'addition' means you should *get ready* to move by facing increasing numbers.
- The operation 'subtraction' means you should *get ready* to move by facing decreasing numbers.
- A 'positive number' means the number of steps to move forward.
- A 'negative number' means the number of steps to move backwards (it doesn't matter which way you face … you move backwards).

What is $5 + -3$ asking you to do? Try it!

Where do you move to? You should be at 2.

5	+	−3
Stand at 5	Get ready to move facing increasing numbers	Take 3 steps backwards

So $5 + -3$ is equivalent to calculating $5 - 3 = 2$.

What is $-3 - -4$ asking you to do? Try it!

Where do you move to? You should be at 1.

−3	−	−4
Stand at −3	Get ready to move facing decreasing numbers	Take 4 steps backwards

So $-3 - -4$ is equivalent to calculating $-3 + 4 = 1$.

Try some other moves and write down equivalent calculations.

When you add a negative number, you move to the left on the number line.
Adding a negative number is the same as subtracting a positive number.

Worked example

$5 + (-3) = ?$

Solution

$5 + (-3) = 5 - 3$
$= 2$

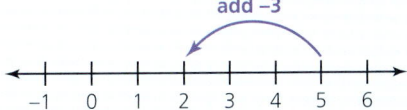

Try this by moving on the number line you used in the activity on page 92.

When you subtract a negative number, you move to the right on the number line. ← *This is what you found by doing the activity on page 92.*
Subtracting a negative number is the same as adding a positive number.

Worked example

$-4 - (-3) = ?$

Solution

$-4 - (-3) = -4 + 3$
$= -1$

Try this by moving on the number line you used in the activity on page 92.

When two signs are written next to each other, you can replace them with a single sign. ← *This is what you found by doing the activity on page 92.*

+ + ⟶ +
+ − ⟶ −
− + ⟶ −
− − ⟶ +

Activity

① **a** Use a copy of this grid to play a game of 'Boxes' with a partner.
Start with a score of 0.
Take it in turns to join any two adjacent dots with a vertical or horizontal line.
If your line completes a square, then add the number in the square to your score.
The winner is the person with the highest score when all the squares have been completed.

b If one person completed all of the boxes in the grid, what would their score be?

+3	0	−1	+1	+2	−2
0	−3	−1	+1	0	−2
−1	−2	+1	−2	+1	−3
+3	0	−3	+5	−1	−2
+1	0	+1	−2	+1	+1
+4	−7	+2	−1	0	+1

Worked example

Work these out without using a calculator, then check your answers on a calculator.

a $3 - (-8) = ?$
b $-7 + (-2) = ?$
c $9 - (+3) = ?$
d $-5 + (+4) = ?$

On some calculators, the negative sign must be written using the (-) key.

Solution

a $3 - (-8) = 3 + 8 = 11$
b $-7 + (-2) = -7 - 2 = -9$
c $9 - (+3) = 9 - 3 = 6$
d $-5 + (+4) = -5 + 4 = -1$

Note

Think about how you moved on the number line for the activity earlier. Try out the moves for these examples!

6.2 Now try these

Band 1 questions

1 Work out these calculations.
   a $3 + 6$ b $7 + 2$ c $4 + 9$ d $1 + 8$
   e $8 - 2$ f $7 - 3$ g $9 - 1$ h $6 - 6$

2 Work out these calculations.
   a $3 - 4$ b $2 - 5$ c $1 - 6$ d $7 - 9$
   e $4 - 10$ f $5 - 8$ g $11 - 13$ h $15 - 23$

3 Work out these calculations.
   a $5 + (-2)$ b $7 + (-5)$ c $9 + (-1)$ d $4 + (-4)$
   e $1 + (-5)$ f $3 + (-4)$ g $8 + (-10)$ h $7 + (-12)$

4 Work out these calculations.
   a $7 - (-1)$ b $4 - (-3)$ c $8 - (-2)$ d $9 - (-7)$

5 Copy and complete these number patterns.

   a $5 - 3 = 2$ b $2 - 2 = 0$ c $1 - 4 = \square$ d $-3 + 0 = -3$
      $5 - 2 = 3$ $2 - 3 = \square$ $1 - 3 = \square$ $-3 + 1 = \square$
      $5 - 1 = \square$ $2 - 4 = \square$ $1 - 2 = \square$ $-3 + 2 = \square$
      $5 - 0 = 5$ $2 - 5 = \square$ $1 - 1 = \square$ $-3 + 3 = \square$
      $5 - (-1) = \square$ $2 - 6 = -4$ $1 - 0 = \square$ $-3 + 4 = \square$
      $5 - (-2) = \square$ $2 - 7 = \square$ $1 - (-1) = \square$ $-3 + 5 = \square$
      $5 - (-3) = 8$ $2 - 8 = -6$ $1 - (-2) = \square$ $-3 + 6 = \square$

6 Work out these calculations.
   a $-3 + 7$ b $-1 + 4$ c $-2 + 9$ d $-5 + 6$
   e $-5 + 2$ f $-9 + 3$ g $-7 + 6$ h $-9 + 9$

7 Work out these calculations.
   a $-6 - 3$ b $-7 - 9$ c $-8 - 2$ d $-5 - 5$
   e $-4 + (-2)$ f $-8 + (-1)$ g $-4 + (-7)$ h $-2 + (-6)$

8 Work out these calculations.
   a $-8 - (-4)$ b $-5 - (-2)$ c $-9 - (-3)$ d $-17 - (-11)$
   e $-4 - (-9)$ f $-5 - (-8)$ g $-2 - (-7)$ h $-12 - (-19)$

6 Negative numbers

9 What number is:
- **a** 5 more than 6
- **b** 2 more than 2
- **c** 4 less than 7
- **d** 8 less than 8
- **e** 3 less than 2
- **f** 4 less than 1
- **g** 4 more than -2
- **h** 9 more than -1
- **i** 6 less than -3
- **j** 3 less than -8
- **k** 7 less than -6
- **l** 4 more than -7?

10 Work out these calculations.
- **a** $(-1) + 7$
- **b** $(-3) + 6$
- **c** $3 - 4$
- **d** $5 - 8$
- **e** $(-5) + 7$
- **f** $5 + -7$
- **g** $(-1) + -2$
- **h** $(-4) - (-3)$
- **i** $(-1) - (-1)$
- **j** $(-1) + (-1)$
- **k** $(-3) + 4$
- **l** $4 + (-2)$
- **m** $(-3) + 1$
- **n** $1 - (-3)$
- **o** $5 + -4$
- **p** $(-6) + 9$
- **q** $2 + (-3)$
- **r** $(-2) - 2$
- **s** $2 - (-3)$
- **t** $2 - (-4)$
- **u** $(-5) + 8$
- **v** $(-3) + 2$
- **w** $(-2) + (-1)$
- **x** $1 - (-1)$
- **y** $(-1) - 1$
- **z** $6 + (-5)$

Match your answers to the letters to complete the message below.

'There are three types of mathematician in the world: those ...'

A	C	D	E	F	H	I	L	N	O	P	S	T	U	W
2	-3	4	5	-6	3	-5	7	-2	-1	-7	-4	1	0	6

Band 2 questions

11 Which of these numbers is furthest from -4?

$-19 \qquad 6 \qquad 24 \qquad -55 \qquad -39 \qquad 41 \qquad 48$

12 Copy and complete this number wall.
The number in each brick is found by adding the numbers in the two bricks below it.

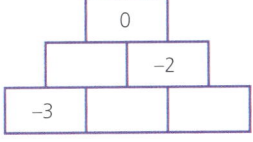

13 Calculate the difference between each pair of temperatures.

a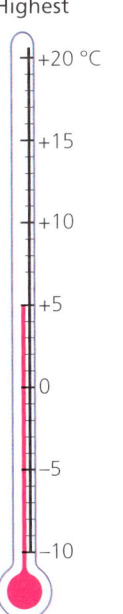

b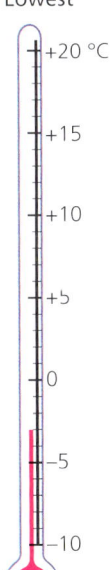

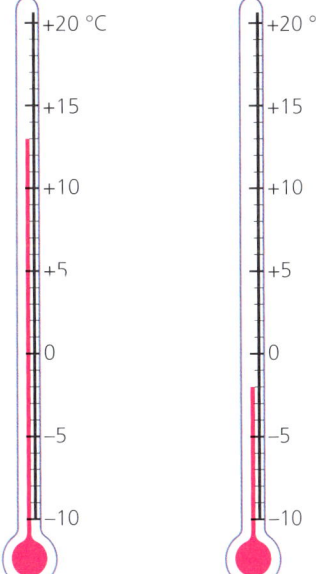

- **c** Highest $-1\,°C$, lowest $-4\,°C$
- **d** Highest $+20\,°C$, lowest $-5\,°C$
- **e** Highest $+2\,°C$, lowest $-6\,°C$
- **f** Highest $-3\,°C$, lowest $-10\,°C$

14 Each of these problems can be worked out using an addition sum.

For each problem, write down the addition and then solve the problem.

Example: The temperature is $-5\,°C$. What will it be after a rise of $8\,°C$?

$(-5) + (+8) = 3$. The temperature will be $3\,°C$.

a I have €25 in my savings bank. I pay a bill for €34. How much money do I have?

b A lift starts in the basement (floor -1) and goes up four floors. What floor is it now on?

c A seagull dives 20 m from a height of 15 m above the water.

How far below the surface does it dive?

d The temperature inside my shed is $10\,°C$ more than outside.

The outside temperature is $-4\,°C$.

What is the inside temperature?

15 Mr Jones has £330 in his bank account.

He receives a cheque for £80 from an insurance company.

He has an electricity bill of £75 to pay.

Mr Jones records his finances on this account sheet.

Description	Credits (+)	Debits (−)	Balance
Starting balance			330
Insurance	80		410
Electricity		75	?

a Copy the account sheet and fill in the missing balance.

b Extend your copy of the account sheet. Write these items in the correct column.

- Payment for suit £145
- Payment to electrician £140
- Weekly wage £650
- Payment for car loan £230
- Payment for shopping £53
- Sale of motorbike £550

Work out the balance each time.

Band 3 questions

16 Two numbers are placed on a number line.

They are both the same distance from -5, and are 42 apart from each other.

What are the two numbers?

17 Use your calculator to work these out.
- a 14.4 − 29.7
- b (−8) − 11.2
- c 117 − (−351)
- d (−67.8) − 32.2
- e 42.8 − (−22.9)
- f 30 + (14.8 − 15.2)
- g (−14.1) − (−26.7)
- h (−6) − 62.7

18 Find the next term in each sequence.
Describe the sequence.
- a 5, 2, −1, −4, −7, …
- b 2, 0, −2, −4, −6, …
- c −20, −19, −17, −14, −10, …
- d −32, −16, −8, −4, −2, …
- e −26, −24, −20, −14, −6, …
- f −13, −12, −10, −7, −3, …

19 Look at this puzzle.
The total for each line of three is 2.

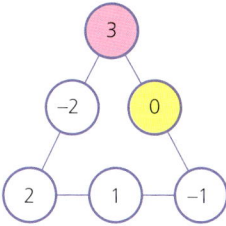

Copy these puzzles and choose numbers to complete them.

a The total for each row of three is 2.

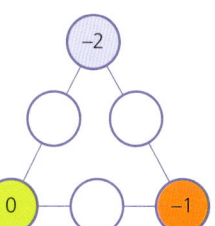

b The total for each row of three is −1.

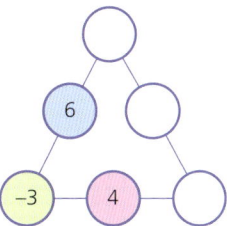

c The total for each row of three is −3.

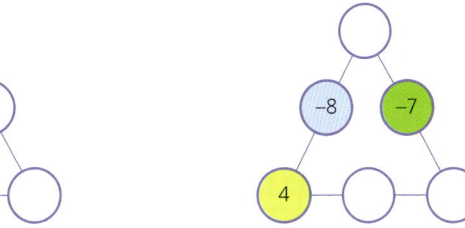

20 Copy and complete these addition grids.

a
+	7	3	−4	10
−1				
4				
−8				
−12				

b
+	5	−6		−4
	−3			
9			6	
		−13		
				−2

21 In a magic square, you get the same answer when you add up the numbers in any row, column or diagonal.
Copy and complete this magic square.

7			4
		2	
	−3	−2	3
−5	6		−8

Strategic competence

㉒ Copy and complete this number wall.
The number in each brick is found by adding the numbers in the two bricks below it.

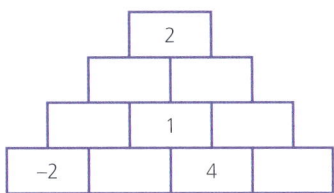

6.3 Multiplying and dividing negative numbers

Skill checker

① a Work out $(-3)+(-3)+(-3)+(-3)$.
  b Write the above calculation as a multiplication.

② a How many -4s must be added to make -20?
  b Write the above question as a division.

Conceptual understanding

Activity

Use a number line on the floor like you did in the activity on page 92.
Calculate 3×-2.
Think of 3×-2 as three lots of -2.
Stand at -2. How should you move?
Did you finish at -6?
Calculate -3×-2.
What does -3×-2 ask you to do?
Remember: Walking backwards facing decreasing numbers means you move in a positive direction.
Did you finish at 6?

Discussion activity

Why does a negative number divided by a negative number give a positive answer?
Why does a negative number divided by a positive number give a negative answer?

When you multiply or divide two numbers with the **same sign**, the answer is **positive**.

$7 \times 3 = 21$ $\qquad\qquad$ $12 \div 4 = 3$
$(-7) \times (-3) = 21$ $\qquad$ $(-12) \div (-4) = 3$

When you multiply or divide two numbers with **different signs**, the answer is **negative**.

$(-3) \times 2 = -6$ $\qquad\qquad$ $15 \div (-3) = -5$
$3 \times (-2) = -6$ $\qquad\qquad$ $(-15) \div 3 = -5$

6 Negative numbers

Worked example

a $4 \times (-2) = ?$
b $-4 \times (-3) = ?$
c $-10 \div 2 = ?$
d $-9 \div (-3) = ?$

Solution

Ignore the signs at first, and work out the calculation. Then decide on the sign of the answer.

a $4 \times 2 = 8$, so $4 \times (-2) = -8$ ← Different signs, so the answer is negative.

b $4 \times 3 = 12$, so $-4 \times (-3) = 12$ ← Same signs, so the answer is positive.

c $10 \div 2 = 5$, so $-10 \div 2 = -5$ ← Different signs, so the answer is negative.

d $9 \div 3 = 3$, so $-9 \div (-3) = 3$ ← Same signs, so the answer is positive.

Note: Think about your number line activity.

Worked example

Marc multiplies a number by itself.
The answer is 36.

a Write down a number that Marc could have used.

Seren multiplies a different number by itself and she also gets the answer 36.

b If Marc used the number given as an answer in part **a**, what number could Seren have used?
c What is the square root of 36?

Solution

a 6
b −6
c 6 ← There are two solutions to $x^2 = 36$, that is, $x = 6$ and $x = -6$. But $\sqrt{36}$ is only 6 (not −6). The symbol $\sqrt{}$ means 'the positive square root of ...'

6.3 Now try these

Band 1 questions

1 Work out these calculations.
   a $2 \times (-3)$
   b $5 \times (-1)$
   c $7 \times (-4)$
   d $8 \times (-2)$
   e $5 \div (-1)$
   f $8 \div (-2)$
   g $9 \div (-3)$
   h $12 \div (-2)$

2 Work out these calculations.
   a $(-2) \times 3$
   b $(-3) \times 4$
   c $(-6) \times 5$
   d $(-9) \times 1$
   e $(-15) \div 3$
   f $(-18) \div 2$
   g $(-49) \div 7$
   h $(-24) \div 3$

3 Work out these calculations.
   a $(-2) \times (-3)$
   b $(-1) \times (-8)$
   c $(-5) \times (-7)$
   d $(-9) \times (-9)$
   e $(-16) \div (-4)$
   f $(-14) \div (-2)$
   g $(-45) \div (-9)$
   h $(-24) \div (-2)$

④ Copy and complete these patterns.

a $5 \times 3 = \square$
   $5 \times 2 = \square$
   $5 \times 1 = \square$
   $5 \times \square = 0$
   $5 \times (-1) = \square$
   $5 \times \square = -10$
   $5 \times (-3) = \square$

b $4 \times 3 = 12$
   $4 \times 2 = \square$
   $4 \times \square = 4$
   $4 \times \square = 0$
   $4 \times (-1) = \square$
   $4 \times (-2) = -8$
   $4 \times \square = -12$

c $(-6) \times 3 = \square$
   $(-6) \times \square = -12$
   $(-6) \times 1 = \square$
   $(-6) \times 0 = \square$
   $(-6) \times \square = 6$
   $(-6) \times (-2) = \square$
   $(-6) \times \square = 18$

⑤ Copy and complete this multiplication grid.

×	-5	-2	0	1	3
-8					-24
-4		8			
-1				-1	
2					
7					

⑥ Work out these calculations.

a $(+2) \times (-5)$
b $(+3) \times (-7)$
c $(-6) \times 5$
d $(-9) \times 4$
e $(-5) \times (-1)$
f $(-8) \times (-9)$
g $0 \times (-7)$
h 6×0
i $(-4) \times 3$
j $5 \times (-6)$
k $(-7) \times (-7)$
l $(-1)^2$

⑦ Work out these calculations.

a $(-8) \div 2$
b $(-10) \div 5$
c $(-6) \div (+3)$
d $8 \div (-4)$
e $16 \div (-8)$
f $(-24) \div (-3)$
g $(-30) \div (-6)$
h $(-42) \div (-6)$
i $42 \div (-3)$
j $48 \div 3$
k $(-26) \div (-13)$
l $(+20) \div (+5)$
m $36 \div 3$

Band 2 questions

⑧ Work out $(-3) + (-3) + (-3) + (-3) + (-3)$.
Why does this give the same answer as $(-3) \times 5$?

 Work out these calculations.

a $\left(-\frac{2}{3}\right) + \frac{4}{7}$
b $\left(-\frac{2}{3}\right) - \frac{4}{7}$
c $\left(-\frac{2}{3}\right) \times \frac{4}{7}$
d $\left(-\frac{2}{3}\right) \div -\frac{4}{7}$

⑩ Work out these calculations.

a 7×-34.6
b $34.7 + (-5.98)$
c $-8.37 \div -0.3$
d 4.6×-2.3
e $-29.2 - (-290.61)$
f $-10.68 \div -1.2$

⑪ Use a calculator to work out the following.

a $-234.8 \times -\frac{4}{7}$
b $-0.897 - \frac{4}{11}$
c $-12.45 \div \frac{7}{9}$

Write your answers correct to one decimal place.

6 Negative numbers

12 Work out these calculations.
  a $(-1) \times (-2) \times (-3)$
  b $(-3) \times 4 \times (-2)$
  c $(-3) \times (-3) \times (-3)$
  d $(-2)^3$
  e $(-10) \times (-10) \times (-10) \times (-10)$
  f $(-1)^4$

13 Work out these calculations.
  a $(-1)^2$
  b $(-2)^2$
  c $(-3)^2$
  d What can you say about the square of any negative number?

14 Work out these calculations.
  a $(-1)^3$
  b $(-2)^3$
  c $(-3)^3$
  d What can you say about the cube of any negative number?

15 Copy and complete these calculations.
  a $(-3) \times \square = -12$
  b $5 \times \square = -40$
  c $(-20) \div \square = 2$
  d $36 \div \square = -4$
  e $\square \div 7 = -4$
  f $\square \times -9 = -27$

16 a Write down a number which, when multiplied by itself, gives an answer of 25.
  b Write down another number which, when multiplied by itself, gives an answer of 25.

17 Sam multiplies two integers together.
  He gets the answer -15.
  a Write down a pair of integers he could use.
  b How many different pairs of integers can be multiplied to give the answer -15?

18 Bethan answers these three questions.
  a $7^2 = ?$ b $(-7)^2 = ?$ c $\sqrt{49} = ?$
  Her answers are shown here in blue.
  Copy Bethan's work and mark it.
  Write down the correct answer next to any of Bethan's wrong answers.

  $7^2 = 49$
  $(-7)^2 = 49$
  $\sqrt{49} = \pm 7$

Band 3 questions

19 Copy and complete this multiplying magic square.
  The numbers in each row, column or diagonal have the same product.

−2		
		−4
12	−1	−18

20 Jemma notices that $(-1)^3$ and $(-1)^5$ both give the same answer of -1.

Are there are any other powers of -1 that would give an answer of -1?

Explain your answer.

21 Copy and complete this number wall.

The number in each brick is found by multiplying the numbers in the two bricks below it.

22 Find the missing number in this calculation.

$2 \times -3 \times \square \times 5 \times -1 = -210$

23 Copy and complete this multiplication grid.

Each white cell is the result of multiplying the purple column and row headers.

×		−3	−1	
	70			
−7	49			
			−5	
−4				−20
	−21		3	

24 The numbers in each pair of blue circles are multiplied to find the numbers in the circles between them.

Copy and complete each diagram.

a 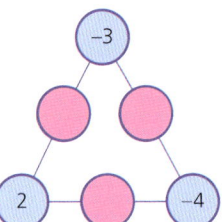 b 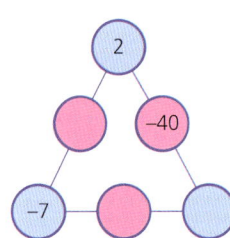 c 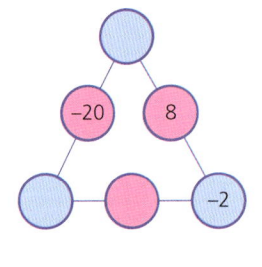 d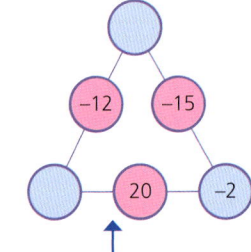

There are two different solutions to part **d**. Can you find both of them?

Key words

Here is a list of the key words you met in this chapter.

Directed   Greater than   Inequality   Less than
Negative   Number line   Positive   Sign

Use the glossary at the back of this book to check any you are unsure about.

6 Negative numbers

Review exercise: negative numbers

Band 1 questions

1 Draw a number line from -10 to 10.
  a Place the numbers $7, -3, 0, 6, -5, -6, -9$ on the line.
  b Which two numbers are the same distance from 0 on your line?

2 What number is:
  a 3 more than 8
  b 5 more than 4
  c 3 less than 9
  d 4 less than 4
  e 2 less than 1
  f 6 less than 0
  g 7 more than -3
  h 8 more than -4
  i 7 less than -2
  j 4 less than -10
  k 6 less than -6
  l 2 more than -8?

3 Work out these calculations.
  a $(-4) \times 5$
  b $4 \times (-5)$
  c $(-4) \times (-5)$

4 For each pair of numbers, identify the highest.
  a 7 and 3
  b -6 and 2
  c -8 and 0
  d -4 and -6

5 Work out these calculations.
  a $(-2) + 7$
  b $(-3) + 8$
  c $(-6) - 8$
  d $(-9) - 7$
  e $6 + (-8)$
  f $8 + 3$
  g $9 - (-1)$
  h $16 - (-3)$
  i $5 - 11$
  j $(-7) + (-8)$
  k $-7 - (-9)$
  l $9 + (-8)$
  m $17 - 13$
  n $(-11) + 5$
  o $-11 - (-4)$

6 Work out these calculations.
  a $18 \div (-3)$
  b $(-12) \div (-3)$
  c $(-10) \div 5$
  d $(-20) \times (-4)$
  e $(-6) \times 4$
  f $30 \times (-2)$

Band 2 questions

7 Which of these numbers is furthest from 0?

  $43 \quad -7 \quad 51 \quad 7 \quad -44 \quad -29 \quad 22$

8 Copy and complete this number wall.
  The number in each brick is found by adding the numbers in the two bricks below it.

9 Work out
  a $(-4) \times (-3) \times (-2)$
  b $(-1) \times 5 \times (-3)$
  c $(-4) \times (-4) \times (-4)$
  d $(-3)^3$
  e $(-10) \times (-10) \times (-10) \times (-10) \times (-10)$
  f $(-1)^6$

10 Insert a suitable sign between these pairs of calculations. ← Choose from =, ⟨, ⟩.
  a $2 \times (-4)$ and $(-3) \times (-2)$
  b $(-5) \times 0$ and $(-4) \times 3$
  c $3 \times (-2)$ and $(-12) \div (-2)$
  d $(-27) \div 9$ and $6 \div (-2)$
  e $(-4) \times (-3)$ and $(-7) \times 2$
  f $(-1)^3$ and $(-1)^4$

11 Calculate the difference between each pair of temperatures.
  a $8\,°C$ and $5\,°C$
  b $-3\,°C$ and $7\,°C$
  c $-5\,°C$ and $-9\,°C$
  d $-11\,°C$ and $13\,°C$
  e $17\,°C$ and $-19\,°C$
  f $-27\,°C$ and $-6\,°C$

12 Work out these calculations.

a 8^2 　　　　b $(-8)^2$ 　　　　c $\sqrt{64}$

13 Which of these numbers is furthest from -9?

| 37 | -14 | 23 | 53 | -38 | 11 | -70 |

Band 3 questions

14 Two numbers are placed on a number line.

They are both the same distance from 18, and are 64 apart from each other.

What are the two numbers?

15 Find the next term in each sequence.

Describe the sequence.

a $8, 4, 0, -4, -8, \ldots$

b $9, 4, -1, -6, -11, \ldots$

c $-23, -21, -18, -14, -9, \ldots$

d $-1, -2, -4, -8, -16, \ldots$

e $1, -3, 9, -27, 81, \ldots$

16 How many different pairs of integers can be multiplied to give the answer -16?

17 The total for each row of three circles is -3.

Copy and complete each diagram.

a 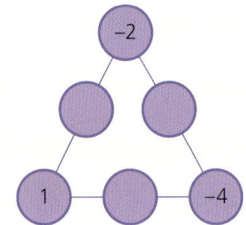　　　b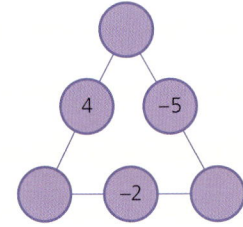

18 Copy and complete this number wall.

The number in each brick is found by multiplying the numbers in the two bricks below it.

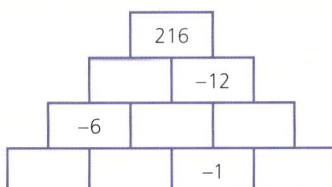

19 Copy and complete this addition grid.

Each white cell is the result of adding the column and row headers in dark blue.

+			-9	3	
			-5		
-5	-13				
				-4	
		4			2
-6		-4			

7 Fractions

Coming up...
- Adding, subtracting, multiplying and dividing fractions
- Interpreting fractions as operators
- Using a calculator to calculate results accurately

Fraction matching

Copy and cut out these fractions on paper or cardboard.

$\frac{1}{2}$	$\frac{1}{3}$	$\frac{2}{3}$	$\frac{1}{4}$	$\frac{3}{4}$	$\frac{1}{5}$	$\frac{2}{5}$	$\frac{3}{5}$	$\frac{4}{5}$	$\frac{1}{6}$	$\frac{5}{6}$	$\frac{1}{7}$	$\frac{2}{7}$	$\frac{3}{7}$	$\frac{4}{7}$	$\frac{5}{7}$	$\frac{6}{7}$
$\frac{2}{4}$	$\frac{2}{6}$	$\frac{4}{6}$	$\frac{2}{8}$	$\frac{6}{8}$	$\frac{2}{10}$	$\frac{4}{10}$	$\frac{6}{10}$	$\frac{8}{10}$	$\frac{2}{12}$	$\frac{10}{12}$	$\frac{2}{14}$	$\frac{4}{14}$	$\frac{6}{14}$	$\frac{8}{14}$	$\frac{10}{14}$	$\frac{12}{14}$
$\frac{3}{6}$	$\frac{3}{9}$	$\frac{6}{9}$	$\frac{3}{12}$	$\frac{9}{12}$	$\frac{3}{15}$	$\frac{6}{15}$	$\frac{9}{15}$	$\frac{12}{15}$	$\frac{3}{18}$	$\frac{15}{18}$	$\frac{3}{21}$	$\frac{6}{21}$	$\frac{9}{21}$	$\frac{12}{21}$	$\frac{15}{21}$	$\frac{18}{21}$
$\frac{4}{8}$	$\frac{4}{12}$	$\frac{8}{12}$	$\frac{4}{16}$	$\frac{12}{16}$	$\frac{4}{20}$	$\frac{8}{20}$	$\frac{12}{20}$	$\frac{16}{20}$	$\frac{4}{24}$	$\frac{20}{24}$	$\frac{4}{28}$	$\frac{8}{28}$	$\frac{12}{28}$	$\frac{16}{28}$	$\frac{20}{28}$	$\frac{24}{28}$

Shuffle the fractions and then lay them out face down.

Take turns to turn over any two fractions.

If the two fractions are equivalent, then the player turning them over removes them from the game and keeps them.

If the two fractions are not equivalent, then turn them face down again.

When all the fractions have been removed from the game, the winner is the one with the most.

7.1 Fractions review

Skill checker

① a The number on the top of a fraction is the n _____ .
   b The number on the b _____ of a fraction is the denominator.
② a Find three multiples of 4.
   b Find three multiples of 5.
   c Find three numbers which are multiples of both 4 and 5.
   d Find the lowest possible multiple of both 4 and 5.

▶ Equivalent fractions

Any fraction can be written in many ways. For example, $\frac{6}{8} = \frac{9}{12} = \frac{12}{16} = \frac{15}{20} = \ldots$ ← *These are all equivalent fractions.*

If a fraction cannot be written using smaller numbers then it is a simplified fraction. ← *A simplified fraction is said to be written in its lowest terms.*

The simplest way to write the above fractions is $\frac{3}{4}$.

Worked example

a Simplify the fraction $\frac{18}{30}$. ← *Type $\frac{18}{30}$ into your calculator. If you then press = your calculator will automatically write the fraction in its lowest terms.*

b Find the missing numerator: $\frac{2}{3} = \frac{?}{15}$.

Solution

a $\frac{18}{30} = \frac{9}{15} = \frac{3}{5}$ ← *18 and 30 are both multiples of 2, so divide them both by 2. 9 and 15 are both multiples of 3, so divide them both by 3. Alternatively, divide both 18 and 30 by 6.*

b $\frac{2}{3} = \frac{2 \times 5}{15} = \frac{10}{15}$ ← *3 has been multiplied by 5 so do the same to the numerator.*

▶ Adding and subtracting fractions

Fractions can only be added or subtracted if they have the same denominators.

If they are different then find equivalent fractions with the same denominators.

Worked example

a $\frac{2}{5} + \frac{1}{4}$

b $\frac{7}{8} - \frac{1}{4}$

Solution

a $\frac{2}{5} + \frac{1}{4} = \frac{8}{20} + \frac{5}{20}$ ← *20 is a common denominator because $4 \times 5 = 20$.*

$= \frac{13}{20}$

b $\frac{7}{8} - \frac{1}{4} = \frac{7}{8} - \frac{2}{8}$ ← *32 could be used as a common denominator, but 8 is smaller.*

$= \frac{5}{8}$

7 Fractions

▶ Finding a fraction of an amount

You can find a fraction of an amount by dividing by the denominator and multiplying by the numerator.

Worked example

Find $\frac{2}{3}$ of £42.

Solution

$\frac{2}{3}$ of £42 = 42 ÷ 3 × 2
$\phantom{\frac{2}{3} \text{ of £42}}$ = 14 × 2
$\phantom{\frac{2}{3} \text{ of £42}}$ = £28

▶ Reciprocals

The reciprocal is what to multiply a number by to make 1.

To find the reciprocal of a fraction you turn the fraction upside down.

Worked example

Write down the reciprocals of these numbers.

a 7   b $\frac{1}{4}$ c $\frac{3}{8}$ d −6   e 1

Solution

a $\frac{1}{7}$ b 4   c $\frac{8}{3}$ d $-\frac{1}{6}$ e 1

▶ Dividing an integer by a fraction

Change a division question into a multiplication by replacing the second fraction with its reciprocal.

Worked example

Calculate $40 \div \frac{2}{5}$.

Solution

$40 \div \frac{2}{5} = 40 \times \frac{5}{2}$ *$\frac{5}{2}$ is the reciprocal of $\frac{2}{5}$.*
$\phantom{40 \div \frac{2}{5}} = 40 \div 2 \times 5$
$\phantom{40 \div \frac{2}{5}} = 20 \times 5$
$\phantom{40 \div \frac{2}{5}} = 100$

▶ Decimal equivalents

You can find equivalent fractions and decimals using tenths, hundredths, thousandths, and so on.

Worked example

a Write 0.48 as a fraction.

b Write $\frac{127}{200}$ as a decimal.

c Write $\frac{3}{8}$ as a decimal.

107

Solution

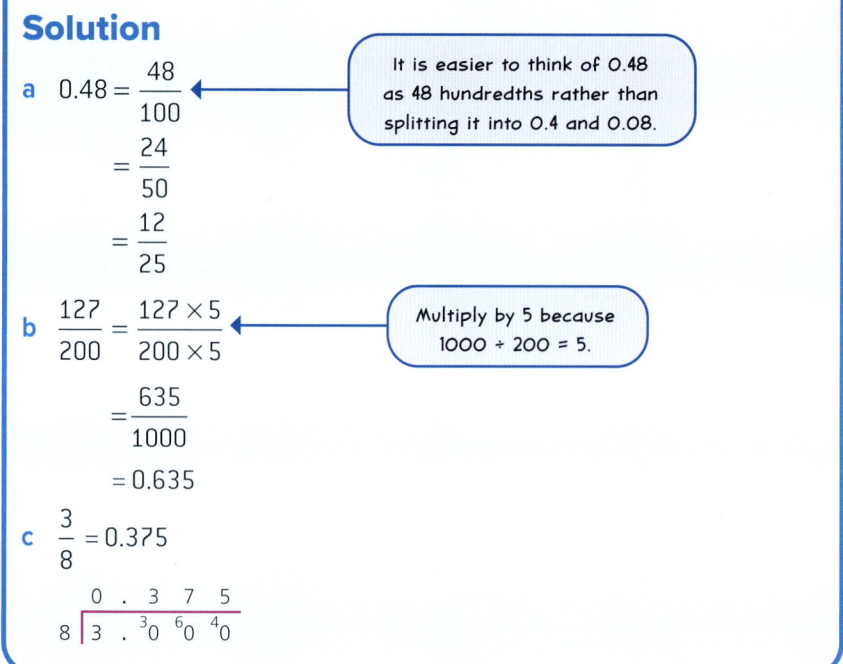

a $0.48 = \dfrac{48}{100}$

$= \dfrac{24}{50}$

$= \dfrac{12}{25}$

It is easier to think of 0.48 as 48 hundredths rather than splitting it into 0.4 and 0.08.

b $\dfrac{127}{200} = \dfrac{127 \times 5}{200 \times 5}$

Multiply by 5 because $1000 \div 200 = 5$.

$= \dfrac{635}{1000}$

$= 0.635$

c $\dfrac{3}{8} = 0.375$

$\;0\,.\,3\,\;7\,\;5$
$8\,|\,\overline{3\,.\,{}^3 0\;{}^6 0\;{}^4 0}$

▶ Percentage equivalents

You can find equivalent fractions and percentages using hundredths.

Worked example

a Write 83.1% as a fraction.

b Write $\dfrac{3}{5}$ as a percentage.

Solution

a $83.1\% = \dfrac{83.1}{100}$

$= \dfrac{831}{1000}$

b $\dfrac{3}{5} = \dfrac{60}{100}$

Alternatively, write $\dfrac{3}{5}$ as a decimal (0.6) and multiply by 100.

$= 60\%$

▶ Ratios

You can use fractions to solve ratio problems.

Worked example

Lewis and Saul share 55 sweets in the ratio 3 : 8.

a What fraction of the total is the smallest share?

b How many sweets are in the largest share?

Solution

a Smallest share is $\dfrac{3}{3+8} = \dfrac{3}{11}$.

b $\dfrac{8}{11} \times 55 = 55 \div 11 \times 8$

$= 5 \times 8$

$= 40$ sweets

7.1 Now try these

Band 1 questions

1 Simplify these fractions. ← Check your answers by typing each one into your calculator and then pressing =.

   a $\dfrac{2}{4}$ **b** $\dfrac{9}{15}$ **c** $\dfrac{12}{16}$ **d** $\dfrac{12}{18}$

2 Find the missing numbers.

   a $\dfrac{1}{2} = \dfrac{3}{\square}$ **b** $\dfrac{3}{4} = \dfrac{\square}{20}$ **c** $\dfrac{\square}{3} = \dfrac{8}{24}$ **d** $\dfrac{\square}{\square} = \dfrac{10}{25}$

3 a Match the equivalent fractions.

$\dfrac{3}{8}$ $\dfrac{18}{21}$ $\dfrac{2}{7}$ $\dfrac{5}{8}$ $\dfrac{25}{55}$ $\dfrac{3}{5}$ $\dfrac{42}{54}$ $\dfrac{9}{24}$

$\dfrac{27}{45}$ $\dfrac{8}{32}$ $\dfrac{4}{14}$ $\dfrac{5}{11}$ $\dfrac{6}{7}$ $\dfrac{7}{9}$ $\dfrac{1}{4}$

   b Which is the odd one out?

4 Work out the following fractions.

   a $\dfrac{1}{2} + \dfrac{1}{3}$ **b** $\dfrac{2}{3} + \dfrac{1}{4}$ **c** $\dfrac{4}{5} - \dfrac{3}{10}$ **d** $\dfrac{4}{9} - \dfrac{5}{12}$

5 Work out the following.

   a $\dfrac{1}{4}$ of £24   **b** $\dfrac{1}{3}$ of 72 kg   **c** $\dfrac{2}{3}$ of 45 m   **d** $\dfrac{9}{10}$ of €90

6 Write down the reciprocals of these fractions.

   a 2 **b** $\dfrac{1}{8}$ **c** $\dfrac{9}{10}$ **d** -11

   e 1 **f** $-\dfrac{5}{4}$

7 Work out the following.

   a $12 \div \dfrac{1}{2}$ **b** $7 \div \dfrac{1}{3}$ **c** $16 \div \dfrac{2}{3}$ **d** $24 \div \dfrac{3}{4}$

Logical reasoning

8 a Match each fraction with a percentage and a decimal.
b Which fraction is the highest?
   How do you know?
c Which fraction is the lowest?
   How do you know?

$\frac{1}{2}$ $\frac{2}{5}$ 0.75 20%

0.5 25%

$\frac{3}{5}$ $\frac{1}{4}$ 0.6 40%

0.25 50%

$\frac{1}{5}$ $\frac{3}{4}$ 0.4 75%

0.2 60%

Fluency

9 Share £36 in the ratio:
a 5 : 7
b 11 : 1
c 5 : 4
d 13 : 5.

Band 2 questions

Strategic competence

10 Copy and complete this statement.
$\frac{1}{2}$ of 60 = $\frac{1}{4}$ of ☐.

11 Jac, Alys and Rhian share £48.
Jac gets $\frac{1}{4}$ and Alys gets $\frac{1}{6}$.
How much money is left for Rhian?

Logical reasoning

12 Ceredig correctly works out $\frac{2}{9} + \frac{1}{6}$.
His working is shown here.

$$\frac{2}{9} + \frac{1}{6} = \frac{12}{54} + \frac{9}{54}$$

$$= \frac{21}{54}$$

$$= \frac{7}{18}$$

Anna says she can do it using simpler fractions.
Show Anna's method.

Fluency

13 a Write each pair of fractions with the same denominator.
i $\frac{1}{4}$ and $\frac{3}{8}$
ii $\frac{4}{5}$ and $\frac{5}{6}$
iii $\frac{3}{10}$ and $\frac{4}{11}$
iv $\frac{5}{6}$ and $\frac{7}{9}$

b Write a suitable inequality sign between each pair of fractions.

Use ⟨ or ⟩.

14 Write these decimals as fractions in their lowest terms.

A fraction in its lowest terms is a fully simplified fraction.

a 0.8
b 0.41
c 0.237
d 0.568

7 Fractions

15 Write these percentages as simplified fractions.
  a 75%   b 90%   c 16%   d 23.6%

Band 3 questions

16 Here are Bryn's end-of-term test results.
  a Which is Bryn's best subject?
  b Which is his worst subject?

17 Dyfed is explaining how to work out 50% of £64.
  He says 'Divide 64 by half'.
  Is he correct?
  Explain your answer.

18 Find a pair of numbers, a and b, which complete this statement correctly.
  $\frac{1}{3}$ of $a = \frac{1}{4}$ of b.

19 Dafydd, Gareth and Harri share some money.
  Dafydd gets $\frac{1}{3}$ and Gareth gets $\frac{1}{4}$.
  What fraction is left for Harri?

20 Copy and complete this inequality with another fraction.
  $\frac{1}{2} < \frac{\square}{\square} < \frac{2}{3}$

21 $\frac{3}{4}$ of $a = \frac{2}{3}$ of b.
  Find the simplified ratio $a : b$.

7.2 Multiplying fractions

Fractions can be multiplied by multiplying the numerators and the denominators.

Skill checker

① Work out the following.
  a $\frac{1}{2}$ of 6   b $\frac{1}{3}$ of 6   c $\frac{2}{3}$ of 6   d $\frac{1}{4}$ of 12

② Copy this rectangle.
  Shade $\frac{1}{3}$ of the rectangle blue.
  Shade $\frac{1}{4}$ of the remainder of the rectangle red.
  What fraction has not been shaded?

Look for common multiples in the numerators and denominators that can be cancelled to make the calculation easier.

Worked example

a $\dfrac{4}{5} \times \dfrac{2}{3}$

b $\dfrac{5}{14} \times \dfrac{3}{20} \times \dfrac{7}{18}$

Solution

a $\dfrac{4}{5} \times \dfrac{2}{3} = \dfrac{4 \times 2}{5 \times 3}$ ← When using a calculator look out for the fraction key.

$= \dfrac{8}{15}$

b $\dfrac{5}{14} \times \dfrac{3}{20} \times \dfrac{7}{18} = \dfrac{\cancel{5}^1}{14} \times \dfrac{\cancel{3}^1}{\cancel{20}_4} \times \dfrac{\cancel{7}^1}{\cancel{18}_6}$ ← 5 and 20 are both multiples of 5. 3 and 18 are both multiples of 3. 7 and 14 are both multiples of 7.

$= \dfrac{1 \times 1 \times 1}{2 \times 4 \times 6}$

$= \dfrac{1}{48}$

Activity

This diagram shows that $\dfrac{5}{8}$ of $\dfrac{3}{5}$ is $\dfrac{3 \times 5}{5 \times 8}$. ← The diagram also demonstrates that $\dfrac{3}{5}$ of $\dfrac{5}{8}$ is $\dfrac{15}{40}$.

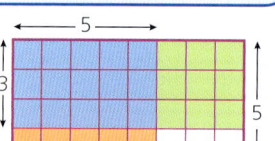

The word 'of' can often be replaced by a multiplication sign.

a Use the above to calculate the simplified answer to $\dfrac{5}{8} \times \dfrac{3}{5}$.

b Can you draw a similar diagram to illustrate the multiplication $\dfrac{3}{7} \times \dfrac{4}{5}$?

c Use your diagram to find the answer to $\dfrac{3}{7} \times \dfrac{4}{5}$.

7 Fractions

7.2 Now try these

Band 1 questions

1 Copy this diagram.

Shade $\frac{1}{3}$ of $\frac{4}{5}$ of the rectangle.

2 Work out these fractions.

a $\frac{1}{2} \times \frac{1}{3}$ b $\frac{1}{4} \times \frac{3}{5}$ c $\frac{2}{7} \times \frac{4}{5}$ d $\frac{3}{8} \times \frac{5}{7}$

3 Work out these fractions and write your answers in simplified form.

a $\frac{1}{4} \times \frac{2}{9}$ b $\frac{5}{6} \times \frac{7}{10}$ c $\frac{3}{8} \times \frac{4}{9}$ d $\frac{6}{7} \times \frac{7}{9}$

4 Cancel these fractions before multiplying.

a $\frac{2}{5} \times \frac{1}{4}$ b $\frac{3}{8} \times \frac{4}{9}$ c $\frac{3}{10} \times \frac{5}{12}$ d $\frac{7}{9} \times \frac{3}{14}$

e $\frac{5}{18} \times \frac{6}{25}$ f $\frac{8}{27} \times \frac{9}{32}$

5 Work out these fraction calculations.

a $\frac{1}{4}$ of $\frac{1}{5}$ of a tin of 120 sweets

b $\frac{2}{3}$ of $\frac{2}{5}$ of a lottery win of £3 million

c $\frac{2}{3}$ of $\frac{2}{7}$ of 42 tonnes of sand

d $\frac{5}{6}$ of $\frac{5}{8}$ of a 96-hectare field of wheat

6 Delroy has a market garden. It is 12 acres in size.

He grows potatoes on $\frac{2}{3}$ of it.

He grows peas on $\frac{3}{4}$ of the rest and asparagus on the remaining area.

a Draw a diagram showing this information.

b What is $\frac{3}{4} \times \frac{1}{3}$ of 12?

c What area does Delroy use for asparagus?

Band 2 questions

7 Lowri is working out $\frac{17}{25} \times \frac{19}{21}$ and $\frac{18}{25} \times \frac{20}{21}$.

She says she can do one of them easily but will use a calculator for the other.

Which one is easy to do?

Explain your answer.

Fluency

8 Work out these fractions.

a $\dfrac{2}{15} \times \dfrac{5}{12} \times \dfrac{8}{9}$

b $\dfrac{3}{8} \times \dfrac{5}{9} \times \dfrac{16}{25}$

c $\dfrac{7}{8} \times \dfrac{12}{21} \times \dfrac{16}{20}$

d $\dfrac{15}{33} \times \dfrac{14}{25} \times \dfrac{11}{21}$

e $\dfrac{12}{45} \times \dfrac{15}{81} \times \dfrac{27}{30}$

f $\dfrac{54}{33} \times \dfrac{49}{56} \times \dfrac{11}{63}$

9 Work out these fractions.

a $-\dfrac{3}{4} \times \dfrac{5}{6}$

b $\dfrac{2}{5} \times -\dfrac{9}{10}$

c $-\dfrac{4}{7} \times -\dfrac{14}{15}$

d $-\dfrac{3}{4} \times \dfrac{8}{15} \times -\dfrac{10}{11}$

e $\dfrac{1}{4} \times -\dfrac{16}{21} \times \dfrac{7}{20}$

f $-\dfrac{21}{32} \times -\dfrac{12}{49} \times -\dfrac{14}{15}$

Remember

positive × positive = positive
positive × negative = negative
negative × positive = negative
negative × negative = positive

10 Use your calculator to work out these fractions.

a $\dfrac{17}{32} \times \dfrac{13}{16}$

b $\dfrac{43}{67} \times \dfrac{78}{92}$

c $\dfrac{23}{98} \times \dfrac{43}{99}$

d $\dfrac{17}{42} \times \dfrac{19}{37} \times \dfrac{78}{95}$

e $-\dfrac{31}{45} \times \dfrac{42}{53} \times -\dfrac{41}{78}$

f $-\dfrac{31}{78} \times -\dfrac{23}{58} \times -\dfrac{21}{77}$

11 Ieuan ordered a lorry load of sand.

Nadir took $\dfrac{3}{4}$ of it.

Ieuan used $\dfrac{1}{3}$ of what was left to build his patio.

The rest was used to build a wall.

What fraction of the lorry load was used to build the wall?

12 A wholesaler sold 5000 bottles of vinegar to a distributor.

The distributor sold $\dfrac{3}{4}$ of the bottles to Mr Williams.

Mr Williams sold $\dfrac{3}{5}$ of these bottles within the first week.

How many bottles did Mr Williams have left?

Strategic competence

13 Work out the area of each of these rectangles.

a $\frac{1}{5}$ m, $\frac{1}{2}$ m

b 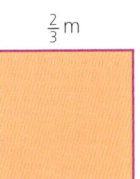 $\frac{2}{3}$ m, $\frac{2}{3}$ m

c 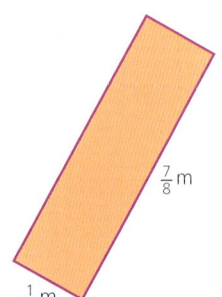 $\frac{7}{8}$ m, $\frac{1}{4}$ m

Band 3 questions

14 A car has $\dfrac{3}{5}$ of a full tank of petrol.

A quarter of this petrol is used on a journey.

What fraction of the petrol tank now needs to be filled to have a full tank?

15 Work out the following fractions.

a $\left(\dfrac{1}{2}\right)^4$ b $\left(-\dfrac{1}{3}\right)^5$ c $\left(\dfrac{2}{3}\right)^3$ d $\left(-\dfrac{1}{2}\right)^6$

16 Megan says that $\left(\dfrac{3}{4}\right)^{10}$ is bigger than $\left(\dfrac{3}{4}\right)^9$ because its power is bigger.
Is she correct?
Explain your answer.

17 Work out $\left(\dfrac{3}{4}\right)^3 \times \left(-\dfrac{4}{5}\right)^3$.

18 Work out the following fractions.

a $\dfrac{1}{2} \times \dfrac{2}{3}$ b $\dfrac{1}{2} \times \dfrac{2}{3} \times \dfrac{3}{4}$ c $\dfrac{1}{2} \times \dfrac{2}{3} \times \dfrac{3}{4} \times \ldots \dfrac{48}{49} \times \dfrac{49}{50}$

7.3 Dividing fractions

Skill checker

1 Work out the following fractions.

a $\dfrac{1}{4} \times \dfrac{1}{3}$ b $\dfrac{1}{2} \times \dfrac{3}{5}$ c $\dfrac{2}{5} \times \dfrac{4}{7}$ d $\dfrac{3}{4} \times \dfrac{8}{9}$

2 Write down the reciprocals of these fractions.

a 5 b $\dfrac{1}{3}$ c $\dfrac{2}{5}$ d $\dfrac{3}{7}$

Change a division into a multiplication by replacing the second fraction with its reciprocal.

Worked example

a $\dfrac{2}{9} \div \dfrac{5}{7}$ b $\dfrac{2}{3} \div 5$ ← It may be useful to remember that 5 can be written as $\dfrac{5}{1}$.

Solution

a $\dfrac{2}{9} \div \dfrac{5}{7} = \dfrac{2}{9} \times \dfrac{7}{5}$ b $\dfrac{2}{3} \div 5 = \dfrac{2}{3} \times \dfrac{1}{5}$

$= \dfrac{2 \times 7}{9 \times 5}$ $= \dfrac{2 \times 1}{3 \times 5}$

$= \dfrac{14}{45}$ $= \dfrac{2}{15}$

7.3 Now try these

Band 1 questions

1 Work out these fractions.

a $3 \div \frac{1}{2}$ b $4 \div \frac{1}{3}$ c $7 \div \frac{1}{4}$ d $8 \div \frac{2}{3}$

2 Work out these fractions.

a $\frac{1}{2} \div 3$ b $\frac{1}{3} \div 5$ c $\frac{1}{4} \div 7$ d $\frac{3}{5} \div 4$

3 Work out these fractions.

a $\frac{1}{4} \div \frac{1}{3}$ b $\frac{1}{5} \div \frac{1}{2}$ c $\frac{1}{7} \div \frac{1}{3}$ d $\frac{1}{10} \div \frac{1}{5}$

4 Work out these fractions.

a $\frac{1}{2} \div \frac{2}{3}$ b $\frac{2}{3} \div \frac{3}{7}$ c $\frac{4}{9} \div \frac{3}{5}$ d $\frac{2}{9} \div \frac{8}{15}$

Band 2 questions

5 Gwilym eats $\frac{4}{7}$ of a watermelon.

a What fraction is left?

He then shares the rest of the watermelon equally amongst three of his friends.

b What fraction of the whole watermelon does each friend receive?

6 Llinos is preparing for a party.

She estimates that each person will eat $\frac{1}{3}$ of a pizza.

How many people can she feed with five pizzas?

7 Work out these fractions.

a $-\frac{1}{3} \div \frac{8}{9}$ b $\frac{2}{7} \div \left(-\frac{9}{10}\right)$ c $-\frac{4}{11} \div \frac{2}{5}$ d $\frac{3}{7} \div \left(-\frac{14}{15}\right)$

8 Use your calculator to work out these fractions.

a $\frac{11}{27} \div \frac{35}{52}$ b $\frac{19}{39} \div \frac{59}{62}$ c $-\frac{14}{37} \div \frac{29}{53}$ d $\frac{41}{97} \div \left(-\frac{83}{89}\right)$

Band 3 questions

9 Bronwen eats $\frac{1}{3}$ of her birthday cake.

She shares the rest of it equally between seven friends.

What fraction of the cake does each person receive?

10 For each rectangle, find the missing length.

a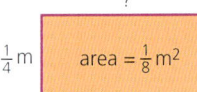

b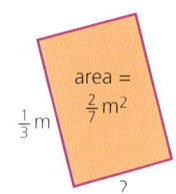

c area = $\frac{3}{11}$ m², side $\frac{2}{5}$ m, ?

7 Fractions

11 Copy and complete these calculations.

a $\square \times -\dfrac{3}{4} = \dfrac{9}{14}$

b $-\dfrac{8}{9} \times \square = -\dfrac{2}{3}$

c $-\dfrac{5}{7} \times \square \times -\dfrac{3}{4} = -\dfrac{9}{14}$

12 Tegan reads $\dfrac{2}{7}$ of a book on the first day.

She then reads $\dfrac{5}{56}$ of the book on each day after.

How many days does it take her to read the whole book?

13 Copy and complete this number wall.

The number in each brick is found by multiplying the numbers in the two bricks below it.

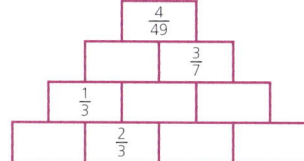

14 Copy and complete this multiplication grid.

×		$-\dfrac{1}{4}$	$-\dfrac{1}{5}$		
$-\dfrac{2}{3}$			$-\dfrac{1}{9}$		
	$\dfrac{3}{8}$		$\dfrac{1}{10}$		
		$-\dfrac{1}{16}$			
$\dfrac{1}{3}$				$\dfrac{1}{6}$	
			$\dfrac{2}{15}$		

Key words

Here is a list of the key words you met in this chapter.

Decimal Denominator Equivalent Integer
Numerator Reciprocal

Use the glossary at the back of this book to check any you are unsure about.

Review exercise: fractions

Band 1 questions

1 Simplify these fractions.

   a $\dfrac{3}{9}$ b $\dfrac{14}{49}$ c $\dfrac{18}{81}$ d $\dfrac{162}{600}$

2 Find the missing numbers.

   a $\dfrac{1}{3}=\dfrac{7}{\square}$ b $\dfrac{3}{5}=\dfrac{\square}{30}$ c $\dfrac{\square}{4}=\dfrac{18}{24}$ d $\dfrac{2}{\square}=\dfrac{10}{35}$

3 Work out these fractions.

   a $\dfrac{2}{5}+\dfrac{1}{4}$ b $\dfrac{3}{8}+\dfrac{5}{11}$ c $\dfrac{9}{14}-\dfrac{3}{10}$ d $\dfrac{8}{21}-\dfrac{3}{14}$

4 Work out these fractions and leave your answers in simplified form.

   a $\dfrac{2}{5}\times\dfrac{1}{8}$ b $\dfrac{1}{6}\times\dfrac{9}{11}$ c $\dfrac{5}{16}\times\dfrac{6}{25}$ d $\dfrac{12}{35}\times\dfrac{14}{15}$

5 Work out these fractions.

   a $\dfrac{1}{3}\div\dfrac{4}{5}$ b $\dfrac{2}{5}\div\dfrac{3}{11}$ c $\dfrac{3}{8}\div\dfrac{9}{16}$ d $\dfrac{5}{8}\div\dfrac{25}{28}$

6 Share £40 in the ratio:

   a 5 : 3   b 3 : 7   c 2 : 3   d 17 : 3.

Band 2 questions

7 Find the missing numbers.

   a $\dfrac{6}{8}=\dfrac{9}{\square}$ b $\dfrac{15}{18}=\dfrac{\square}{30}$ c $\dfrac{\square}{24}=\dfrac{21}{28}$ d $\dfrac{20}{\square}=\dfrac{15}{81}$

8 Work out these fractions. 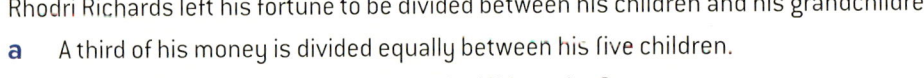 *Don't forget to use BIDMAS.*

   a $\dfrac{2}{3}\times\dfrac{4}{5}+\dfrac{1}{4}$ b $\dfrac{1}{5}\div\dfrac{4}{7}-\dfrac{1}{9}$ c $\dfrac{1}{4}+\dfrac{2}{3}\times\dfrac{9}{10}$ d $\dfrac{9}{10}-\dfrac{4}{9}\div\dfrac{8}{11}$

9 Rhodri Richards left his fortune to be divided between his children and his grandchildren.

   a A third of his money is divided equally between his five children.

       What fraction of his fortune does each child receive?

   b The remainder is divided equally between his eleven grandchildren.

       What fraction of his fortune does each grandchild receive?

   c His fortune was worth £1.65 million. Calculate the amount that each child and grandchild receives.

10 Write these decimals as fractions in their lowest terms.

   a 0.3   b 0.45   c 0.415   d 0.048

11 Write these fractions as decimals.

   a $\dfrac{3}{4}$ b $\dfrac{7}{10}$ c $\dfrac{17}{20}$ d $\dfrac{39}{40}$

12 Use a calculator to work out these fractions.

   a $\dfrac{23}{37}\times\dfrac{12}{17}+\dfrac{5}{19}$ b $\dfrac{13}{15}\div\dfrac{24}{17}+\dfrac{8}{19}$ c $\dfrac{9}{14}-\dfrac{2}{13}\times\dfrac{29}{30}$

7 Fractions

Band 3 questions

13 Two rectangles are placed next to each other.

Calculate:

a   the perimeter of the combined shape

b   the area of the combined shape.

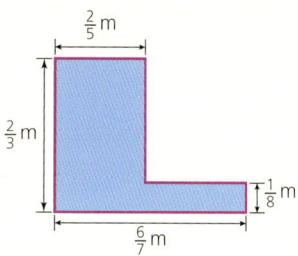

14 Copy and complete this multiplication grid.

×	$\frac{1}{2}$	$\frac{3}{7}$	$\frac{7}{9}$		$\frac{1}{5}$		$\frac{1}{3}$
$\frac{1}{2}$							
$\frac{3}{4}$							
$\frac{2}{5}$							
$\frac{5}{8}$							
			$\frac{2}{3}$				
$\frac{5}{16}$			$\frac{3}{16}$				
					$\frac{1}{15}$	$\frac{1}{4}$	$\frac{1}{9}$

15 Work out these fractions.

a $-\dfrac{3}{5} \times \dfrac{7}{9} - \dfrac{4}{15}$

b $-\dfrac{3}{4} \div \left(-\dfrac{5}{8}\right) + \dfrac{7}{10}$

c $-\dfrac{3}{8} + \dfrac{5}{8} \times -\dfrac{3}{7}$

16 Copy and complete this inequality with another fraction.

$\dfrac{2}{3} < \square < \dfrac{3}{4}$

17 A rectangle and a triangle are placed next to each other.

Calculate:

a   the perimeter of the combined shape

b   the area of the combined shape.

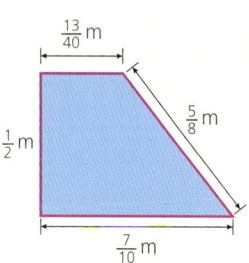

18 Copy and complete this number wall.

The number in each brick is found by adding the numbers in the two bricks below it.

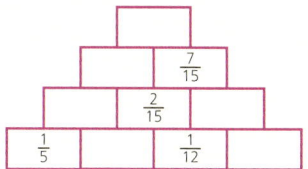

8 Expressions and formulas

Coming up...

- Recognising and using relationships between operations
- Using and interpreting algebraic notation
- Substituting numerical values into formulas and expressions
- Simplifying and manipulating algebraic expressions
- Understanding and using standard mathematical formulas
- Rearranging formulas to change the subject
- Using algebraic methods to solve linear equations in one variable
- Expanding brackets and factorising algebraic expressions

Number crossword

Copy and complete this cross-number grid.

Across

1. $12 + 3 \times 10^2$
3. $(3+7)^2 \times 6 - 8$
5. $11 \times (7 + 5)$
6. 12^2
8. $(64 \times 6) + 26$
10. 37×3
12. $5^4 + 99$
14. 29×8
15. 11×13
16. $108 \times 8 - 10$

Down

1. $3 \times 10^2 + 1$
2. $15^2 - 11$
3. $5 \times 10^2 + 2 \times 10 + 4$
4. $15^2 - 5^2$
7. $19 \times 23 - 6^2$
9. $2^6 \times 3$
10. $11^2 + 70$
11. 41×3
12. $7 \times 8 \times 13$
13. A multiple of (10 across)

8.1 Working with letter symbols review

Skill checker

① a $7 + 2 \times 3$ b $13 - 3^2$ c $(5 + 3)^2$ d $\dfrac{15 - 3}{1 + 5}$

② Evaluate these expressions when $p = 8$ and $q = 4$.
   a $5p$ b $p - q$ c p^2 d $\dfrac{p}{q}$

③ a $-8 + 3$ b $7 \times (-2)$ c $-10 - (-2)$ d $-20 \div (-4)$

▶ Using formulas

A **formula** is a rule for working something out.

Worked example

This formula converts temperatures measured in Celsius to Fahrenheit.
$$F = 1.8C + 32$$
where C is the temperature in Celsius, and F is the temperature in Fahrenheit.

Use the formula to convert -5 degrees Celsius into Fahrenheit.

It is important to define your variables carefully.

Solution

$F = 1.8 \times -5 + 32$
$= -9 + 32$
$= 23°$

Cross-curricular activity

In your science lesson, investigate the temperature ranges of the planets in the solar system. What are the temperatures in Celsius and Fahrenheit?

▶ Simplifying expressions

Separate terms can sometimes be combined to simplify an expression.

Worked example

Simplify the following expressions.

a $3x + 7x^2 - 2x$

b $2p \times 5q \times 3r$

Solution

a $3x + 7x^2 - 2x = 7x^2 + 3x - 2x$

You can write the terms in any order, but you must keep the signs in front of each term.

$= 7x^2 + 1x$

The x terms can be combined but an x^2 term can only be combined with other x^2 terms.

$= 7x^2 + x$

When simplifying, the final expression should not include any coefficients of 1 as they are unnecessary.

b $2p \times 5q \times 3r = 2 \times 5 \times 3 \times p \times q \times r$

When finding a product, you can multiply in any order.

$= 30pqr$

Omit multiplication signs and write the number at the front of a product.

8.1 Now try these

Band 1 questions

1. Evaluate these expressions when $a = 7$ and $b = 3$.
   a. $a + b$
   b. $3a - 2b$
   c. $a^2 + 5b$
   d. $\dfrac{a - b}{2}$

2. Simplify these expressions.
   a. $7p + 3p - 6p$
   b. $5x^2 - x^2$
   c. $3g \times 7h$

3. Use the formula $v = u + at$ to find the value of v when:
   a. $u = 4, a = 2, t = 0.5$
   b. $u = 8, a = -3, t = 2$
   c. $u = -7, a = 6, t = \dfrac{1}{3}$.

4. Simplify these expressions.
   a. $3a + 8c - 2a + 6c$
   b. $2x^2 + 3x + x^2 - 8x$
   c. $7ac + 3ab - 2ac + bc$
   d. $2w^2 \times 3w \times 5y$

Band 2 questions

5. In these grids, two expressions are added to give the expression above.
   Copy the grids and find the missing simplified expressions.

   a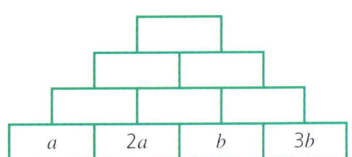

   b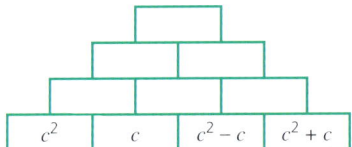

6. A rectangle has a length of $7x$ centimetres and a width of $3x$ centimetres.
   Write a simplified expression for the:
   a. perimeter
   b. area.

7. A taxi firm charges customers £5 for each journey plus £2 per kilometre.
   a. Calculate the cost of a journey of 7 kilometres.
   b. Write an expression for the cost of a journey of n kilometres.
   c. Write a formula for the cost of a journey.
      Define your variables.

Band 3 questions

8. In these grids, two expressions are added to give the expression above.
   Copy the grids and find the missing simplified expressions.

   a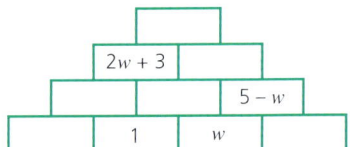

   b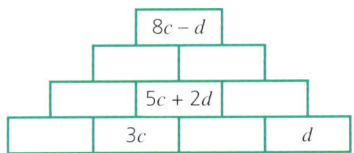

9. Using the formula $y = mx + c$ find the value of:
   a. c when $y = 14, m = 3$ and $x = 2$
   b. m when $y = 20, x = -3$ and $c = 5$
   c. x when $y = 31, m = -5$ and $c = -4$
   d. y when $x = 0.7, m = -6$ and $c = \dfrac{2}{5}$.

10 Meinir is going ice-skating with her two parents and three young cousins. She uses this formula to work out the cost.

Cost (£) = 5a + 3c

a What do a and c stand for?
b How much does it cost for this group of people to skate?

AFONFFORDD ICE RINK

Adults (Oedolion) £5
Children (Plant)   £3

11 Simplify these expressions.
a $\frac{a}{7} + \frac{3a}{7}$
b $\frac{5b}{6} + \frac{b}{6}$
c $\frac{7c}{4} + \frac{2c}{4}$
d $\frac{2d}{5} - \frac{d}{5}$
e $\frac{7e}{4} - \frac{e}{4}$
f $\frac{3f}{4} \times g$
g $\frac{3h}{4} \times \frac{h}{3}$
h $\frac{4k}{5} \div \frac{k}{m}$

12 Evaluate these expressions when $n = 3$.
a $2n + 5$
b $2(2n + 5)$
c n^2
d $3n^2$
e $n^2 + 1$
f $20 - n^2$
g n^3
h $2n^3$
i $1 + 2n^3$

13 Look at Catrin's maths homework:

When $x = 3$, then:
a $4x^2 = 4 \times 3^2 = 12^2 = 144$ ✗
b $10 - x^2 = 10 - 3^2 = 7^2 = 49$ ✗
c $5(x - 2) = 5(3 - 2) = 15 - 2 = 13$ ✗

What mistakes has she made?
What should the answers be?

14 Evaluate $\frac{x + y}{2x}$ when:
a $x = 3$ and $y = 9$
b $x = 20$ and $y = 60$
c $x = 1$ and $y = 7$
d $x = 7$ and $y = 0$.

15 In a newsagent's shop:
• The price of any football magazine is £f.
• The price of any music magazine is £m.

$5 - f$ $f + m$ $1 - m$ $5f + 5m$ $f - m$ $5f + m$ $5f$ $f - 5$ $m - 1$

Choose the correct expression from the box to match each of the following statements.
a The cost of five football magazines
b The cost of one football and one music magazine
c The change from £5 when you buy one football magazine
d The cost of a music magazine when its price is reduced by £1
e How much more a football magazine costs than a music magazine.

8.2 Expanding brackets

Skill checker

① a $8 \times (5 + 7)$ b $8 \times 5 + 8 \times 7$

② a $7 \times (8 - 3)$ b $7 \times 8 - 7 \times 3$

③ a $\frac{1}{2} + \frac{1}{3}$ b $\frac{1}{2} - \frac{1}{3}$ c $\frac{1}{2} \times \frac{1}{3}$ d $\frac{1}{2} \div \frac{1}{3}$

▶ Expanding brackets

Multiplying out brackets is sometimes referred to as 'expanding brackets'.

Worked example

Expand the brackets in each of these expressions.

a $7(a + 2)$ b $m(4m + 2n - p)$

Solution

a $7(a + 2) = 7a + 14$ *Multiply each term in the bracket by 7.*

b $m(4m + 2n - p) = 4m^2 + 2mn - mp$ *Multiply each term in the bracket by m.*

Worked example

Expand and simplify $3(2x + 7) - 2(x - 4)$.

Solution

$3(2x + 7) - 2(x - 4) = 6x + 21 - 2x + 8$

$= 4x + 29$

Be careful when expanding a bracket that follows a negative sign.
$-2 \times -4 = +8$

8.2 Now try these

Band 1 questions

❶ Which expression is not equal to the others?

$2a + 4b$ $3b + a + 2b + a - b$ $2a + b - 5b$ $2(a + 2b)$

❷ Expand the brackets in each of these expressions.

a $3(a + 4)$ b $2(b - 5)$ c $7(c + 4)$ d $3(2d - 5)$

❸ Match each coloured expression to a white expression.

A $2a + 2$ 2 $2(a + 1)$ C $12d - 4$ 4 $4(3d - 1)$ 5 $3(a + b)$

B $3a + 3b$

1 $3(5a - 8)$ 3 $12(1 + 2b - 3c)$ D $12 + 24b - 36c$ E $15a - 24$

4 Expand the brackets in these expressions.
 a $7(e + 2f)$
 b $9(7g - 2h + 8)$
 c $6(m^2 + 3m - 5)$

5 Expand and simplify these expressions.
 a $3(n + 4) + 2(5n - 1)$
 b $7(2p - 8) - 5(p - 6)$

6 Expand the brackets in these expressions.
 a $3r(4r - 7)$
 b $7u(u + 6 - 2v)$
 c $-3w(4x - 5y)$

Band 2 questions

7 Expand and simplify these expressions.
 a $2(3r - 7) + 3r(4r + 1)$
 b $6(8 - 3u) - 4u(5 - 2u)$

8 Expand and simplify $3y(x - 2y) + 5x(2y - 7) + y(3y - y^2)$.

9 Romesh evaluates $4 \times (3 + 6)$.
 His working is shown here.

 $4 \times (3 + 6)$
 $= 4 \times 9$
 $= 36$

 a Describe another way to evaluate this expression.
 b Which method is quicker?

10 a Write an expression using a bracket for the area of this rectangle.
 b Expand the brackets.
 c i Find the area of the rectangle when $p = 3$.
    ii Use your answer to check your expressions in parts **a** and **b**.

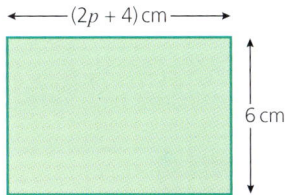

$(2p + 4)$ cm
6 cm

Band 3 questions

11

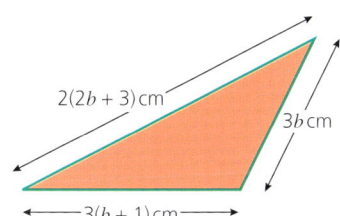

$2(2b + 3)$ cm
$3b$ cm
$3(b + 1)$ cm

 a Write an expression for the perimeter of this triangle.
 b Expand the brackets and simplify your answer.
 c Find the perimeter of the triangle when $b = 2$.
 d Is it possible for b to equal 1? Explain your answer.

12 Expand and simplify these expressions.
 a $\frac{1}{2}(4x + 2) + 5$
 b $\frac{2}{3}(6x - 6 + 4)$

13 Expand and simplify $\frac{3}{4}(8p - 4r) + \frac{1}{3}(9p + 6r)$.

Curriculum for Wales Mastering Mathematics: Book 2

Logical reasoning

⑭ Mari has got both of her homework questions wrong.
Find her mistakes and write out the correct solutions.

a $2n(3n + 4)$
   $= 5n^2 + 6n$ ✗

b $3(2n - 5) - 4(n - 2)$
   $= 6n - 15 - 4n - 8$
   $= 2n - 23$ ✗

⑮ Eira and Jac have simplified $2(5x + 6) - 3(2 - 4x)$.
Eira says, 'The answer is $6 - 2x$.'
Jac says, 'No, it's $14x + 22$'

a Substitute $x = 2$ into all three expressions to show who is definitely wrong.
b Expand and simplify $2(5x + 6) - 3(2 - 4x)$.
   Are either of them right?

⑯ Try this 'mind reading' trick on a friend.

> Think of the number of your birthday month.
> Now add 2, multiply by 5 and subtract 7.
> Then multiply by 4 and add 9.
> Multiply the answer by 5 and add on the number of the day you were born.
> What is your answer?

You can work out your friend's birthday by subtracting 105.

a How does subtracting 105 help?
b Write down an expression for each step of the puzzle to show why it works.

8.3 Factorising expressions

Skill checker

① Write down all the factors of 24.
② What type of number has exactly two factors?
③ Expand $3(2x+5y)$.
④ Find the highest common factor of 18, 12 and 24.

▶ Factorising expressions

Factorising expressions is the reverse of expanding (multiplying out) brackets.

Conceptual understanding

Worked example

Factorise fully each of these expressions.

a $8x - 6$
b $12w^2 + 15w$

> The word 'fully' is used to mean that all factors must be removed. Always assume 'factorise' means 'fully factorise'.

Solution

a $8x - 6 = 2(4x - 3)$
b $12w^2 + 15w = 3w(4w + 5)$

> 2 is the highest factor of both 8 and 6, so 2 is written in front of the bracket.
> Divide each term by 2 to find the terms inside the bracket.

> Each term divides by 3 and by w, so the factor removed is $3w$.
> $12w^2 + 15w = 3(4w^2 + 5w)$ would be partially factorised, but not fully factorised because the factor of w also needs to be taken out.

8.3 Now try these

Band 1 questions

1 Copy and complete these factorisations.
   a $3x + 6y = 3(x + \square)$
   b $10v - 15w = 5(\square - 3w)$

2 Copy and complete these factorisations.
   a $12r - 15u = \square(4r - 5u)$
   b $8w + 6x - 10y = \square(4w + 3x - 5y)$

3 Factorise these expressions.
   a $2a + 6$
   b $3b - 9$
   c $6c + 3$

4 Factorise these expressions.
   a $14d - 21$
   b $6e + 9$
   c $12f - 16$

5 Factorise these expressions.
   a $8g + 14h$
   b $9i + 12j - 15$
   c $16k + 20m - 8$

6 Factorise these expressions.
   a $n^2 + 7n$
   b $2m^2 - 5m$
   c $p^3 + 8p$

Band 2 questions

7 Factorise these expressions.
   a $a^3 + 5a^2$
   b $14b^3 - 6b^2$
   c $6c^3 + 8c^2 - 2c$

8 Factorise these expressions.
   a $d^2 + 8d - cd$
   b $e^2f + ef^2$
   c $8m^2n + 6mnp$

9 Rhian factorises $9r^2 + 6r$.
   She writes her answer as $3(3r^2 + 2r)$.
   Is Rhian's factorisation correct? Explain your answer.

10 Factorise $7a^2 + 3ab - 6a - 8ab + 3a^2 - 9a$.

11 Look at the expressions below. Some are in factorised form, others in expanded form.
   Find the matching pairs.
   Write them in two columns, with the expanded form on the left and its factorised form on the right beside it.
   One of the expressions cannot be factorised.

   $x^2 + xy$ $x(x+1)$ $xy + y^2$ $x(x-y)$ $x^3 + x^2 + x$ $x^2 + 1$ $x^2 + x$ $x(x+y)$ $x(x^2 + x + 1)$ $y(x+y)$ $x^2 - xy$

Band 3 questions

12 Factorise $g^3h^2i + 4g^2hi^2 - 7gh^3i$.

13 In this wall, two expressions are added to give the expression above.
   Copy the wall and find the missing expressions.
   Factorise your answers where possible.

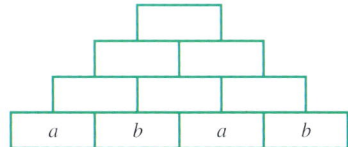

14 Factorise these expressions.
   a $9(x + 2y) + 4(3x - y)$
   b $9(3x + 2y) - 7(x - y)$

Logical reasoning

⑮ Two small rectangles are placed side by side to make a big rectangle.

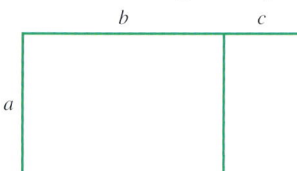

Write factorised expressions for the big rectangle's:

a perimeter

b area.

8.4 Rearranging formulas

Skill checker

① Write down the inverse of each of these operations.

The inverse of an operation is the reverse of the operation.

a $+3$ b $\times 6$ c -7 d $\div 2$

② Use the formula $C = 5n + 8$ to find C when $n = 7$.

③ Use the formula $y = 3x - 2$ to find x when $y = 4$.

▶ Rearranging formulas

Sometimes you need to rearrange a formula to change its **subject**.

Write the formula as a number machine, and then reverse the number machine.

Note

The subject of the formula is the letter which is written on its own on one side of the equal sign.

Conceptual understanding

Worked example

Rearrange this formula to make n the subject.

$$W = 10n + 8$$

Solution

Write the formula as a number machine.

$n \longrightarrow \boxed{\times 10} \longrightarrow \boxed{+8} \longrightarrow W$

Reverse the number machine.

$n \longleftarrow \boxed{\div 10} \longleftarrow \boxed{-8} \longleftarrow W$

Then rewrite it as a formula.

$n = \dfrac{W - 8}{10}$ ← *The subject is usually written on the left.*

Alternatively, rearrange a formula by treating it as an equation and solving it for the required subject.

$W = 10n + 8$ ← *Here, the subject is W.*

$W - 8 = 10n$

$\dfrac{W - 8}{10} = n$

$n = \dfrac{W - 8}{10}$ ← *The subject is usually written on the left.*

8 Expressions and formulas

8.4 Now try these

Band 1 questions

1 Work out the rules for these adding and subtracting number machines.
  a 2 ⟶ ☐ ⟶ 5
  b 5 ⟶ ☐ ⟶ 2
  c 18 ⟶ ☐ ⟶ 25
  d 25 ⟶ ☐ ⟶ 18

2 Teilo is going on holiday in Europe and wants to exchange his spending money into euros before he travels.
  a Copy and complete this number machine.

  Pounds (£) ⟶ [?] ⟶ Euros (€)

  Today's exchange rate is:
  1 pound sterling (£) = 1.15 euro (€)

  It converts British pounds into euros.
  b Teilo wants to exchange £140.
     Use the number machine to find out how many euros he will receive.
  c Reverse the number machine.
  d Carys is returning from a holiday in Germany.
     She finds an exchange shop offering to buy her euros back for the same exchange rate, 1 pound sterling (£) for 1.15 euro (€).
     She has €23 to exchange back into pounds.
     Use your reversed number machine to find out how many pounds she'll get back.

3 Jabir has a pay-as-you-go phone tariff.
  All calls cost 10p per minute.
  He tops up with £20.
  a How much does he pay for a 6-minute call?
  b Copy and complete this number machine to work out how much Jabir pays for his calls.

  ☐ ⟶ [?] ⟶ ☐

  c Jabir has £4.60 credit left.
     Reverse your number machine to work out how many minutes of calls he can make.

4 Reverse these number machines.
  a a ⟶ [+3] ⟶ b
  b c ⟶ [−2] ⟶ d

5 Reverse these number machines.
  a e ⟶ [×4] ⟶ f
  b g ⟶ [÷7] ⟶ h

6 a Write this number machine as a formula.
     m ⟶ [+8] ⟶ p
  b Reverse the number machine.
  c Write the reversed number machine as a formula with m as the subject.

7 a Write this formula as a number machine.
     $v = w − 9$
  b Reverse the number machine.
  c Write the reversed number machine as a formula with w as the subject.

Band 2 questions

8 Which letter is the subject in each of these formulas?
  a $v = u + at$
  b $mx + c = y$
  c $p = 3r + s$
  d $v = w − y^2$

Fluency

9 Rewrite these formulas with n as the subject.
   a $a = n + 4$ b $b = n - 7$ c $c = n + p$ d $d = n - r$

10 Rewrite these formulas with m as the subject.
   a $e = 3m$ b $f = \dfrac{m}{2}$ c $g = mx$ d $h = \dfrac{m}{w}$

Strategic competence

11 Which of these formulas is not a correct rearrangement of the others?

$a = bc \qquad b = \dfrac{a}{c} \qquad cb = a \qquad c = \dfrac{b}{a}$

Fluency

12 a Write this formula as a number machine.
   $y = 3x + 2$
   b Reverse the number machine.
   c Write the reversed number machine as a formula with x as the subject.

13 a Copy these number machines.
   i $v \longrightarrow \boxed{\times 2} \longrightarrow \boxed{+ 5} \longrightarrow w$
   ii $v \longrightarrow \boxed{\times 5} \longrightarrow \boxed{+ 2} \longrightarrow w$
   iii $v \longrightarrow \boxed{+ 2} \longrightarrow \boxed{\times 5} \longrightarrow w$
   iv $v \longrightarrow \boxed{\div 2} \longrightarrow \boxed{+ 5} \longrightarrow w$

   Do these formulas match the number machines? Label your diagrams with any that match.

   $w = \dfrac{1}{2}v + 5 \qquad w = 5v + 2$

   b Reverse each number machine, and write it as an algebraic formula.

14 Write these formulas with x as the subject.
   a $y = 8x - 7$ b $y = 3x + 7$ c $y = \dfrac{x}{2} + 1$

Band 3 questions

15 Rearrange these formulas to make x the subject.
   a $y = mx + 2$ b $y = 5x + c$ c $y = mx + c$

16 Rearrange $v = u + at$ to make:
   a u the subject   b a the subject   c t the subject.

17 Rearrange $a = 3c + 2e$ to make:
   a c the subject   b e the subject.

18 Rearrange $f = gh - mn$ to make:
   a g the subject   b m the subject.

Key words

Here is a list of the key words you met in this chapter.

Coefficient   Equation   Expand brackets   Expression
Factorise   Formula   Simplify   Solve
Subject   Substitute Term   Unknown
Variable

Use the glossary at the back of this book to check any you are unsure about.

Review exercise: expressions and formulas

Band 1 questions

1 Evaluate these expressions when $p = 7$ and $q = 3$.
 a $2p + q$
 b $2pq^2$
 c $3p - 5q$
 d $p^2 + q^2$

2 Simplify these expressions.
 a $8a + 7a - 3a$
 b $5b^2 + 3b^2$
 c $3c \times 4cd$

3 Expand these expressions.
 a $3(f + 6)$
 b $4(2g - 7)$
 c $-5(3h - 4i)$

4 Factorise these expressions.
 a $4j + 6$
 b $3m^2 + 4m$
 c $24n + 40$

5 Reverse these number machines.
 a $m \longrightarrow \boxed{-8} \longrightarrow n$
 b $a \longrightarrow \boxed{\times 7} \longrightarrow c$

6 a Use the formula $C = 5b - 2$ to find the value of C when:
   i $b = 2$
   ii $b = 8$
   iii $b = 12$
   iv $b = 20$.
 b The formula refers to this notice about paperback books.
   i What does b stand for?
   ii What does C stand for?

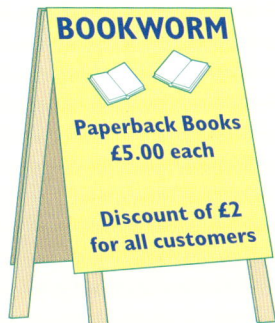

BOOKWORM
Paperback Books
£5.00 each

Discount of £2 for all customers

Band 2 questions

7 Write simplified expressions for the:
 a perimeter
 b area
of this rectangle.

(rectangle labelled $8w$ and $5w$)

8 Ceri has $2m + 5$ sweets and Caitlyn has $2(m + 2)$ sweets.
Who has more sweets?
Explain your answer.

9 Using the formula $D = b^2 - 4ac$ find the value of:
 a D when $b = 5, a = 2, c = -7$
 b c when $D = 25, b = 7, a = 2$.

10 Niall expands this expression $3(2c - 7)$.
Here is his working.

> $3(2c - 7)$
> $= 6c - 7$

Explain the mistake he has made.

11 Factorise these expressions.

  a $3a^2 - 9a$ b $8b^2 + 10b^3 - bc$ c $4g^2h - 6gh^3$

12 Rewrite these formulas with w as the subject.

  a $y = w - 8$ b $d = m + w$ c $f = 9w$ d $a = \dfrac{w}{c}$

Band 3 questions

13 Nerys expands and simplifies this expression.
$5(2a + c) - 3(a - 2c)$
Her working is shown here.

$5(2a + c) - 3(a - 2c)$
$= 10a + 5c - 3a - 6c$
$= 7a - c$

Explain her error.

14 Factorise $2(5a + 4c) - 5(6c - 7a) + 10a$.

15 Rearrange $E = I(R + r)$ to make R the subject.

16 In this number wall, two numbers are added to give the number above.

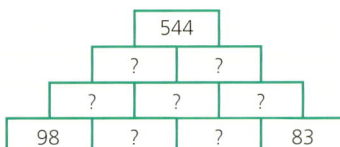

Anwen says there is not enough information to find the missing numbers.

  a Is she right?

Shibo says he can find one of the missing numbers.

  b Is he right? Explain your answer.

Hint
Let the two middle bricks on the bottom row be a and b. Then write the other bricks in terms of a and b.

Consolidation 3: Chapters 6–8

Band 1 questions

1 Find:
   a $\frac{1}{2}$ of 6 km
   b $\frac{2}{3}$ of £60
   c $\frac{1}{4}$ of 12 kg
   d $\frac{3}{5}$ of 20 litres
   e $\frac{3}{4}$ of 640
   f $\frac{3}{10}$ of 1 kilobyte.

2 Work out these calculations.
   a -4×7
   b $-8 \times (-9)$
   c $32 \div (-8)$
   d $(-3)^2$

3 Work out these fractions.
   a $\frac{1}{3} + \frac{1}{3}$
   b $\frac{1}{5} + \frac{2}{5}$
   c $\frac{2}{7} + \frac{4}{7}$
   d $\frac{2}{9} + \frac{5}{9}$
   e $\frac{5}{8} - \frac{3}{8}$
   f $\frac{5}{7} - \frac{2}{7}$
   g $\frac{3}{4} - \frac{1}{4}$
   h $\frac{5}{9} - \frac{2}{9}$

4 Write these fractions in their lowest terms.
   a $\frac{6}{12}$
   b $\frac{14}{18}$
   c $\frac{60}{80}$
   d $\frac{28}{91}$

5 Draw a number line from -10 to 10.
   a Place the numbers 9, -8, -7, 3, -5, 5, 0 on the line.
   b Which two numbers are the same distance from 0 on your line?

6 What number is:
   a i   4 more than 5   b i   3 more than 3   c i   8 more than 7   d i   10 more than 3
       ii 4 less than 5   ii 3 less than 3   ii 8 less than 7   ii 10 less than 3?

7 Share £28 in the ratio:
   a $3 : 1$
   b $1 : 6$
   c $3 : 4$
   d $11 : 3$.

8 Copy and complete these number patterns.
   a $3 - 2 = 1$
       $3 - 1 = 2$
       $3 - 0 = \square$
       $3 - (-1) = 4$
       $3 - (-2) = \square$
       $3 - (-3) = \square$
       $3 - (-4) = 7$
   b $6 - 4 = 2$
       $6 - 5 = \square$
       $6 - 6 = \square$
       $6 - 7 = \square$
       $6 - 8 = -2$
       $6 - 9 = \square$
       $6 - 10 = -4$
   c $4 - 4 = \square$
       $4 - 3 = \square$
       $4 - 2 = \square$
       $4 - 1 = \square$
       $4 - 0 = \square$
       $4 - (-1) = \square$
       $4 - (-2) = \square$

9 Work out the following.
   a $(-5) + 7$
   b $(-3) - 2$
   c $8 - 11$
   d $6 + 4$
   e $6 - 9$
   f $(-3) + 5$
   g $(-2) + 3 + 1$
   h $4 - 6 - 3$
   i $(-5) - 2 + 3 + 1$

10 Cancel these fractions before multiplying.
   a $\frac{5}{8} \times \frac{2}{3}$
   b $\frac{3}{16} \times \frac{4}{5}$
   c $\frac{7}{9} \times \frac{6}{14}$
   d $\frac{11}{12} \times \frac{2}{9}$
   e $\frac{5}{14} \times \frac{7}{8}$
   f $\frac{2}{30} \times \frac{13}{14}$

Fluency

11 Find the value of $2(l + w)$ when:
  a $l = 3$ and $w = 7$
  b $l = 5$ and $w = 5$
  c $l = 10$ and $w = 2$
  d $l = 4$ and $w = 0$
  e $l = 2$ and $w = 2$
  f $l = 10$ and $w = 12$
  g $l = 5$ and $w = 9$
  h $l = 8$ and $w = 5$
  i $l = 18$ and $w = 20$
  j $l = 25$ and $w = 50$.

12 Expand the brackets in these expressions.
  a $5(2 + c)$
  b $3(x - 6)$
  c $7(p + q)$
  d $4(2d + 3)$
  e $4(b + 8)$
  f $8(2r - 1)$
  g $3(1 - k)$
  h $3(1 + 2m - 3n)$
  i $5(2a - 3b)$
  j $12(2 + 2d + 2e)$
  k $2(6u - 8v - 4w - 10x)$
  l $6(2y - x - z)$

13 Factorise these expressions fully.
  a $3x + 6$
  b $2x + 10$
  c $10x + 2$
  d $10x - 2$
  e $5x + 15$
  f $15f - 45$
  g $30s - 40$
  h $12b + 6a$
  i $48 - 24a$

Strategic competence

14 One afternoon last winter the temperature in Haverfordwest was 4 °C at 2 p.m.
Between 2 p.m. and 2 a.m. the temperature fell by 12 °C.
What was the temperature at 2 a.m.?

15 Copy and complete.
  a $\frac{1}{2} + \square = 1$
  b $\frac{3}{4} + \square = 1$
  c $\frac{1}{3} + \square = 1$
  d $\frac{2}{5} + \frac{1}{5} + \square = 1$

16 Gethin's freezer was set at −28 °C.
He turned it up to −18 °C.
By how many degrees did he increase the temperature?

Band 2 questions

Fluency

17 Write each set of numbers in order, starting with the lowest.
  a 7, 0, 3, 10, −5, 100
  b −7, 0, −3, −10, 5, −100
  c 7, 0, −3, 10, −5, 100
  d −7, 0, 3, −10, 5, −100

18 Write these numbers as fractions in their lowest terms.
  a 60%
  b 0.45
  c 11.8%
  d 0.225

19 Copy and complete the table.

Temperature in morning (°C)	Change in temperature (°C)	Temperature in afternoon (°C)
+4	Up 2	
+3	Down 4	
−1	Up 5	
−2	Down 3	
−10	Up 8	
−3	Down 1	
0	Down 4	
+1		+5
−2		−3

20 Insert < or > between each pair of numbers.
  a 17 and 25
  b 8 and −9
  c −14 and −33

Consolidation 3

21 These are the temperatures in four cities recorded on 18 January.
Cardiff: 5 °C   Moscow: −9 °C   Rome: 7 °C   New York: −3 °C
  a What was the difference between the temperature in Cardiff and the temperature in Moscow?
  b How many degrees warmer was it in Rome than in New York?
  c How many degrees colder was it in Moscow than in New York?
  d On 19 January the temperature in New York fell by 5 °C compared with the previous day.
     What was the temperature in New York on 19 January?

22 The coldest temperature ever recorded was −89 °C, in Antarctica in 1988.
The hottest temperature ever recorded was 56.7 °C in Death Valley, California, in 1913.
What is the difference between these two temperatures?

23 Work out these fractions.
  a $\frac{1}{2} \times \frac{1}{4}$
  b $\frac{2}{5} \times -\frac{1}{3}$
  c $-\frac{3}{8} \times -\frac{2}{5}$
  d $\frac{4}{9} \times \frac{7}{10}$
  e $-\frac{1}{3} \div \frac{3}{4}$
  f $\frac{2}{5} \div -\frac{6}{7}$
  g $-\frac{3}{10} \div -\frac{5}{6}$
  h $-\frac{2}{15} \div \frac{8}{9}$

24 Wai Peng is reading a book 400 pages long. On Monday he reads $\frac{1}{2}$ of it.
On Wednesday he reads $\frac{1}{2}$ of what now remains. On Thursday he reads $\frac{1}{2}$ of what now remains.
  a How many pages has he still to read on Friday?
  b What fraction is this of the pages in the book?
  c What is $\frac{1}{2} \times \frac{1}{2} \times \frac{1}{2} \times \frac{1}{2}$?

25 Copy and complete.
  a $\frac{5}{12} + \frac{7}{12}$
  b $\frac{3}{7} + \left(-\frac{2}{7}\right)$
  c $\frac{10}{19} + \left(-\frac{3}{19}\right)$
  d $\frac{2}{5} - \frac{1}{10}$
  e $\frac{2}{5} + \frac{3}{10}$
  f $\frac{3}{4} + \left(-\frac{5}{12}\right)$
  g $1 - \frac{2}{5}$
  h $\frac{1}{3} + \frac{1}{4} + \frac{1}{5}$

26 Find the values of the following.
  a $3(x + y)$ when:
     i $x = 2$ and $y = 8$ ii $x = 4$ and $y = 1$ iii $x = 2$ and $y = 0$ iv $x = 5$ and $y = -1$.
  b $5(x - y)$ when:
     i $x = 10$ and $y = 8$ ii $x = 10$ and $y = 9$ iii $x = 10$ and $y = 0$ iv $x = 5$ and $y = 5$.
  c $(x + y) \times (x - y)$ when:
     i $x = 5$ and $y = 2$ ii $x = 4$ and $y = 3$ iii $x = 6$ and $y = 0$ iv $x = 4$ and $y = 1$.
  d $(a + b + c) \div 3$ when:
     i $a = 5, b = 6$ and $c = 7$ ii $a = 1, b = 1$ and $c = 1$ iii $a = 2, b = 1$ and $c = 0$ iv $a = 9, b = 6$ and $c = -6$.

27 Expand the brackets and simplify these expressions.
  a $5(b - 2) + 4(2 - b)$
  b $3(d + 2) - 2(d - 1)$

28 Expand the brackets in these expressions.
  a $3(a - b)$
  b $5(3x + 8y)$
  c $x(x - 1)$
  d $x(x + 5)$
  e $2x(x + 5)$
  f $2x(3x + 5)$
  g $x(x + y)$
  h $4x(x - y)$
  i $4x(5x - 7y)$
  j $x(x^2 - x - 1)$
  k $3x^2(2x + 5)$
  l $3x^2(4x - 11y)$

29 Factorise these expressions fully.
  a $5x + 15$
  b $14a + 21b$
  c $xy + xz$
  d $5xy + 10xz$
  e $3x - ax$
  f $ax^2 + bx^2$
  g $x^2 - xy$
  h $x^2 - 5xy$
  i $5y + y^2$
  j $x^3 + x^2 - x$
  k $x^4 + x^2$
  l $6x^4 - 9x^2$

Strategic competence

30 Copy these calculations. Fill in the missing numbers.
- a $8 - \square = 3$
- b $5 + \square = -9$
- c $-7 + \square = -4$
- d $8 + \square = 3$
- e $6 - \square = 8$
- f $-4 + \square = 1$
- g $\square + 7 = 2$
- h $\square - 7 = -10$
- i $\square + 2 = -8$
- j $-1 - \square = -6$
- k $-15 + \square = -9$
- l $\square + 4 = -5$

31 The area of this rectangle is $(9x + 6)$ cm².

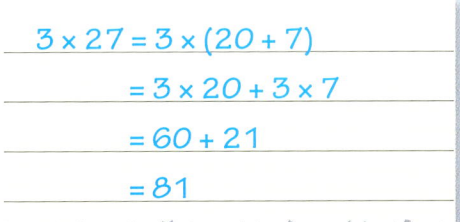

- a Factorise $9x + 6$ fully.
- b What is the height of the rectangle?
- c What is the length of the rectangle when $x = 2$?
- d What is the area of the rectangle when $x = 2$?

32 Seren is using brackets to help her with multiplication.
Work out these using the same method as Seren.
- a 5×23
- b 6×32
- c 7×47
- d 8×84

$3 \times 27 = 3 \times (20 + 7)$
$= 3 \times 20 + 3 \times 7$
$= 60 + 21$
$= 81$

33 Four-fifths of the trees in a park are deciduous.
One-third of these deciduous trees are oak trees.
What fraction of the trees in the park are oak trees?

Band 3 questions

Fluency

34 Use your calculator to work these out.
- a $236.7 - 378.95$
- b $-45 - (-789)$
- c 34.7×98.32
- d $\frac{32}{73} \times \frac{14}{61}$
- e $\frac{32}{73} + \frac{14}{61}$
- f $-\frac{32}{73} \times \left(\frac{14}{61} - \frac{11}{34}\right)$

Strategic competence

35 Two numbers are placed on a number line.
They are both the same distance from 0 and are 30 apart from each other.
What are the two numbers?

36 Siôn gave Rhys $\frac{2}{5}$ of his sweets.
Rhys gave a quarter of these sweets to his sister, Bethan.
What fraction of Siôn's sweets did each person get?

Fluency

37 These are the midday temperatures during one week on Yr Wyddfa.
Work out the mean temperature for the week.

Sunday −10°C
Monday −7°C
Tuesday −5°C
Wednesday 0°C
Thursday 2°C
Friday 2°C
Saturday −3°C

38 Rearrange $y = mx + c$ to make:
- a c the subject
- b x the subject
- c m the subject.

Strategic competence

39 These five numbers are all the same distance from one another.

Copy the number line and fill in the gaps.

Remember
To calculate the mean, you add up all the numbers and then divide by how many there are.

9 Equations

Coming up...
- Solving equations review
- Recognising an inequality
- Solving equations with an unknown on both sides
- Solving equations involving brackets
- Solving word problems

Ink splats!
Aki's pen has leaked all over his maths homework.
Find the missing number under the ink splats.

1. $12 - \blacksquare = 8$
2. $\blacksquare + 7 = 22$
3. $9 - \blacksquare = 4$
4. $\blacksquare \times 4 = 12$
5. $\blacksquare \div 3 = 5$
6. $2 \times \blacksquare + 3 = 11$
7. $9 - \blacksquare \times 4 = 1$
8. $\blacksquare - (-13) = 13$
9. $18 \div \blacksquare = 3$
10. $\blacksquare - 2 \times 4 = 7$
11. $18 - 6 \times \blacksquare = 0$
12. $12 - 2 \times (\blacksquare) = 22$

9.1 Solving equations review

Skill checker

1. Solve these equations.
   - a $\quad a + 6 = 10$
   - b $\quad 3 + b = 9$
   - c $\quad c + 4 = 12$
   - d $\quad d - 2 = 9$
   - e $\quad e - 5 = 4$
   - f $\quad 15 - f = 12$
   - g $\quad 10 - g = 4$
   - h $\quad 4h = 12$
   - i $\quad 5i = 20$

2. Find the value of each expression when $x = 6$.
   - a $\quad x + 3$
   - b $\quad 15 - x$
   - c $\quad 3x$
   - d $\quad 10 + 2x$
   - e $\quad \dfrac{12}{x}$
   - f $\quad \dfrac{x}{3}$

3. Work out these calculations.
   - a $\quad 4 + (-2)$
   - b $\quad 4 - (-2)$
   - c $\quad -4 \times (-2)$
   - d $\quad 4 \div (-2)$
   - e $\quad -4 \div (-2)$
   - f $\quad -4 \times (-2)$
   - g $\quad -4 + (-2)$
   - h $\quad -4 - (-2)$
   - i $\quad -2 - (-4)$

Remember
$\dfrac{12}{x}$ means '$12 \div x$'.

Note
See Chapter 6 for a reminder of negative numbers.

▶ Solving equations with an unknown on one side

Remember that an **equation** says that one expression is equal to another.

For example,

$$4x - 1 = 2 \text{ or } 20 - 2y = 24.$$

x and y are called 'unknowns' (variables).

The solution doesn't have to be a positive whole number – it could be negative or a fraction or both!

Solving an equation means finding the value of the **unknown** that makes the equation true – this is the **solution** to the equation.

You can use the **balance method** to solve an equation.

Apply the same operation to both sides of the equation to keep it balanced.

Worked example

Solve these equations.

a $4x - 1 = 2$

b $20 - 2y = 24$

Note

A function machine could be used to solve the equation in example a.

How could a function machine be used in example b?

Solution

a
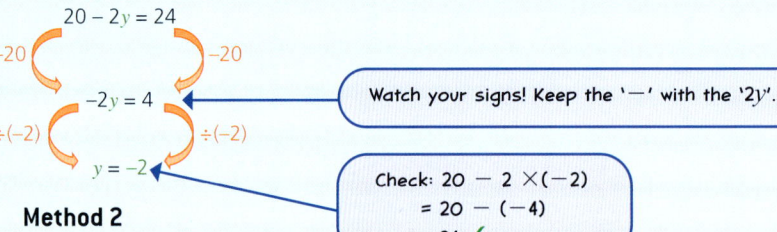

So $x = \dfrac{3}{4}$.

Imagine a pair of old-fashioned scales – to keep them balanced you have to add the same amount to BOTH sides.

You can check your answer by substituting it back into the equation.

$3 \times \dfrac{3}{4} - 1$
$= 3 - 1$
$= 2$ ✓

b **Method 1**

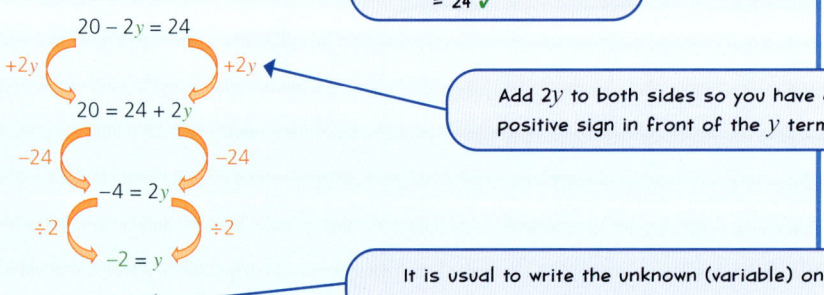

Watch your signs! Keep the '−' with the '2y'.

Check: $20 - 2 \times (-2)$
$= 20 - (-4)$
$= 24$ ✓

Method 2

$20 - 2y = 24$

$+2y +2y$

$20 = 24 + 2y$

$-24 -24$

$-4 = 2y$

$\div 2 \div 2$

$-2 = y$

So $y = -2$.

Add $2y$ to both sides so you have a positive sign in front of the y term.

It is usual to write the unknown (variable) on the left-hand side.

Some equations involve fractions. You can clear the fraction by multiplying both sides of the equation by the denominator (bottom) of the fraction.

If there is more than one fraction, you multiply by the common denominator of **ALL** the fractions.

9 Equations

Worked example

Solve these equations.

a $\dfrac{12}{a} = 3$ b $\dfrac{b}{2} + 7 = 6$

Solution

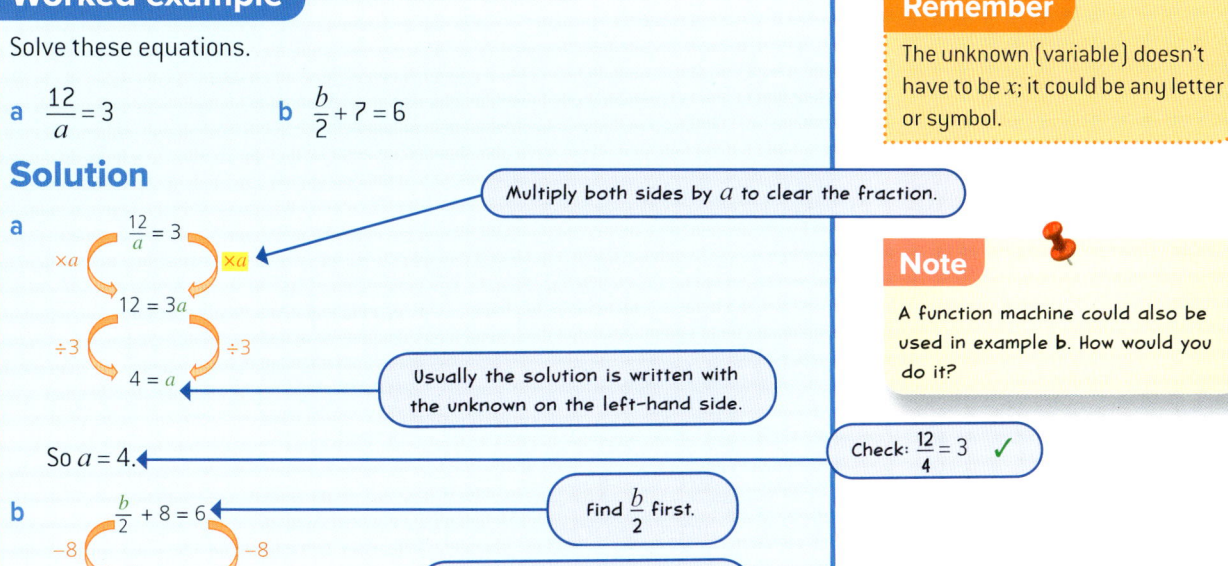

Multiply both sides by a to clear the fraction.

Usually the solution is written with the unknown on the left-hand side.

So $a = 4$.

Check: $\dfrac{12}{4} = 3$ ✓

Find $\dfrac{b}{2}$ first.

Multiply both sides by 2 to clear the fraction.

Check: $\dfrac{-4}{2} + 8$
$= -2 + 8$
$= 6$ ✓

So $x = -4$.

Remember
The unknown (variable) doesn't have to be x; it could be any letter or symbol.

Note
A function machine could also be used in example **b**. How would you do it?

Activity

Osian has got his Maths homework wrong.

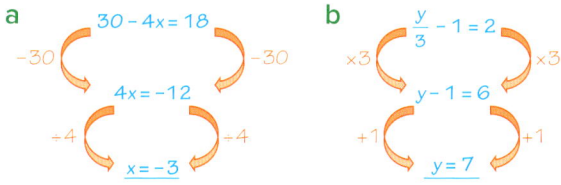

① How can you show Osian that he has the wrong answers without solving the equation?
② Where has Osian gone wrong?
③ Work out the correct solutions to Osian's equations.

Some problems can be solved by first writing down an equation and then solving it.

Worked example

In the diagram, all angles are in degrees.
The two angles form a right angle.

a Write down an equation in x.
b Work out the size of each angle.

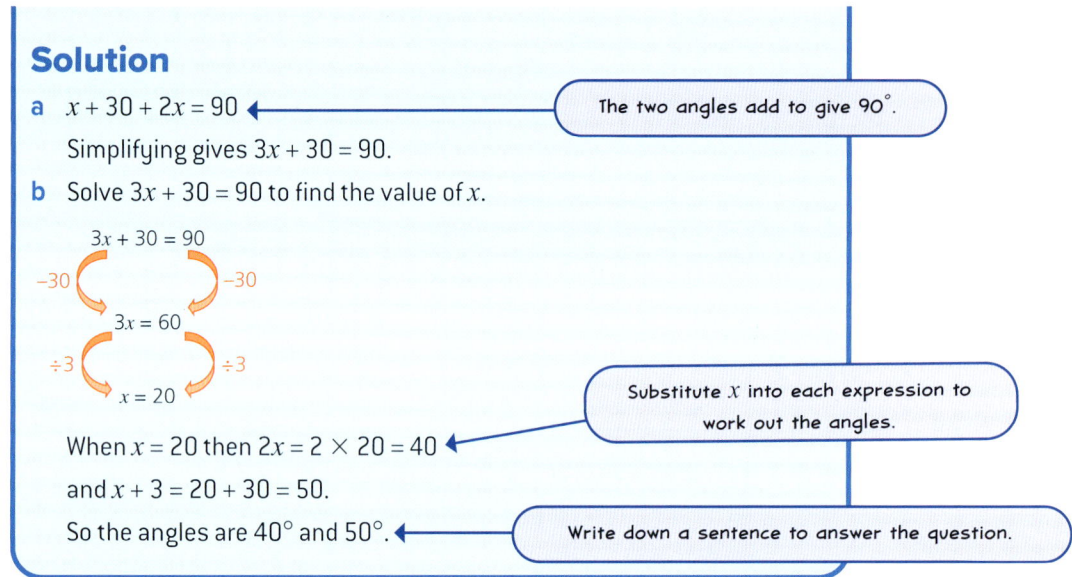

Solution

a $x + 30 + 2x = 90$ ← The two angles add to give 90°.

Simplifying gives $3x + 30 = 90$.

b Solve $3x + 30 = 90$ to find the value of x.

$3x + 30 = 90$
$-30 \quad -30$
$3x = 60$
$\div 3 \quad \div 3$
$x = 20$

When $x = 20$ then $2x = 2 \times 20 = 40$
and $x + 3 = 20 + 30 = 50$.

So the angles are 40° and 50°. ← Write down a sentence to answer the question.

(Substitute x into each expression to work out the angles.)

▶ Inequalities

An inequality says that one expression is **not equal** to another.

It uses an inequality symbol in place of an equals sign (=).

For example, Jasmyn thinks of a number.

She says 'My number is **more than 10**.'

Jasmyn writes this as an inequality:

$n > 10$

> **Remember**
>
> $>$ means 'greater than' (more than).
> $<$ means 'less than'.
> $\geq$ means 'greater than or equal to'.
> $\leq$ means 'less than or equal to'.

Communication using symbols

Discussion activity

Can you think of a way to help remember what these symbols mean?

Activity

Match together these statements with their inequalities.

n add 2 is less than or equal to 4	n is greater than 4	$2n > 4$	$n + 2 \geq 4$
n is less than or equal to 4	n is greater than or equal to 4	$\frac{n}{2} < 4$	$n - 2 \geq 4$
Twice n is greater than 4	n is less than 4	$n < 4$	$n > 4$
Half n is less than 4	Half n is greater than 4	$\frac{n}{2} > 4$	$n \geq 4$
2 more than n is greater than or equal to 4	2 less than n is greater than 4	$n \leq 4$	$n + 2 \leq 4$

9.1 Now try these

Band 1 questions

1. Solve these equations.
   - a $a - 2 = 7$
   - b $b + 7 = 9$
   - c $5 + c = 16$
   - d $d - 1 = 2$
   - e $8 - e = 5$
   - f $17 - f = 1$

2. Solve these equations.
   - a $5p = 15$
   - b $3q = 18$
   - c $24 = 4r$
   - d $\frac{s}{6} = 2$
   - e $2 = \frac{t}{7}$
   - f $3 = \frac{u}{2}$

3. Solve these equations. *(Show how to check one of your solutions.)*
   Show clearly each step of your working.
   - a $5r + 3 = 8$
   - b $3 + 2s = 21$
   - c $5t - 2 = 3$
   - d $2 + 4u = 34$
   - e $15v - 3 = 27$
   - f $9w - 20 = 52$
   - g $7x - 10 = 25$
   - h $8 + 4y = 8$
   - i $7z - 45 = 25$

4. Solve the equations to find the number which should go in each box.
   - a $2 \times \square = 4.2$
   - b $\square + 6.2 = 11$
   - c $5.4 = 2.1 + \square$
   - d $\frac{\square}{4} = 3.2$
   - e $10 - \square = 12$
   - f $18 = 15 - \square$

5. Solve the equations to find the number which should go in each box.
   - a $2 \times \square - 4.6 = 10$
   - b $3 \times \square - 0.5 = 1$
   - c $4.2 + \frac{\square}{2} = 7.2$

Band 2 questions

6. Solve these equations. *(Show how to check one of your solutions.)*
   Show clearly each step of your working.
   - a $3a + 8 = 2$
   - b $2b + 5 = 10$
   - c $5c - 4 = 18$
   - d $\frac{d}{2} + 5 = 3$
   - e $\frac{e}{2} + 2 = 3.2$
   - f $\frac{f}{2} - 5 = -6$

7. Solve these equations.
   - a $-4j = 12$
   - b $10 - 2k = 7$
   - c $7 - 4l = 15$
   - d $6 - \frac{m}{3} = 7$
   - e $15 - 3n = 1.2$
   - f $5 = 4 - \frac{p}{4}$

8. Decide whether each of these is an expression, formula, equation or inequality.
   Make a copy of the table and place a tick in the correct column for each one.

	Expression	Formula	Equation	Inequality
$2b + 4 = 10$				
$P = 2l + 2w$				
$2l + 2w$				
$6 + a < 10$				

9 Look at this algebra wall.

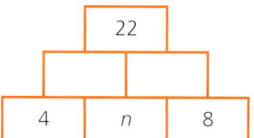

The number in each brick is the **sum** of the numbers in the two bricks beneath it.
   a   Make a copy of the algebra wall.
       Fill in expressions for the numbers in the bricks in the middle row.
   b   Write down an equation for n.
   c   Solve your equation to find the value of n.

10 The symbols ☺, ☺ and ☹ represent numbers.
   ☺ + ☺ + ☺ = ☺   ☹ × ☺ = ☺   ☹ + ☹ = ☺
   a   Show that ☹ = 3.
   b   Work out the value of each symbol.

11 The lengths of the sides of the triangle are in centimetres.
   The perimeter of the triangle is 17 cm.
   a   Find a simplified expression for the perimeter of the triangle.
   b   Write down an equation for the perimeter of the triangle.
   c   Solve your equation and work out the length of each side of the triangle.

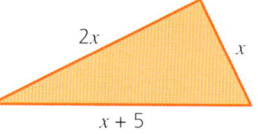

Band 3 questions

12 a   Josh's teacher writes the equation $a + b = c$ on the board.
       Josh says, 'That means that $b - c = a$ is also true.'
       Show that Josh is wrong.
   b   Given $a + b = c$, write down which of the following equations are also true.

   $b + a = c$ $c - b = a$ $c - a = b$ $a - c = b$ $c = a + b$

   c   Given $xy = z$, write down two other equations using x, y and z which are also true.

13 Solve these equations.
   a $\dfrac{10}{j} = 2$ b $\dfrac{1}{2}k = 8$ c $\dfrac{2l}{3} = 8$
   d $\dfrac{12}{m} - 4 = 0$ e $\dfrac{8}{5n} = 2$ f $3 - \dfrac{5p}{2} = 6$

14 In the diagram, all angles are in degrees.

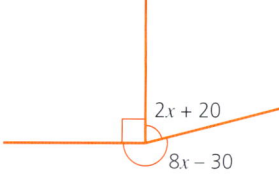

Work out the size of the acute angle.

15 All the lengths on the rectangle are in centimetres.
Work out the area of the rectangle.

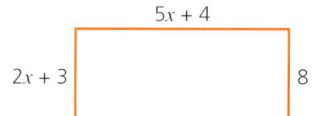

9 Equations

16 Sayeed is x years old.
Tomos is 5 years older than Sayeed.
Rhiannon is twice as old as Sayeed.
The total of their ages is 153.
How old is Sayeed?

17 Look at how Lilia solves the equation $\frac{x}{2} + \frac{x}{3} = 5$.

	$\frac{x}{2} + \frac{x}{3} = 5$
Multiply by 6:	$3x + 2x = 5$
Simplify:	$5x = 5$
Divide by 5:	$x = 1$

a Without solving the equation, show that Lilia has the wrong answer.
b Find and correct Lilia's mistake.
c Solve these equations.
   i $\frac{x}{2} - \frac{x}{3} = 1$
   ii $\frac{x}{4} + \frac{x+2}{3} = 3$
   iii $\frac{2}{x+4} = \frac{1}{x}$

9.2 Solving equations with an unknown on both sides

Skill checker

Play a game of equation 'Four in a line' with a partner.

Arrange the numbers 1–16 randomly in a four-by-four grid like this one.

13	2	12	4
3	14	11	7
6	1	8	16
9	10	15	5

Take it in turns to choose an equation from the ones below.

$5w - 6 = 29$	$-3 = 30 - 3k$	$\frac{p}{4} + 3 = 5$	$2t - 5 = 3$
$28 - 2z = 0$	$6 = 20 - 7y$	$9r - 30 = 60$	$5 - 3x = 2$
$\frac{n}{6} - 2 = 0$	$3b - 15 = 12$	$3f - 20 = 19$	$\frac{24}{a} + 3 = 7$
$60 - 3d = 15$	$5 = \frac{15}{c}$	$2s - 10 = 22$	$4 + 3u = 19$

Both of you should solve the equation and cross out the solution from your grid.
The first person to cross out a row, column or diagonal of four from their grid is the winner!

Solving equations

Look at these scales.

They have some identical parcels and 1 kg weights.

These scales represent the equation:

$3x + 1 = 2x + 5$

To find the mass of one parcel, you need to solve the equation.

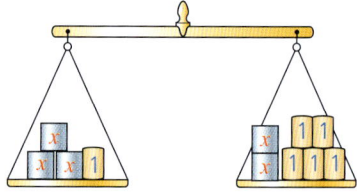

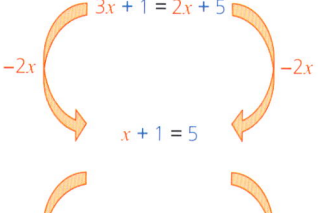

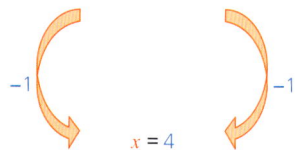

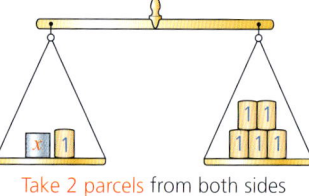

Take 2 parcels from both sides

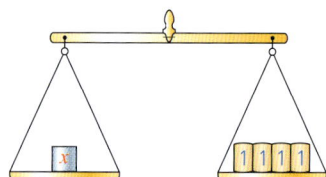

Take 1 kg from both sides

Check:
$3 \times 4 + 1 = 2 \times 4 + 5$
$12 + 1 = 8 + 5$
$13 = 13$ ✓

Both sides of the equation have the same value when $x = 4$. So the solution is correct.

Think carefully about which side you should eliminate x from.

Look for the side with the smallest amount of x, and eliminate x from that side.

Worked example

Solve $4 + 2x = 19 - x$.

Solution

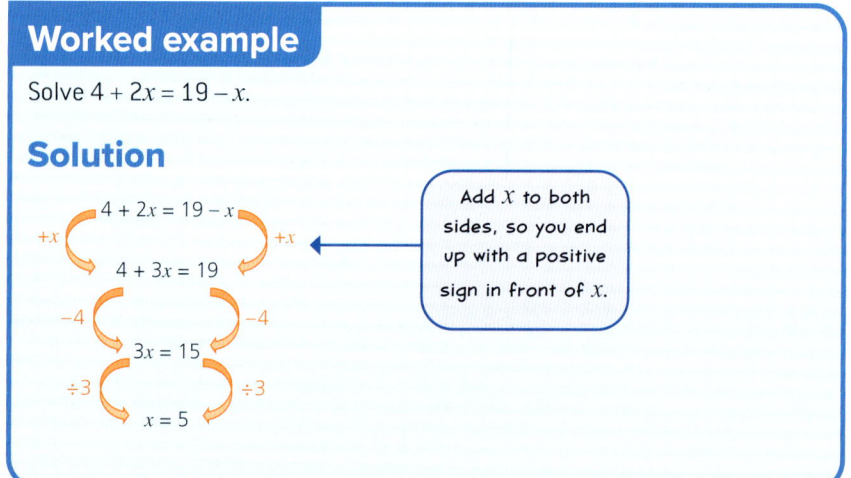

Add x to both sides, so you end up with a positive sign in front of x.

9 Equations

You can use algebra tiles to help you solve equations.

These tiles show the equation:

$3x - 4 = 6 - 2x$

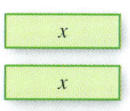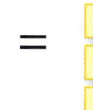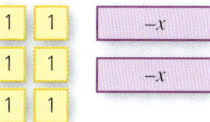

Use algebra tiles to show the equation.

Adding $2x$ to both sides gives:

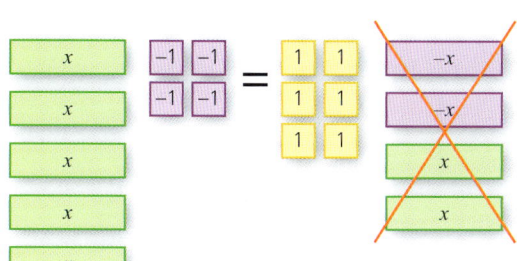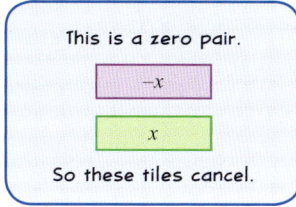

This is a zero pair.

So these tiles cancel.

$5x - 4 = 6$

Adding 4 to both sides gives:

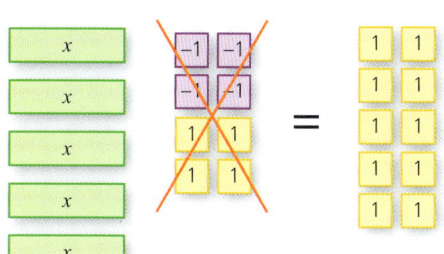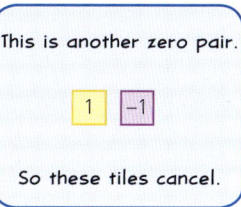

This is another zero pair.

So these tiles cancel.

$5x = 10$

Dividing both sides by 5 gives:

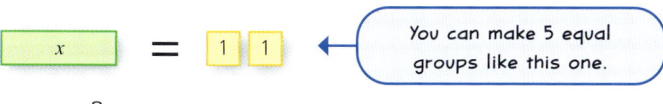

You can make 5 equal groups like this one.

$x = 2$

Activity

Use algebra tiles to solve these equations.

① $5x + 3 = 8$
② $4 - 2x = 6$
③ $4x + 3 = x + 9$
④ $x + 3 = 7 + 3x$
⑤ $6x - 8 = 13 - x$
⑥ $12 - 2x = 2x - 8$

Some word problems can be solved by first writing down an equation and then solving it.

Curriculum for Wales Mastering Mathematics: Book 2

Worked example

Cai and Heulwen do the same amount of exercise at the sports centre.

Cai goes swimming and then spends 1 hour playing football.

Heulwen only goes swimming.

She swims for four times longer than Cai does.

a Work out how long Cai goes swimming for.

b Work out how long Heulwen goes swimming for.

Solution

a Let s = number of minutes Cai spends swimming. *(Start by saying what the letter symbol stands for.)*

Cai swims for s minutes and plays football for an hour.

Heulwen swims for $4 \times s$ minutes.

Write down expressions for the amount of exercise they each do.

Heulwen: $4s$ **Cai: $s + 60$**   *(Watch out! You must work with the same units, so change 1 hour to 60 minutes as s is in minutes.)*

Heulwen's amount of exercise = Cai's amount of exercise

So

$4s = s + 60$ $-s$
$3s = 60$ $\div 3$
$s = 20$

So Cai spends 20 minutes swimming.

Check: $4s = s + 60$
$4 \times 20 = 20 + 60$ ✓

b Heulwen spends 4×20 minutes = 80 minutes swimming.

Activity

① Show how the example above would have been different in each of these three situations.

a s = number of hours Cai spends swimming

b s = number of minutes Heulwen spends swimming

c s = number of hours Heulwen spends swimming

② Which method do you think is easiest?

9.2 Now try these

Band 1 questions

① Complete the working to solve these equations.

a

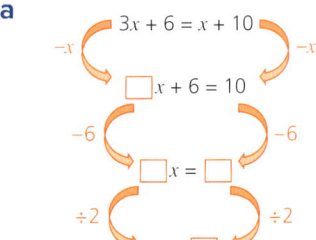

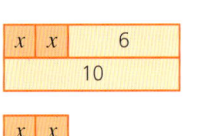

b $6x + 1 = x + 16$

$\square x + 1 = 16$
$\square x = \square$
$\div \square$
$x = \square$

c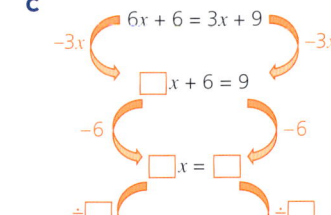

You can use a bar model diagram to help.

9　Equations

② Write down an equation to match each balance problem.

For example, the equation for part **a** is $2x = 12$.

Solve your equations to find x, the weight of one coloured cylinder in each case.

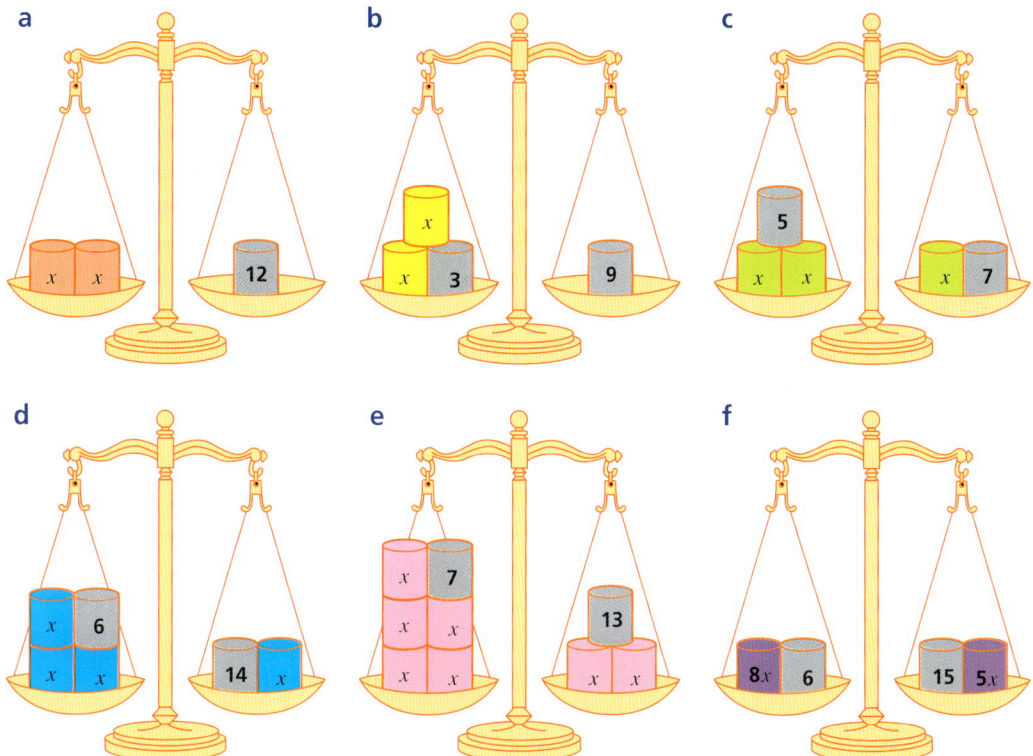

③ In each part, the diagram shows two bars which are the same length.

For each part, work out **i** the value of the letter and **ii** AB, the total length of one bar.

a
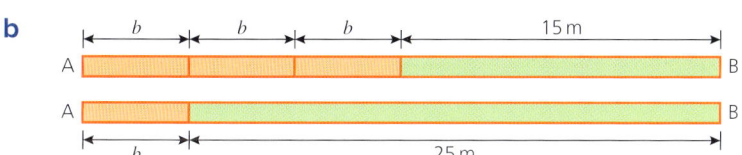

b
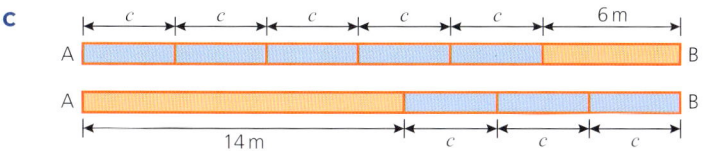

c

Band 2 questions

④ Solve these equations. ← Show how to check one of your solutions.

   a $3x + 5 = 2x + 9$　　b　$4x + 6 = 8 + 3x$　　c　$7 + 3x = 4x + 2$　　d　$3 + 7x = 7 + 3x$

⑤ Solve these equations.

   a $3x - 2 = 4x - 5$　　b　$7x - 8 = 4x + 4$　　c　$5x - 5 = 2x + 4$　　d　$6x - 10 = 4x$

6 Solve these equations.
   a $2a + 7 = 3a + 4$
   b $7b - 5 = 6b + 1$
   c $4c + 3 = 2c + 5$
   d $2d - 4 = 8 - d$
   e $6e + 3 = 12 - 3e$
   f $7f - 2 = 4f + 7$
   g $3 + 2g = 13 - 3g$
   h $10 - h = 16 - 3h$

7 Aled and Bryn are having an argument about this equation.
   $$7x - 8 = 19 - 2x$$
   Aled says, 'I think $x = 4$.'
   Bryn says, 'No, you are wrong. $x = 3$.'
   a How can you check who is right?
   b Who is right?
   c Show how to solve the equation.
      Explain your method at each stage.

8 Priya thinks of a number, x.
   She says, 'I think of a number. I multiply it by 4 and add 2. The answer is four more than twice my number.'
   a Write down expressions for:
      i 'multiply my number by 4 and add 2'
      ii 'four more than twice my number'.
   b Form an equation for x.
   c Solve your equation to find Priya's number.

9 Harri is buying some pencils from the local shop.
   He says, 'With the money in my pocket I can buy 6 pencils and get 32p change or 4 pencils and get 88p change.'
   a Write down an expression (use p for the cost in pence of one pencil) for the cost of buying:
      i six pencils and getting 32p change
      ii four pencils and getting 88p change.
   b Form an equation with your two expressions by making them equal.
   c Solve your equation to find the cost of one pencil.
      Check you have solved the equation correctly.
   d How much money does Harri have?

Band 3 questions

10 Solve these equations.
   a $2.1x + 0.4 = 1.1x + 2.4$
   b $3.1x + 0.4 = 1.1x + 2.4$
   c $2.4x + 3.9 = 9.9 + 0.4x$
   d $2.9x - 0.5 = 3.1 + 0.9x$
   e $10.4x + 12 = 36 - 1.6x$
   f $4.2x + 10 = 9.2x + 15$

11 Pedr is the penguin keeper at Afonffordd Zoo.
   He is designing a new enclosure for the penguins.
   It must fit around the penguin pond.
   Here are his two designs.
   The measurements are in metres.

   i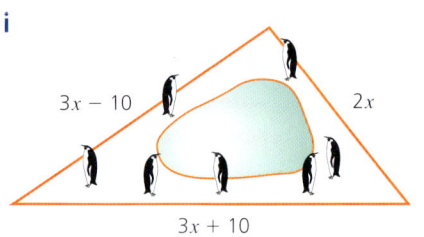

   ii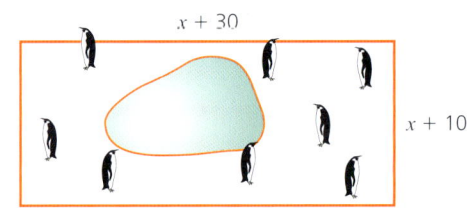

Strategic competence

a Write expressions for the perimeters of the enclosures.
Simplify your expressions.
b Both penguin enclosures use the same length of fencing.
Write an equation for x.
c Solve your equation to find x.
d What are the dimensions of each penguin enclosure?
e How much fencing will Pedr need for each enclosure?

12 Megan says, 'I think of a number. I multiply it by 3 and subtract 6. This gives the same answer as when I subtract my number from 30.'

What is Megan's number?

13 Find the area of this rectangle.

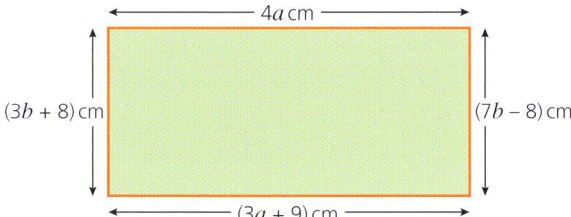

14 Benji has £n.
Tilly has five times as much money as Benji.
Tilly gives £10 to Benji.
Now they both have the same amount of money.

a Use this information to form an equation for n.
b Solve your equation to work out how much money Benji has now.

9.3 Solving equations with brackets

Skill checker

Match together equivalent expressions.

$4(4x+2)$	$2(8x+4)$	$12x-6$	$6(3x+2)$
$3(6x+4)$	$24-8x$	$18x+12$	$8(1+2x)$
$8(3-x)$	$2(12-4x)$	$3(4x-2)$	$6(2x-1)$
$8+16x$	$12+18x$	$4(6-2x)$	$3(2x+6)+6(2x-1)$

▶ Solving an equation with brackets

Some equations involve brackets.

Worked example

Solve the equation $5(3x - 7) = 10$.

Solution

Method 1: Expanding the brackets

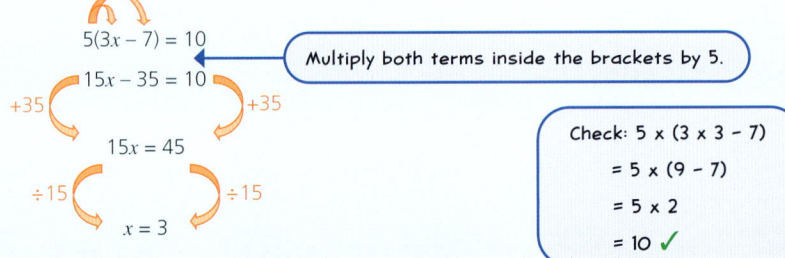

Multiply both terms inside the brackets by 5.

Check: $5 \times (3 \times 3 - 7)$
$= 5 \times (9 - 7)$
$= 5 \times 2$
$= 10$ ✓

Method 2: Dividing by the number outside the bracket

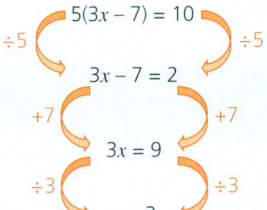

Which method do you think is easiest? Which method would be best for $3(3x - 7) = 10$? Explain why.

Note

See Chapter 8 for a reminder of expanding (multiplying out) brackets.

Remember

The unknown is only on one side, so a function machine could be used to solve this equation.

How would you do it? What are the steps involved and in what order would you write them in the function machine?

Sometimes you need to expand the bracket and then simplify before you can solve the equation.

Worked example

A cafe sells coffee and muffins.
A cup of coffee costs £c. A muffin costs £1 more than a coffee.
Carys buys 6 cups of coffee and 4 muffins. She spends £19.
Work out the cost of a cup of coffee.

Solution

A muffin costs $(c + 1)$ pounds.
Total cost of 6 coffees and 4 muffins is $6c + 4(c + 1) = 19$.
Expand the brackets: $6c + 4c + 4 = 19$
Simplify:

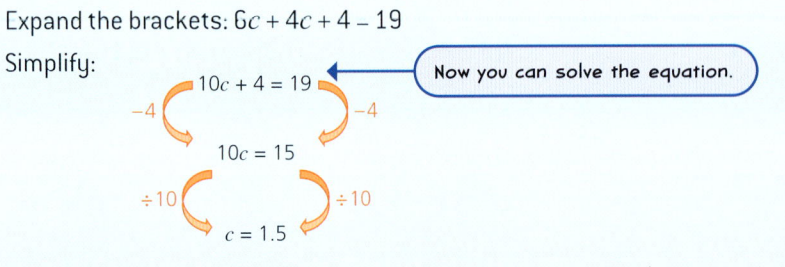

Now you can solve the equation.

A cup of coffee costs £1.50.

Remember

Always use two decimal places for money.

9 Equations

Activity

Choose TWO of the following equations.

$5a - 6 = 14$	$30 = 9 - 7k$	$\dfrac{b}{3} - 4 = 2$
$15 - 3x = 3$	$3(s - 6) = 12$	$5(2p + 3) = 32$
$3(e - 4) + 2(3e + 1) = 35$	$6t - 4 = 4t + 6$	$4(2y - 4) = 10(11 - y)$

Design a poster showing clearly all the steps required to solve each equation.
Show how to use algebra tiles on your poster.
On your poster, show how to check the solution to each equation.

9.3 Now try these

Band 1 questions

1. **a** **i** Expand $2(x + 3) = \Box x + \Box$.
   ii Use your answer to help you solve $2(x + 3) = 10$.
   b **i** Expand $5(x + 2) = \Box x + \Box$.
   ii Use your answer to help you solve $5(x + 2) = 25$.
   c **i** Expand $4(x + 1) = \Box x + \Box$.
   ii Use your answer to help you solve $4(x + 1) = 8$.

2. Look at this rectangle.
   All the lengths are in cm.
   a Complete these expressions for the area of the rectangle.
   i $\Box \times (x + \Box)$
   ii $\Box x + \Box$
   b The total area of the rectangle is 18 cm².
   Use your answers to part **a** to help you solve $3(x + 4) = 18$.

3. Look at this rectangle.
   All the lengths are in cm.
   a Complete these expressions for the area of the rectangle.
   i $\Box \times (x + \Box)$
   ii $\Box x + \Box$
   b The total area of the rectangle is 8 cm².
   Use your answers to part **a** to help you solve $2(x + 1) = 8$.

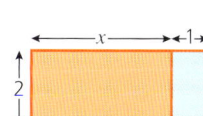

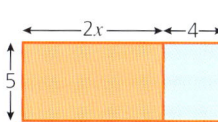

4. **a** Use this rectangle to help you expand $5(2x + 4)$.
   b Use your answer to help you solve $5(2x + 4) = 60$.

Logical reasoning

5 a Aloke and Ruby are solving the equation $3(x + 4) = 18$.

Here is the start of their working.

Aloke
$3(x + 4) = 18$
$3x + 12 = 18$
$3x = 6$
$x =$ _____

Ruby
$3(x + 4) = 18$
$(x + 4) = \dfrac{18}{3}$
$x + 4 = 6$
$x =$ _____

Complete their working.

Explain each step of their working.

b Use both methods to solve these equations.

   i $4(d + 2) = 20$ **ii** $5(h - 3) = 15$ **iii** $5(x + 4) = 20$

   iv $2(3x + 1) = 20$ **v** $3(2m - 4) = 12$

Band 2 questions

Fluency

6 Solve these equations by expanding the brackets first.

Check one of your solutions by substituting back into the original equation.

  a $4(w + 2) = 14$ **b** $5(x - 3) = 11$ **c** $10(7 - y) = 16$ **d** $8 = 5(2z - 4)$

7 Solve these equations by dividing both sides by the number outside the bracket first.

  a $4(q + 3) = 20$ **b** $6(r - 7) = 18$ **c** $2 = 2(3s - 8)$ **d** $5(6 - t) = 15$

Strategic competence

8 Ali is thinking of a number.

She says, 'I think of a number. I double it and subtract 6. When I multiply the result by 3 the answer is 24.'

a Copy and complete the following table to build up an equation for Ali's number.

Instruction	Algebra
I think of a number	n
Double it ...	
Subtract 6 ...	
Multiply the result by 3 ...	
The answer is 24.	

b Solve your equation to work out Ali's number.

c Show how you can check your solution is right.

9 a Find an expression for the perimeter of this rectangle.

b The perimeter of the rectangle is 32 cm.

   Write down an equation for this information.

c Solve your equation to find the value of b.

d What is the length and width of the rectangle?

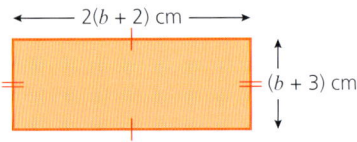

Fluency

10 a Expand the following expressions.

   i $4(3x + 2)$ **ii** $6(2x - 1)$

b Use your answers to part **a** to help you simplify $4(3x + 2) + 6(2x - 1)$.

c Use your answer to part **b** to help you solve $4(3x + 2) + 6(2x - 1) = 122$.

9 Equations

11 Solve these equations.
- a $\quad 7(b+2) - 3b = 18$
- b $\quad 4(3x-2) - 7x = 2$
- c $\quad 2(5a-3) + 8 - 4a = 44$
- d $\quad 3(p+2) + 2p = 36$
- e $\quad 5(q-1) - 4q = 9$

12
- a Write down an expression for the sum of the angles in this triangle. Simplify your expression.
- b Write down an equation for the sum of the angles in this triangle.
- c Use your expression to work out each of the angles in the triangle.

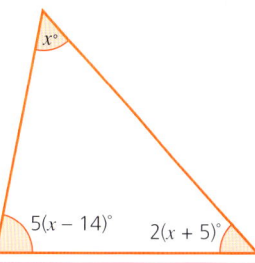

Band 3 questions

13 Solve these equations.
- a $\quad 2(b+3) = 7b+1$
- b $\quad 5(f-1) = 4f+1$
- c $\quad 4(2e-3) = 6e+2$
- d $\quad 4(2a+1) = 5a+7$
- e $\quad 5(3n-2) = 10n$

14 Solve these equations.
- a $\quad 4(k+2) = 24$
- b $\quad 7(6-2r) = 28$
- c $\quad 5t+8 = 2(2t+9)$
- d $\quad 3(n-1) = 3-3n$
- e $\quad 4(g+1) = 3(g+3)$

15 Solve these equations.
- a $\quad 2(p-2) = p-1$
- b $\quad 3(x+2) = 2x+13$
- c $\quad 3(2-3r) + 2 = 2(5-r)$
- d $\quad 3(2h-3) = 3+4h$
- e $\quad 2(5-k) = 4(3-k)$

16 Work out the perimeter of this rectangle.

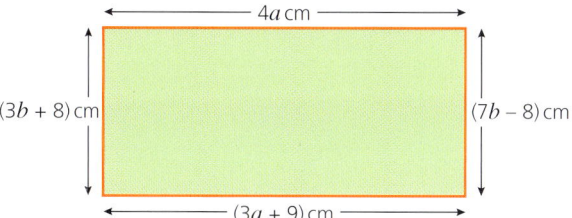

17 How can you solve $5(4x-6) - 2(4x-6) = 6$ without multiplying out any brackets?

18 Look at this two-way function machine.

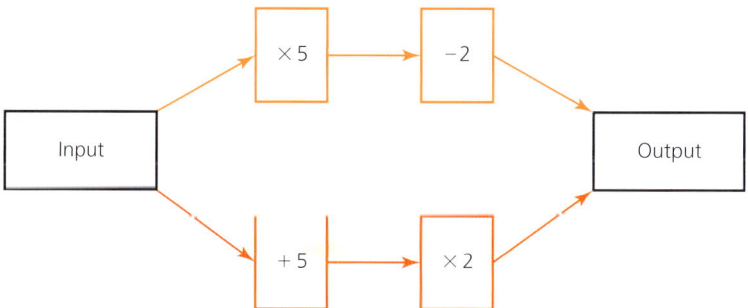

- a Find the input which gives the same output whichever path you take on the function machine.
- b Design your own two-way function machine that gives the same output either way round for:
  - i exactly one input
  - ii two inputs
  - iii all inputs.

Key words

Here is a list of the key words you met in this chapter.

Equation   Expand   Expression   Simplify   Solve   Substitute   Unknown

Use the glossary at the back of this book to check any you are unsure about.

Review exercise: equations

Band 1 questions

1 Solve each of these equations.

Make sure that each side of the equation balances the other.

Show how to check one of your solutions.

- **a** $5a = 15$
- **b** $10 - b = 7$
- **c** $7 = c - 8$
- **d** $\dfrac{d}{4} = 5$
- **e** $19 = 6 + e$
- **f** $3f + 6 = 15$
- **g** $12 + 2g = 16$
- **h** $2 = 7h - 5$
- **i** $10 - 3i = 4$

2 Alun is weighing some tins of soup using scales.

The soup tins are all the same.

Each weighs m kg.

Alun writes:

$$4m + 5 = 2m + 6$$

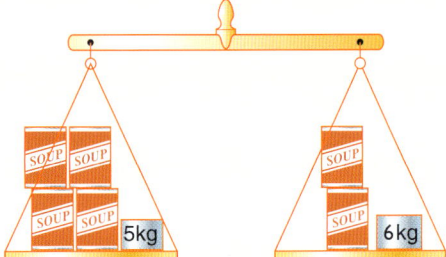

- **a** Solve Alun's equation to find the weight of one tin of soup.
- **b** Check you have solved the equation correctly.
- **c** What is the total weight on the left-hand side of the scales?

3 Ffion is f years old.

Ffion's little sister, Alwena, is 7 years younger than her.

- **a** Write down an expression for Alwena's age.

The sum of their ages is 25.

- **b** How old is Ffion?

4 Ami, Ben and Callum, share £150 between them.

Callum receives £x.

Ami gets twice as much as Callum.

Ben gets £10 more than Ami.

- **a** Complete these expressions for the amount of money that each person receives.

  Callum: x Ami: ☐x Ben: ☐$x +$ ◇

- **b** Form an equation for the total amount of money.
- **c** Solve your equation.

  How much money do they get each?

Band 2 questions

5 Solve these equations.

- **a** $2n + 1 = n + 3$
- **b** $3p - 2 = 2p + 1$
- **c** $1 + 4r = 2r + 5$
- **d** $3b - 2 = 4 + b$
- **e** $10 + 2f = 15 - 3f$
- **f** $3 + 3w = 10 - 4w$

6 Solve these equations.

Show how to check one of your solutions by substituting back into the original equation.

- **a** $5(2a + 1) = 17$
- **b** $4(3 + 2b) = 52$
- **c** $7(12 - 2c) = 28$
- **d** $6(2d - 1) = 3$
- **e** $10(8 - 4e) = 0$
- **f** $4(3 - f) = 16$

7 Janice hires a car for a day.

Her bill comes to £82.40.

Let m stand for the number of miles she drove.

Write down an equation involving m and use it to find how many miles Janice drove.

CAR HIRE
£50 per day
40p per mile

8 Look at this diagram.

- **a** Write an expression in terms of x for the sum of the three angles.
- **b** Form an equation in x and solve it to find the value of x.
- **c** Hence work out the value of each angle.

Angles: $(x + 15)°$, $(x + 5)°$, $2x°$

9 There are 18 children at a party.

There are x children not wearing fancy dress.

There are four more children wearing fancy dress than there are children not wearing fancy dress.

- **a** Write this information as an equation in x.
- **b** Solve your equation to find out how many children are not wearing fancy dress.

10 The length of a rectangle is four times its width.

The width is w cm.

The perimeter is 50 cm.

Find the area of the rectangle.

Band 3 questions

11 Solve these equations.

- **a** $4(2e + 3) = 7e + 2$
- **b** $5(3n - 2) = 10n$
- **c** $2(b + 3) = b + 7$
- **d** $3(x + 1) = 2x + 5$
- **e** $5(x - 4) = 2x + 1$

12 Solve these equations.

- **a** $\dfrac{x}{5} = 4$
- **b** $\dfrac{x}{4} = 5$
- **c** $\dfrac{1}{4}(x + 1) = 5$
- **d** $\dfrac{1}{4}(x - 1) = 5$

13 Solve each of these equations and write down the corresponding letter from the table.

What word do you spell?

7	−7	−$\dfrac{1}{2}$	8	10	−6	0	−4	2
P	E	R	D	G	T	U	A	S

- **a** $\dfrac{3x}{14} = \dfrac{3}{7}$
- **b** $6(x - 2) = 5x - 12$
- **c** $3(x - 2) = 5(x - 4)$
- **d** $3(x + 8) = -x - 4$
- **e** $3(2x + 1) = 0$

14 A coach party of 50 people are going to a zoo.

The total entrance cost for the whole group is £768.

Let c stand for the number of children in the group.

ZOO ENTRANCE
Adults..........£18
Children.......£12

- **a** Write down an expression for the number of adults in the group.
- **b** Write down an equation involving c.
- **c** Solve your equation to find the number of children and the number of adults in the group.

15 Julian says, 'In seven years' time, I will be twice as old as I was three years ago.'

How old is Julian now?

16 Try solving these equations.

a $4(x + 2) + 2x = 6x + 8$

b $6x - 2(2x + 1) = 2x - 2$

c $12x + 9 = 3(4x + 3)$

What happens? Why does this happen?

17 Work out the area of this square.

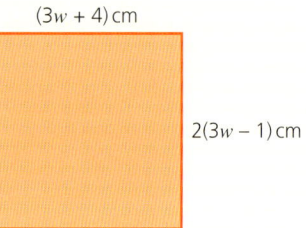

$(3w + 4)$ cm

$2(3w - 1)$ cm

18 A farmer has 40 fewer cows than she has sheep.

She has twice as many hens as she has sheep.

The total number of animals on her farm is six times the number of cows.

How many of each type of animal does she have?

Cross-curricular activity

Create a computer program which can solve equations.

First, focus on equations with unknowns on one side.

10 Working with 2D shapes

Coming up...
- Classifying special types of quadrilateral
- Finding the area and perimeter of shapes made from triangles and rectangles (review)
- Finding the area of trapeziums and parallelograms

Join the dots

You can join dots on a three-by-three square to make different quadrilaterals like this:

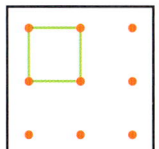 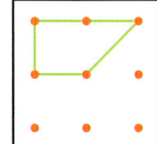

① Use square dotty paper or a 9-pin geoboard and an elastic band to find as many DIFFERENT quadrilaterals as you can.

Write down the name of as many of the quadrilaterals as you can.

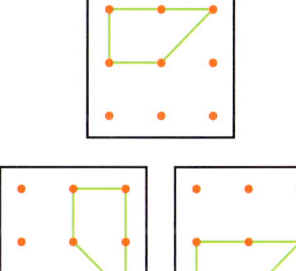

> These shapes are **congruent** (the same shape and size) so you can only include one of them.

② Choose a quadrilateral that you have found.

Describe your quadrilateral to a friend.

Can they can draw your quadrilateral without looking at your shape?

Here are some phrases you might find helpful:

| Equal sides | Opposite sides | Adjacent sides |

> Adjacent means 'next to'.

| Parallel sides | Line of symmetry | Rotational symmetry | Right-angles |

10.1 Types of quadrilateral

Skill checker

① Write down the number of lines of symmetry that these shapes have.

a 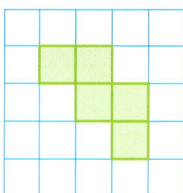   b 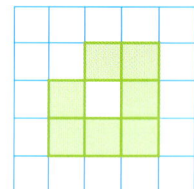   c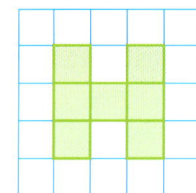

② Make a copy of these shapes on squared paper.
Shade in one more square so that the diagram has rotational symmetry of order 2.

a 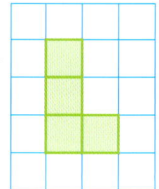   b 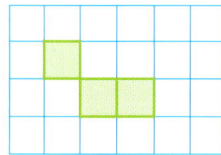   c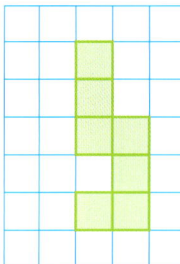

③ Which of these angles is
   a acute
   b obtuse
   c reflex?

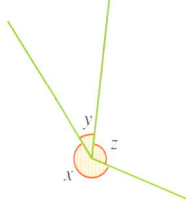

④ Match these triangles with their correct labels.

   isosceles   scalene   equilateral

   right-angled isosceles   right-angled triangle

a 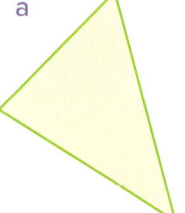   b 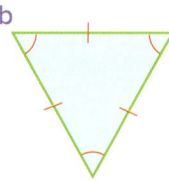   c 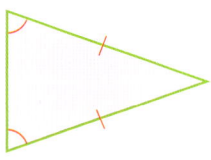   d 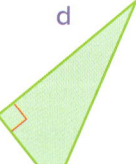   e

▶ Classifying quadrilaterals

A **quadrilateral** is a two-dimensional closed shape with four straight sides.

> **Remember**
> Any four-sided closed shape is a quadrilateral.

There are seven special types of quadrilateral to remember.

Make a copy of this table.

As you work through this section, you will find some other properties of quadrilaterals which you should add to your table.

Shape	Definition: A quadrilateral with...	Other properties
Parallelogram (Paralelogram)	2 pairs of parallel sides	Opposite angles equal
Rhombus (Rhombws)	4 equal sides	2 pairs of parallel sides Opposite angles equal
Rectangle (Petryal)	4 right angles	2 pairs of equal-length opposite sides
Square (Sgwâr)	4 right angles 4 equal sides	

> Equal sides are marked with the same number of dashes.
> Parallel sides are marked with the same number of arrows.
> Equal angles are marked with the same number of arcs.

Trapezium (Trapesiwm) 	1 pair of parallel sides	
Isosceles trapezium (Trapesiwm isosgeles) 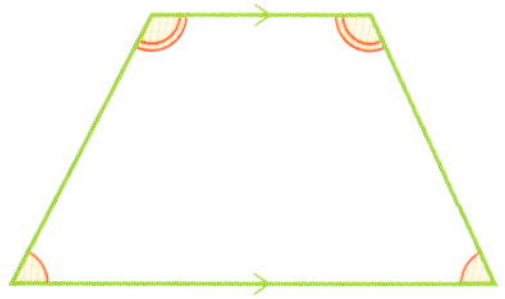	A truncated isosceles triangle (Truncated means the top has been cut off.)	1 pair of parallel sides 2 equal sides 2 pairs of equal angles
Kite (Barcut) 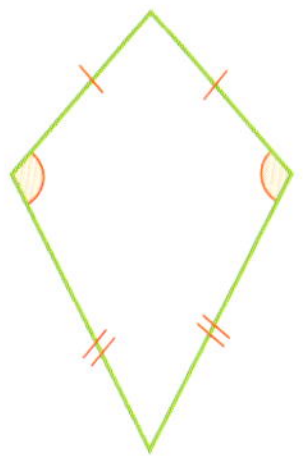	2 pairs of equal adjacent sides (Adjacent means 'next to'.)	1 pair of equal angles
Arrowhead (Blaen saeth) 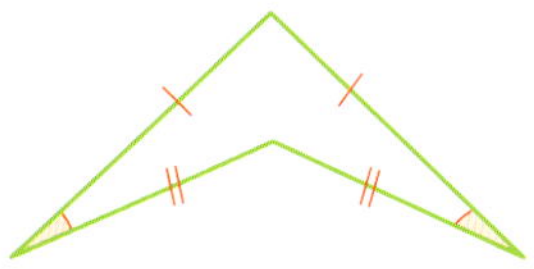	A kite with a reflex angle	2 pairs of equal adjacent sides 1 pair of equal angles 1 reflex angle

Discussion activity

The interesting building below is based on a parallelogram and is located in Hamburg, Germany.

Can you find any other interesting buildings based on quadrilaterals (or any other shape) on the internet? Who are the architects? Find out if they have designed any other famous buildings.

10 Working with 2D shapes

Activity

① a Trace this rhombus onto a piece of paper and cut it out.
   i   Fold your rhombus along BD.
       What do you notice?
   ii Open your shape and fold it along AC.
       What do you notice?
   iii Copy and complete these statements.
       The diagonals of a rhombus are lines of _____.
       AM is the same length as __.
       DM is the same length as __.
  b i   Without opening your shape between folds, fold it along AC and then BD.
         What shape have you made?
     ii Are the diagonals AC and BD equal in length?
     iii What type of angle is AMB?
② Repeat question 1 for the other special quadrilaterals.

In the activity, you found that the diagonals of a rhombus cut each other **exactly in half**.

You also found that the diagonals are perpendicular (at right angles) to each other.

You say that the diagonals **bisect** each other at right angles.

> Bisect means 'cut exactly in half'.

When classifying shapes, the key things to look for are:
- the number of sides
- the number of pairs of equal sides
- whether the diagonals bisect at right angles
- the lines of symmetry
- whether opposite sides are parallel
- the sizes of the angles
- the number of pairs of equal angles.

Remember that the angles in a quadrilateral add up to 360°.

> You proved this in Book 1.

Discussion activity

Quadrilaterals can be found in all manner of places outside the classroom.
See how many different types you can spot around your school and where you live.

Worked example

Work out the sizes of the angles a, b and c.

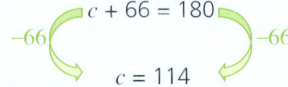

Solution

The shape is a parallelogram as it has two pairs of parallel sides.

A parallelogram has two pairs of equal angles and opposite angles are equal.

So angle a is 66°.

There are two different methods you can use to find b and c.

Method 1:

$$c + 66 = 180$$
$$c = 114$$

(−66 from both sides)

> **Remember**
> Allied angles add up to 180°.

So angle c is 114°.

Opposite angles are equal, so angle b is also 114°.

Method 2:

$$66 + a + b + c = 360$$

You know $a = 66$ and b and c are the same.

So

Replace a with 66 and b with c.

$$66 + 66 + c + c = 360$$
$$132 + 2c = 360$$

Simplify.

$$2c = 228$$
$$c = 114$$

(−132 from both sides, then ÷2)

So angle c is 114°.

Opposite angles are equal, so angle b is also 114°.

10 Working with 2D shapes

10.1 Now try these

Band 1 questions

1 Describe each of these shapes as fully and as accurately as you can.

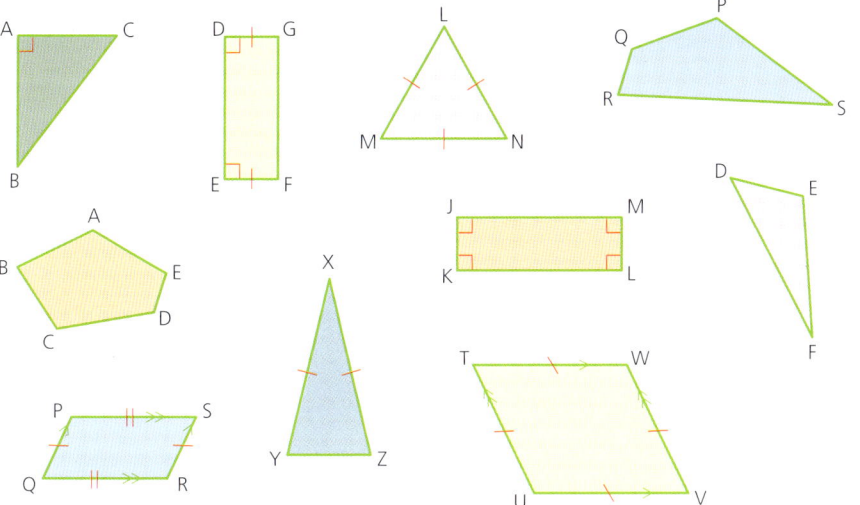

2 Find:
  i the number of lines of symmetry
  ii order of rotational symmetry of each of these shapes.
  a square
  b rectangle
  c isosceles trapezium
  d parallelogram
  e kite
  f rhombus

3 a Draw a quadrilateral with four right angles.
  What is the name of your shape?
  b Draw a different type of quadrilateral with four right angles.
  What is the name of this shape?

4 a Draw a quadrilateral with two pairs of parallel sides.
  What is the name of your shape?
  b Draw and name another type of quadrilateral with two pairs of parallel sides.
  c How many different types of quadrilateral have two pairs of parallel sides?
  Write down the names of these shapes.

Band 2 questions

5 Write down the size of each lettered angle.
Give a reason for each of your answers.

a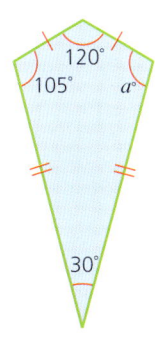
b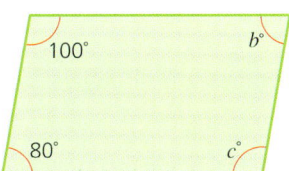
c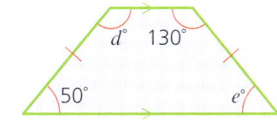

163

6 Copy and complete this table.

The first row has been completed for you.

Diagram and shape name	Opposite sides parallel	Number of pairs of equal sides	Opposite angles equal	Four right angles	Diagonals equal in length	Diagonals bisect each other	Diagonals perpendicular to each other
Rectangle	2 pairs	2 pairs	✔	✔	✔	✔	✘
(square)							
(parallelogram)							
(rhombus)							
(trapezium)							
(isosceles trapezium)							
(kite)							
(arrowhead)							

7 A regular polygon has all sides the same length and all angles the same size.
  a What is the special name for a regular triangle?
  b Is a rhombus a regular quadrilateral? Give a reason for your answer.

8   The diagram shows two sides of a quadrilateral.
    Make four copies of the diagram on squared paper.
    Use them for parts *a* to *d*.
    a   Add two sides to make a parallelogram.
    b   Add two sides to make an isosceles trapezium.
    c   Add two sides to make a trapezium with a right angle.
    d   Add two sides to make a different trapezium.

Band 3 questions

9   Work out the size of each lettered angle.
    Give a reason for each of your answers.
    a 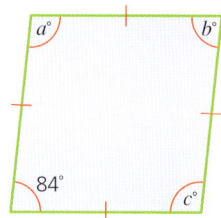   b 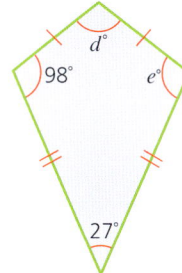   c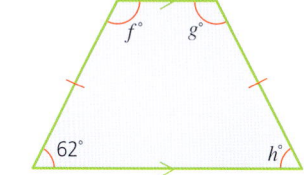

10 a   Jamil says that a rectangle is a special type of parallelogram.
        Is Jamil correct? Give a reason for your answer.
    b   Roshni says that a rhombus is a special type of square.
        Is Roshni correct? Give a reason for your answer.

11 Alwena thinks of a shape.
    She says, 'My shape has four sides and two pairs of equal sides.'
    a   Which shapes could Alwena be thinking of?
    Alwena says, 'It also has one pair of equal angles.'
    b   Which shapes could Alwena be thinking of?
    Alwena says, 'One of its angles is a reflex angle.'
    c   Which shape is Alwena thinking of?

12 Amal has two congruent right-angled triangles.

    Show how she can join her triangles to make:
    a   a rectangle
    b   a kite
    c   a parallelogram.
        Find two different answers to part **c**.

13 Fatima draws a coordinate grid and plots the points A(-1, 4) and C(3, 2) on her grid.
    A and C are the diagonally opposite vertices of a square ABCD.
    Work out the coordinates of B and D.

10.2 Area

Skill checker

1. Work out the area and perimeter of each of these shapes.

   Remember
   Area of rectangle = length × width
   Perimeter = distance around the outside of a shape

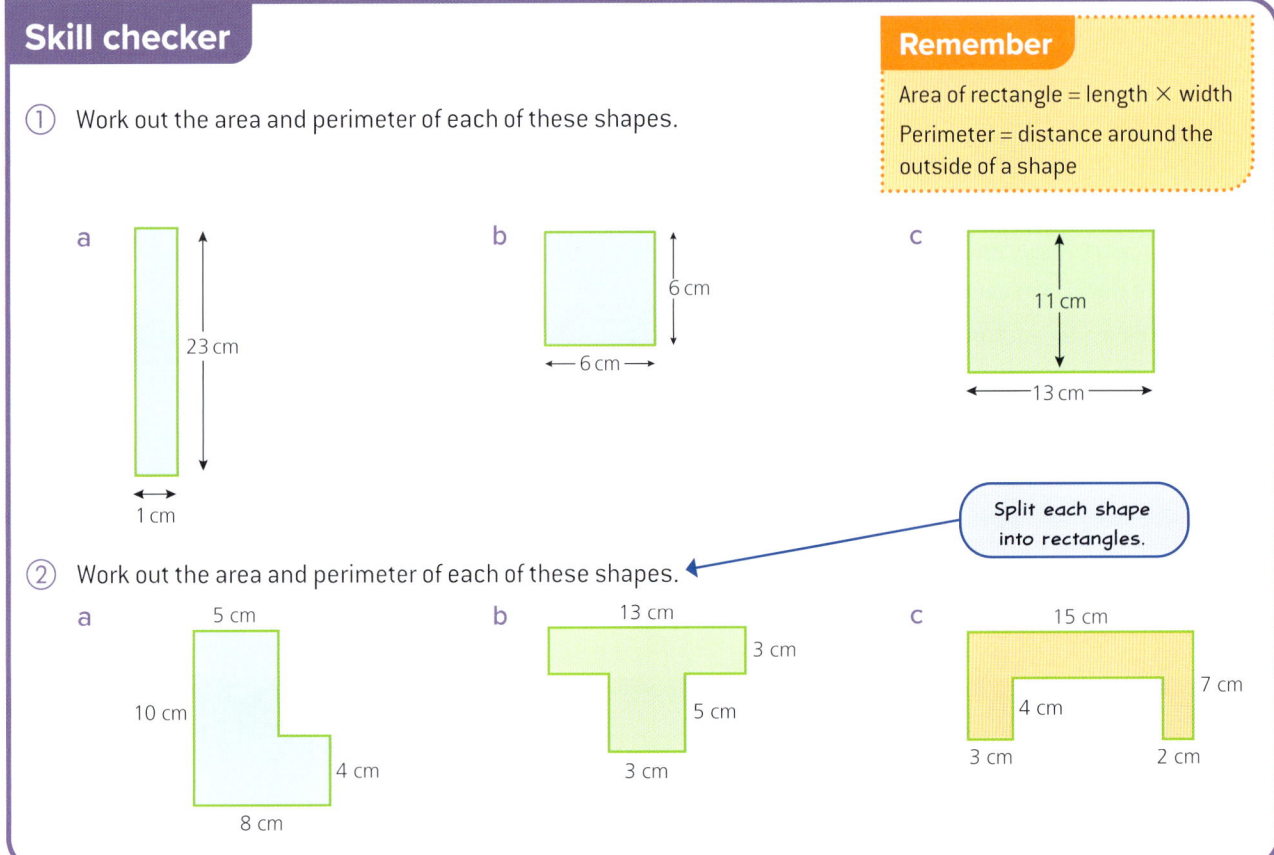

2. Work out the area and perimeter of each of these shapes. *Split each shape into rectangles.*

▶ Area of a parallelogram

Look at this rectangle.

The area of the rectangle = length × width
= 7 cm × 3 cm = 21 cm²

You can cut a triangle from one side of the rectangle and place it on the other side.

You have made a parallelogram.

The area of the parallelogram is the same as the area of the rectangle.

So area of parallelogram = length of base × height
= 7 cm × 3 cm = 21 cm²

This works for any parallelogram.

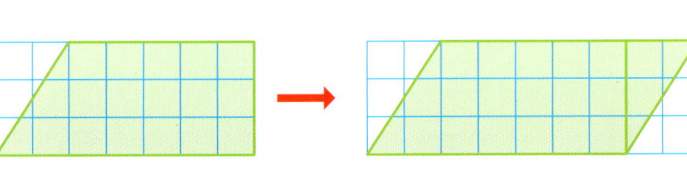

Area of a parallelogram = base × perpendicular height   In symbols this is $A = bh$

Worked example

Work out:
a the perimeter
b the area
of this parallelogram.

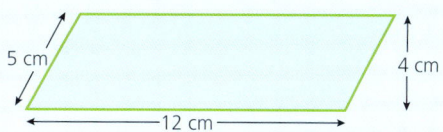

Solution

a Perimeter of parallelogram = 12 cm + 5 cm + 12 cm + 5 cm
   $\qquad\qquad\qquad\qquad\qquad\;\; = 34$ cm

b $A = bh$
   $\;\;\; = 12 \times 4$
   $\;\;\; = 48$

So the area is 48 cm².

Activity

① Draw a parallelogram.
  Show how your parallelogram can be cut in half to make:
  a two congruent triangles
  b two congruent trapeziums.
  Is there more than one way of doing this? Why/why not?
  Is this true for any parallelogram?

② Can you join any pair of congruent
  a triangles
  b trapeziums
  together to form a parallelogram?

▶ Area of a triangle – review

You can split a parallelogram into two triangles.

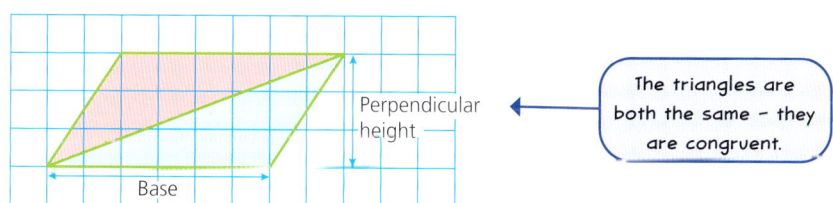

The triangles are both the same – they are congruent.

So the area of a triangle is half the area of a parallelogram.

Area of triangle = $\frac{1}{2}$ × base × perpendicular height

In symbols this is $A = \frac{1}{2} bh$

Worked example

Work out the area of this triangle.

Solution

Rotate the page so you can see which side is the base.

$A = \frac{1}{2}bh$

$= \frac{1}{2} \times 18 \times 8$

$= 72$

> Use 18 cm as the base. You know the height is 8 cm.
> You cannot use 12 cm as the base as you do not know the height of the triangle when it is that way round.

So the area is 72 cm².

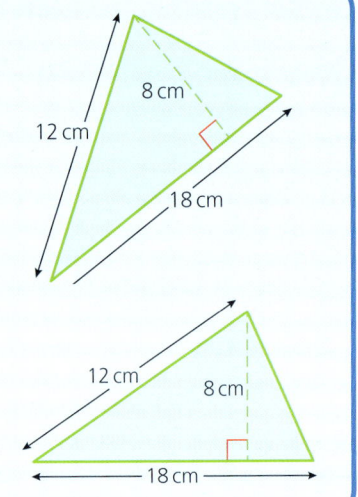

▶ Area of a trapezium

Conceptual understanding

You can split a parallelogram into two trapeziums.

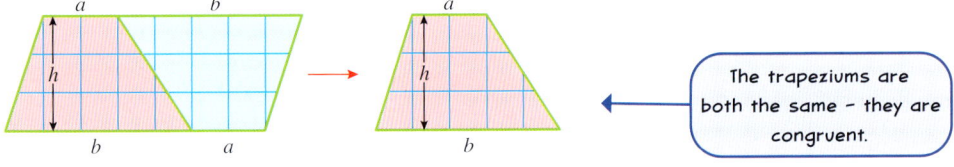

> The trapeziums are both the same – they are congruent.

The area of the whole parallelogram is $(a + b) \times h$.

So the area of each trapezium is $\frac{1}{2}(a + b) \times h$.

> Each trapezium is half a parallelogram.

Communication using symbols

You do not need to write the × signs and you can multiply in any order.

Area of a trapezium $= \frac{1}{2}(a + b)h$

Activity

Conceptual understanding

Romesh looks up the formula for the area of a trapezium on the internet.

He finds three different formulas.

$A = \frac{1}{2}(a+b)h$ $\quad A = \frac{1}{2}h(a+b)$ $\quad A = \frac{(a+b)h}{2}$

Show that all three formulas give the same value for A when $a = 2$, $b = 5$ and $h = 3$.

Why do all three formulas give the same answer?

10 Working with 2D shapes

Worked example

Work out the area of this trapezium.

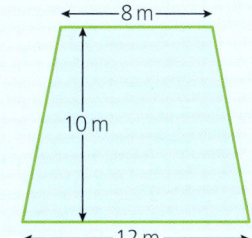

Solution

Parallel sides are 8 m and 12 m, so $a = 8$ and $b = 12$.
Height = 10 m, so $h = 10$.

$A = \frac{1}{2}(a+b)h$ ← This is half the **sum of the parallel** sides multiplied by the **distance between them**.

$= \frac{1}{2}(8 + 12) \times 10$

$= \frac{1}{2} \times 20 \times 10$

$= 100$

So the area is 100 m². ← Make sure you use the right units!

10.2 Now try these

Band 1 questions

1 Work out the area of each of these triangles.

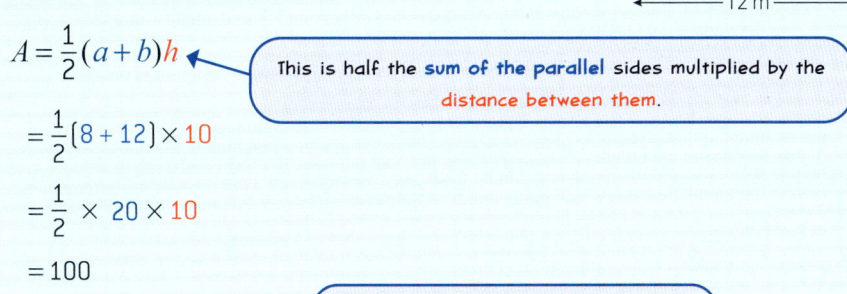

a 4 mm, 12 mm
b 5.4 cm, 4.2 cm
c 3.2 m, 6.1 m

2 Work out the area of these parallelograms.

a 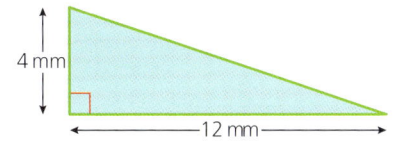 4 cm, 10 cm

b 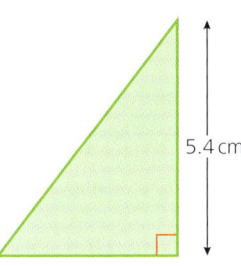 1.9 m, 3.2 m

c Write the formula for the area of this parallelogram using b and h.

3 This is Cai's garden.

a Find the area of the whole garden.
b Find the area of the patio.
c What is the area of the grass?

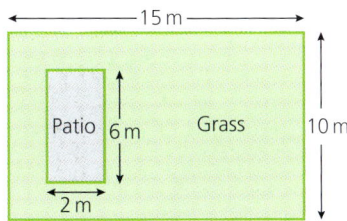

15 m, 10 m, Patio 6 m, Grass, 2 m

169

Fluency

4 Work out the area of this pentagon.

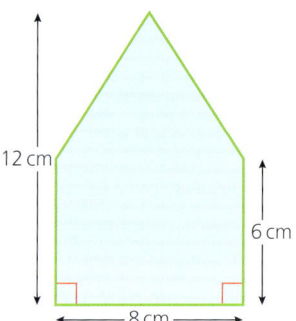

Logical reasoning

5 Look at this shape.

a Find the area of part A.
b Find the area of part B.
c What is the area of the whole shape?
d Amrita wants to work out the perimeter of the whole shape.
  She writes:

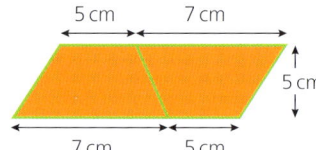

> Perimeter of the whole shape = Perimeter of A + Perimeter of B

Is Amrita right? Explain your answer.

Band 2 questions

6 a Find the area of this parallelogram.

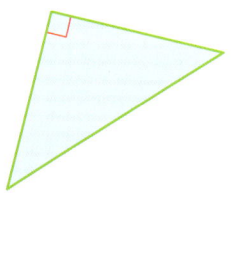

b What is the area of each trapezium in this diagram?

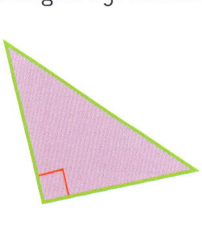

Strategic competence

7 Calculate the area of each of these triangles by measuring them.

a  b c d

e 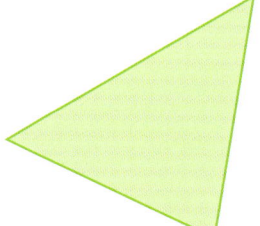 f g  h

10 Working with 2D shapes

8 Find the area of each of these parallelograms.
Make sure you give the units for each of your answers.

a    b 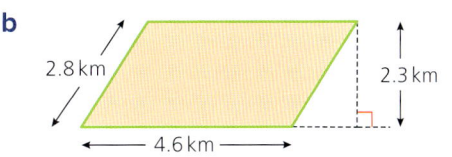   c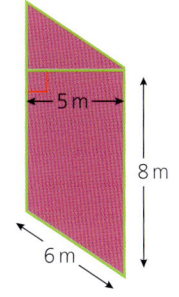

9 This parallelogram has an area of 144 cm².

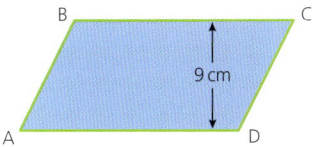

a Work out the length of AD.
b Write down the length of BC.

10 Calculate the area of each of these trapeziums.

a    b 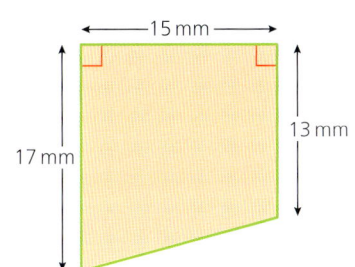   c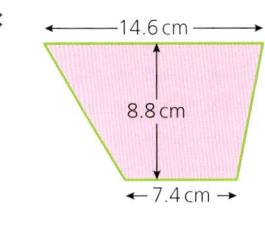

11 Find the area of each of these shapes.
Show the formula you are using.

a    b 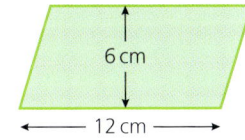   c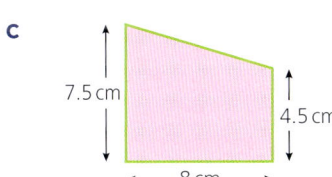

12 Find the area of each of these shapes.

a

b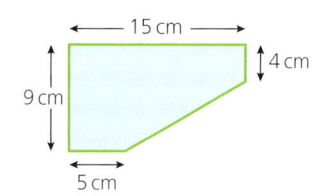

c

13 Catrin has three congruent rectangles.

Inside each rectangle she draws a triangle.

Each triangle has a base length of b cm.

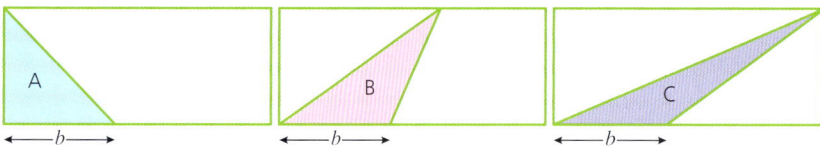

Which triangle has the greatest area?

Give a reason for your answer.

14 Work out the area of the shaded shape.

a 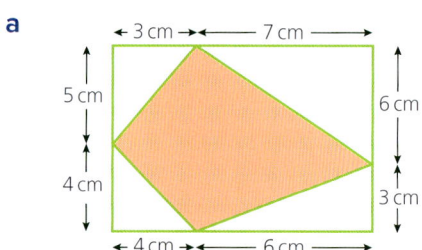   b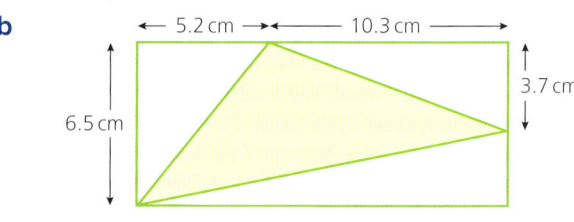

Band 3 questions

15 A square has a side length of 8 cm.

The area of a triangle is the same as the area of the square.

The base length of the triangle is 0.2 m.

Calculate the height of the triangle.

Hint

These shapes can be split into two or more simpler shapes.

10 Working with 2D shapes

16 Use the formula for the area of a trapezium to work out the missing values in this table.

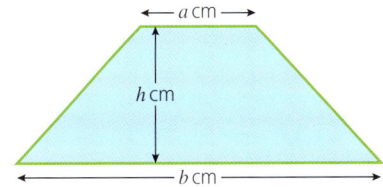

a cm	b cm	h cm	Area
3	4	5	
4.5	12	6	
5	9		28 cm²
7		6	33 cm²
	4.5	8	30 cm²

Hint
Substitute the values into the formula and then solve the equation.

17 This shape is made from a parallelogram and a trapezium.

The area of the parallelogram is 32 cm².

Find the area of the whole shape.

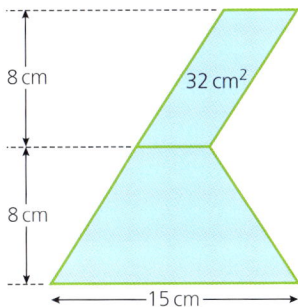

18 This diagram shows two squares. The measurements are in metres.

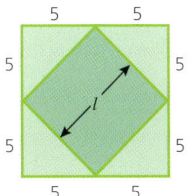

Work out the length l.

Give your answer correct to the nearest centimetre.

19 The length of the diagonals of this rhombus are 18 cm and 24 cm.

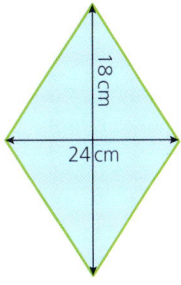

a Work out the area of this rhombus.

b Work out a formula for the area of a rhombus with diagonals of length a cm and b cm.

Key words

Here is a list of the key words you met in this chapter.

Area	Arrowhead	Congruent	Equilateral
Formula	Isosceles	Kite	Parallel
Parallelogram	Perimeter	Perpendicular	Quadrilateral
Rectangle	Rhombus	Right angle	Square
Trapezium			

Use the glossary at the back of this book to check any you are unsure about.

Review exercise: working with 2D shapes

Band 1 questions

1 Look at this pattern.

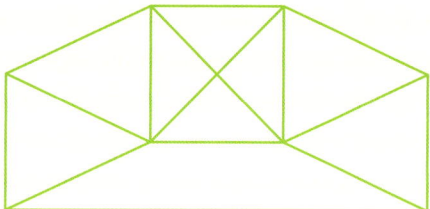

It is made of triangles and quadrilaterals.

Copy and complete this table.

Only count each shape once.

Shape	How many are in the pattern?
Right-angled triangle	
Other isosceles triangle	
Square	1
Other parallelogram	
Other trapezium	
Kite	

2 The diagram shows a children's play area.

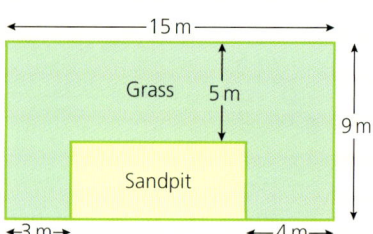

a Work out the area of the sandpit.

b Work out the area of the grass.

3 Calculate the area of each of these shapes.

Make sure you give the correct units with your answers.

a
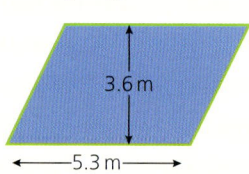

b
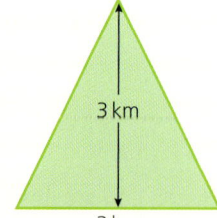

c
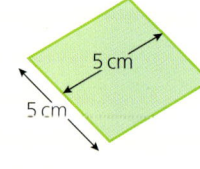

10 Working with 2D shapes

④ Work out the perimeter and area of each of these shapes.

a 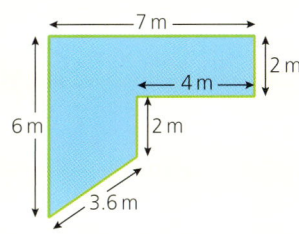 b 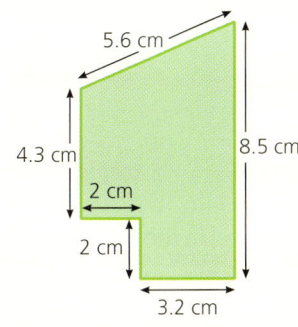 c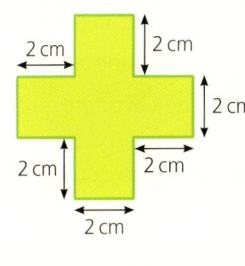

Band 2 questions

⑤ Here is a plan of Megan's garden.

a Fencing costs £1.20 per metre.
Calculate the cost to put a fence around the garden.

b Megan decides to sow the garden with lawn seed.
One packet of seed covers 12.5 m².
How many packets will she need?

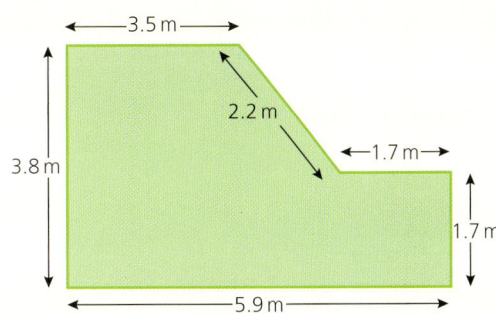

⑥ Work out the area of each of these parallelograms.

a 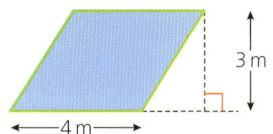 b 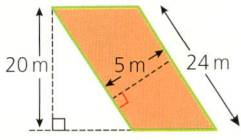 c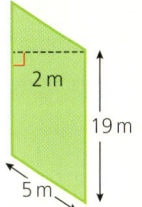

⑦ The area of this parallelogram is 90 cm².

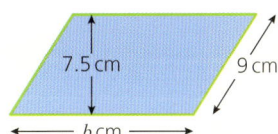

Calculate the perimeter of the parallelogram.

⑧ Look at the diagrams below.

In each case, one vertex of the quadrilateral is missing.

Write down the coordinates of the missing vertex to make:

a a square   b a kite   c a rhombus.

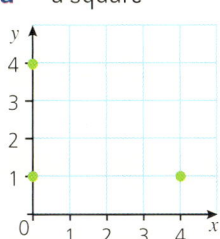 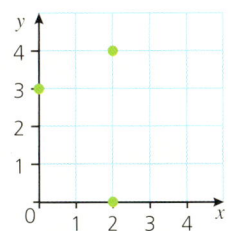 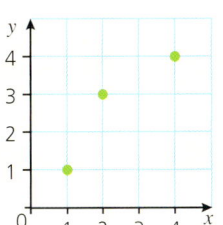

Logical reasoning

9 Which of the special quadrilaterals:
  a can contain a right angle
  b can include a reflex angle?

10 The names of some quadrilaterals are below.

rectangle   rhombus   kite   trapezium   square

Write each quadrilateral in the correct position in the table below.

	Two pairs of parallel sides	One pair of parallel sides	No pairs of parallel sides
Equal-length diagonals			
Diagonals not equal length			

Fluency

11 Work out the area of this trapezium.

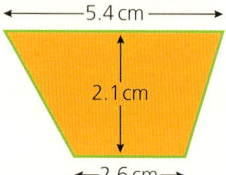

Band 3 questions

Strategic competence

12 Find the area of the parallelogram ABCD with vertices at A(0, 1), B(0, 5), C(6, 7) and D(6, 3).

13 The total area of this parallelogram is 81 cm².

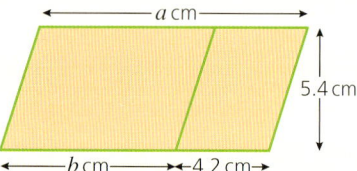

Work out value of a and of b.

14 The area of this trapezium is 37.6 cm².

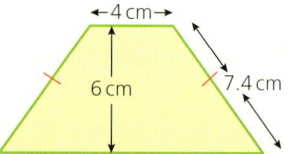

Calculate the perimeter of the trapezium.

Logical reasoning

15 Jasmyn writes down some statements about shapes.

Decide whether each of Jasmyn's statements is true or false.

Explain your reasoning fully.

  a A trapezium with a right angle is a rectangle.
  b Every rhombus is also a kite.
  c An arrowhead cannot have an obtuse angle.

16 The diagram shows a parallelogram inside a trapezium.

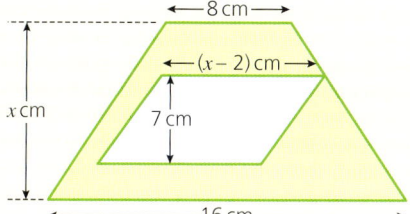

The parallelogram has half the area of the trapezium.

a   Calculate the value of x.
b   Work out the area of the shaded region.

Consolidation 4: Chapters 9–10

Band 1 questions

1 Write down the mathematical name of these shapes. Do you know what they are in Welsh?

a 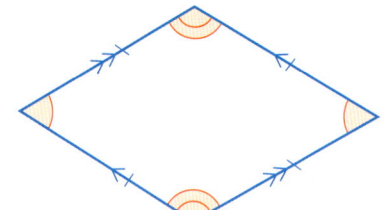   b 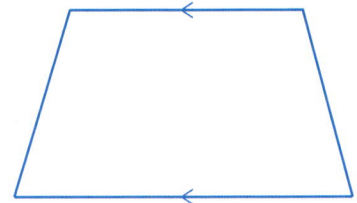   c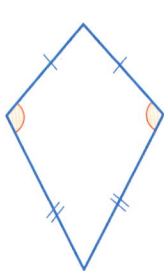

2 Solve these equations.

a $\quad 2a = 6$ b $\quad b - 4 = 6$ c $\quad 12 - c = 4$

d $\quad \dfrac{d}{3} = 2$ e $\quad \dfrac{12}{e} = 4$ f $\quad \dfrac{3}{4}f = 24$

3 Jamal thinks of a number.

He subtracts his number from 20.

Jamal multiplies the result by 3.

His answer is 39.

What is Jamal's number?

4 Jac has some bricks to build a wall.

Each small brick weighs $\dfrac{1}{2}$ kg.

Eighty small bricks and sixty large bricks weigh 160 kg.

What is the weight of one large brick?

5 These shapes have been drawn on centimetre-squared paper.

    i Write down a calculation to work out the area of each shape.

    ii Check your calculation works by counting the squares.

a 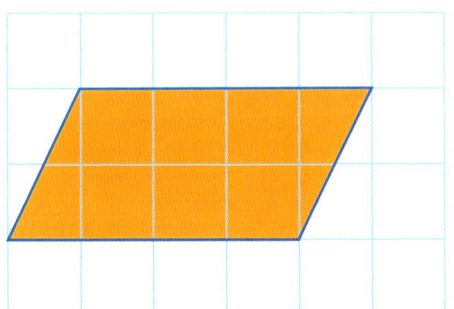   b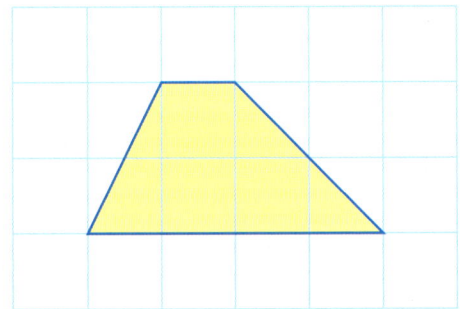

Consolidation 4

6 The diagram shows two ornamental ponds surrounded by concrete.

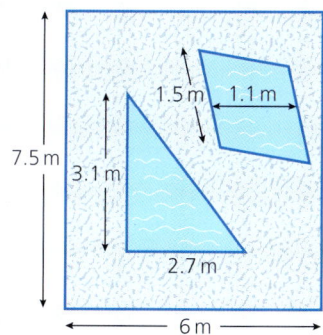

a Calculate the area of concrete.

Concrete costs £2.70 per square metre.

b Calculate the cost of the concrete surrounding the ponds.

Band 2 questions

7 Solve these equations.

a $3x - 4 = 17$

b $8 - 2y = 4$

c $3z + 4 = 16$

d $\frac{x}{2} + 5 = 9$

e $\frac{2z}{3} - 1 = 5$

f $5 - \frac{z}{4} = 3$

> **Remember**
> You can use function machines to solve these types of equations.
> How could you use them to solve the equations in parts **b** and **f**?

8 Nia has solved some equations for her Maths homework.

i Without solving the equations, show that Nia's answers are wrong.

ii Show how to solve the equations correctly.

a $2(x - 4) = 20$

Expand brackets $2x - 4 = 20$

$+4$ ⟶ $2x = 24$ ⟵ $+4$

$\div 2$ ⟶ $x = 12$ ⟵ $\div 2$

b $4x + 6 = 2x - 4$

$-2x$ ⟶ $2x + 6 = 4$ ⟵ $-2x$

-6 ⟶ $2x = -2$ ⟵ -6

$\div 2$ ⟶ $x = -1$ ⟵ $\div 2$

9 List all the types of quadrilateral that have these features.

a Four equal angles

b Four equal sides

c At least one pair of parallel sides

d Two pairs of parallel sides

e One but not two pairs of equal angles

10 Calculate the area of each of these shapes.

a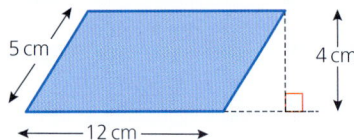

b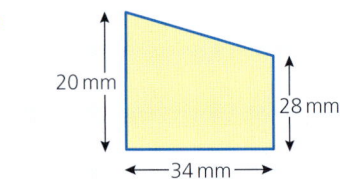

11 A rectangle is $(x + 2)$ cm wide and 4 cm long.

The perimeter of the rectangle is 16 cm.

Work out the area of the rectangle.

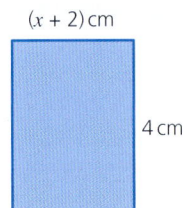

12 Solve these equations.

a $5k - 1 = 2k + 2$

b $15 - x = 23 - 2x$

c $12 - 3c = 18 - 5c$

d $4m - 3 = 5 + 2m$

13 The head of a fish is 6 cm long.

The body of the fish is twice as long as the head and tail together.

In total, the fish is 54 cm long.

How long is the tail?

14 Look at this diagram.

a Write an equation in x.

b Solve your equation to find the value of x.

c Hence find the size of each angle in the triangle.

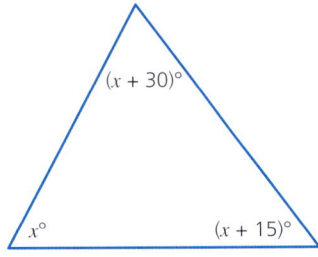

Band 3 questions

15 Solve these equations.

a $12(x + 4) + 4(2x + 3) = 50$

b $12(x + 4) - 4(2x + 3) = 50$

16 Ami, Bethan and Cerys share £620.

Ami gets £30 less than Bethan.

Cerys gets three times as much as Ami.

How much money does each person receive?

17 Kwame has designed a badge.

The design is made from a trapezium and a parallelogram.

The area of the trapezium is the same as the area of the parallelogram.

The area of the parallelogram is 300 mm².

Calculate the height of the badge.

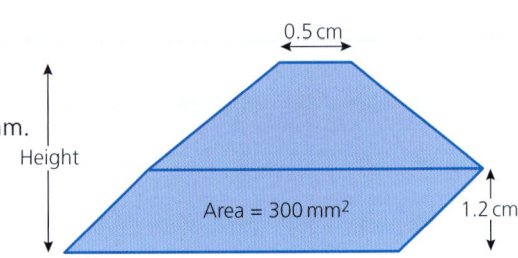

18 This shape is a regular octagon.

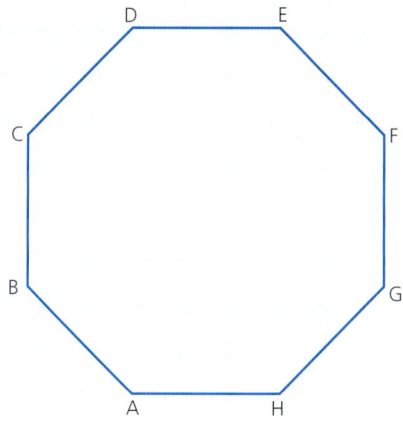

Find ways in which you can use the points and lines on it to make these quadrilaterals:

a   a square
b   a rhombus
c   a trapezium
d   a kite
e   an arrowhead
f   a rectangle that is not a square
g   a parallelogram that is not a rectangle or a rhombus
h   any other quadrilateral.

You may extend the lines that are on the diagram.

You may also make new lines by joining the vertices and you may extend these lines.

19 Osian and Suki are trying to solve this equation

$10n - 18 = 2(7 - 3n)$

Osian says, 'I think $n = 1$.'

Suki says, 'No, $n = 2$.'

a   Without solving the equation, check who is right.
b   Solve the equation, showing your working.

20 Solve these equations.

a $3(x + 2) = 2(x + 3)$
b $5(x + 2) = 2(x + 20)$
c $10(x + 1) = 5(x + 3)$
d $7(x - 1) = 2(x + 4)$
e $3(3g - 8) = 2g - 3$

21 Meinir is buying some bananas on holiday.

She can buy five bananas and have 10 cents left over.

Or she can buy three bananas and have 50 cents left over.

a   Write an equation using this information.
b   Solve your equation to find the cost of one banana.
    Check your answer.
c   How much money does Meinir have?

22 Calculate the size of one of the acute angles in this rhombus.

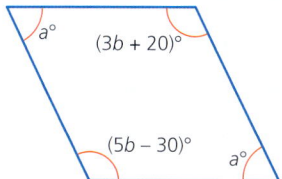

11 Properties of 3D shapes

Coming up...
- Understanding the properties of 3D shapes
- Recognising nets
- Finding the surface area of a cuboid
- Finding the volume of a cuboid

Painted cubes

① The cube shown has dimensions 3 cm × 3 cm × 3 cm and is made up from several smaller cubes measuring 1 cm × 1 cm × 1 cm.

The 3 cm cube is painted orange on the outside.

   a How many 1 cm cubes make up the 3 cm cube?

   b Some of the smaller cubes have one face painted, some have two faces painted and some have three.

      Are there any 1 cm cubes that have no faces painted?

   c Copy and complete this table for the 1 cm cubes.

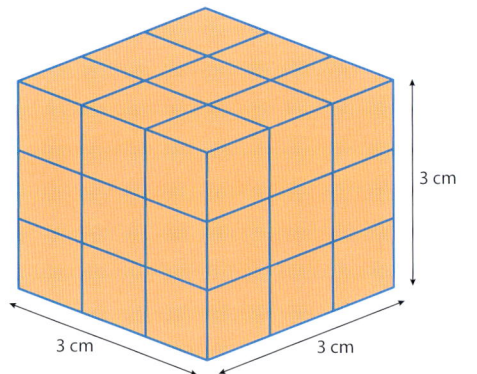

Number of faces painted	0 faces painted	1 face painted	2 faces painted	3 faces painted
Number of cubes				

② Repeat question 1 with a 4 cm cube made from 1 cm cubes.

11.1 Properties of 3D shapes

Skill checker

① Put the correct labels into the correct spaces.

   Face (Wyneb)   Edge (Ymyl)   Vertex (Fertig)

② Can you use a formula?

   a The formula for the area of a rectangle is $A = h \times w$.

      If $h = 2$ cm and $w = 3$ cm, find the value of A in cm².

   b The formula for the volume of a 3D wedge is $V = \frac{1}{2} l \times b \times h$.

      Find the value of V in cm³ if $l = 2$ cm, $b = 3$ cm and $h = 4$ cm.

   c The formula for the height of a particular trapezium is $h = \frac{A}{2w}$.

      Find the height in cm when $A = 30$ cm² and $w = 3$ cm.

   d The formula for the width of a square tube is $w = \frac{V}{h^2}$.

      If $V = 45$ cm³ and $h = 3$ cm, find the width in cm.

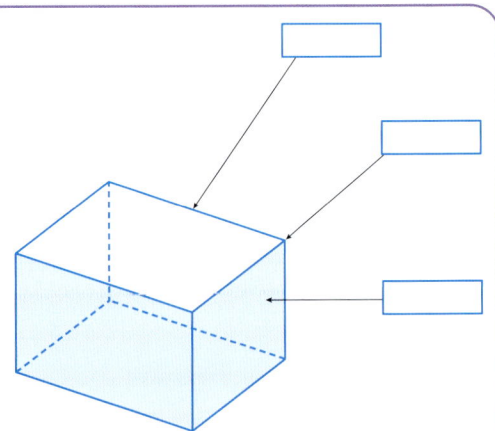

11 Properties of 3D shapes

▶ Prisms

A **prism** is a 3D shape with a constant cross-section and no curved faces. If you cut the shape parallel to the end, the cut will be the same shape as the end.

There are many different types of prism.

The two **solids** below are both prisms.

'Solid' is another word for a 3D shape.

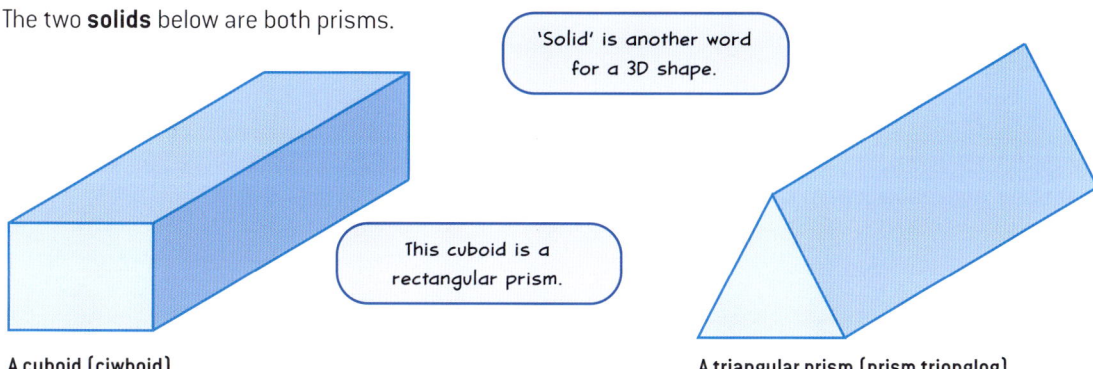

This cuboid is a rectangular prism.

A cuboid (ciwboid)   A triangular prism (prism trionglog)

The cross-section of the cuboid is a rectangle or a square.

The cross-section of the triangular prism is a triangle.

Be aware that a prism may not be very long! The shapes below are both prisms.

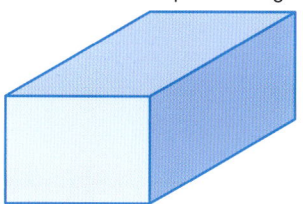 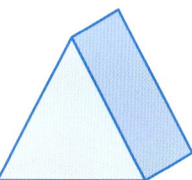 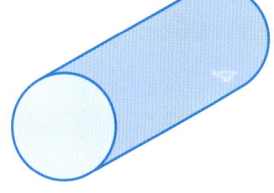

A cuboid (ciwboid)   A triangular prism (prism trionglog)   A cylinder (silindr)

The cross-section of a prism can be any **polygon**.

You will come across a hexagonal prism, an octagonal prism and other types of prism.

The cross-section of the cylinder is a circle, but it is not a prism because it has a curved face.

The Giant's Causeway in Northern Ireland is made up of hundreds of rocks in the shape of hexagonal prisms.

▶ Other 3D shapes

Not all 3D shapes are prisms.

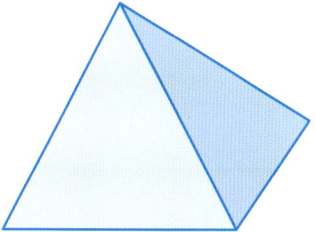  

A square-based pyramid (pyramid sylfaen sgwâr)   A cone (côn)   A sphere (sffêr)

As well as the square-based pyramid, there are other pyramids; for example, a triangle-based pyramid. A cone is a type of pyramid. What other name could you give it?

> A triangle-based pyramid is also called a 'tetrahedron'.

Worked example

This object is a square-based pyramid.

a If you cut the pyramid vertically through points V, E and G, what shape is the cut?

b If you cut the pyramid horizontally, what shape is the cut?

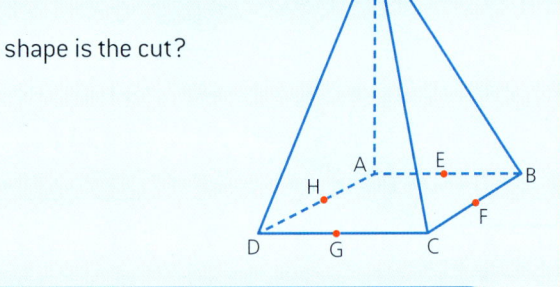

Solution

a The cut is an isosceles triangle.

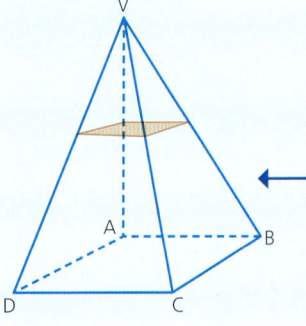

> The isosceles triangle VEG is called a 'plane of symmetry' of the pyramid. It splits the pyramid into two identical pieces. Using the letters on the diagram, identify another plane of symmetry.

b The cut is a square.

> The square is not a plane of symmetry of the pyramid. It splits the pyramid into two very different pieces.

Activity

Euler's Formula is:

$$V + F = E + 2$$

where:

- ▶ V is the number of **vertices** or corners
- ▶ F is the number of **faces**
- ▶ E is the number of **edges**.

Euler's Formula works for all 3D shapes, as long as none of the faces are curved.

Maths in context

Leonhard Euler was an eighteenth century Swiss mathematician. He worked in many areas of mathematics and is thought of as one of the greatest mathematicians ever to have lived.

Find out more about the life and work of Euler and the contributions he made to mathematics.

Do you understand his work? What level of mathematics do you need to reach to understand his theories?

11 Properties of 3D shapes

For example, look at the square-based pyramid below.

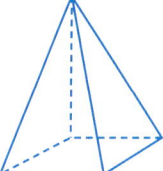

① Copy and complete the statements below.
The pyramid has _____ faces.
The pyramid has _____ vertices.
The pyramid has _____ edges.

The base is a square face. Add on the number of triangular faces.

Count the number of vertices around the square base first. Are there any more?

Count the number of edges around the square base first. How many more are there?

② Copy and complete the two lines below to check that Euler's Formula works for the pyramid:
$F + V =$ ___ + ___ = ___
$E + 2 =$ ___ + 2 = ___
Choose another 3D shape without curved faces and repeat.

If you are asked to find the number of faces, vertices and edges for a 3D shape, Euler's Formula can be a good way to check your answer.

③ Find out more about the life and work of Euler and the contributions he made to mathematics.
Do you understand his work? What level of mathematics do you need to reach to understand his theories?

Cross-curricular activity

The history of mathematics is a fascinating subject in its own right! Find out more in your history lessons.

11.1 Now try these

Band 1 questions

❶ Match these names to the 3D objects below:

cuboid (ciwboid)   **cylinder** (silindr)   **hexagonal prism** (prism hecsagonol)

triangular prism (prism trionglog)   **cube** (ciwb)   **octagonal prism** (prism wythonglog)

a   b   c
d   e   f
g   h   i

❷ Is a sphere a prism? Explain your answer.

Curriculum for Wales Mastering Mathematics: Book 2

Logical reasoning

3 All the pupils in Class 8J are asked to give one fact about 3D shapes. Here are some of their answers. Say whether each one is correct or incorrect and explain your answers.
- a 'A triangular prism has five faces.'
- b 'A square-based pyramid is another name for square prism.'
- c 'Each face of a square-based pyramid is a square or a rectangle.'
- d 'A cylinder has more curved faces than a sphere.'
- e 'A triangular prism has nine vertices.'
- f 'A triangular prism has nine edges.'
- g 'Triangular prism is another name for triangular pyramid.'
- h 'A cuboid will always have a bigger volume than a cube.'
- i 'A cone only has one vertex.'
- j 'A cube is the only 3D shape with 6 faces.'

Band 2 questions

Fluency

4 Complete the table below. Can you remember their names in Welsh?

	Name of shape	Faces	Vertices	Edges
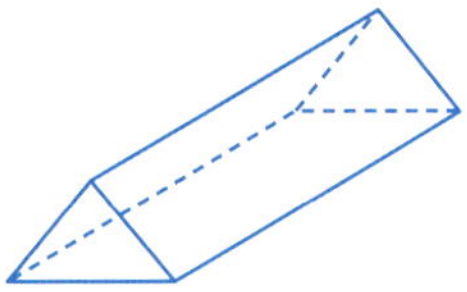				
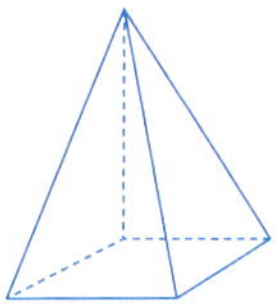				
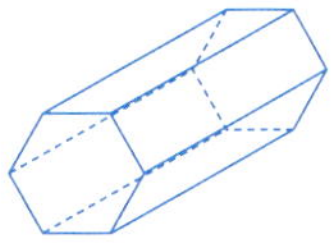				
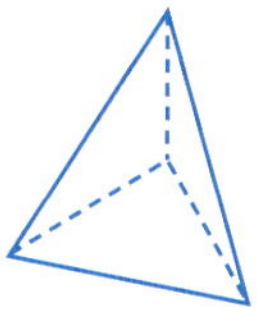				

5 The picture below shows a cylinder.

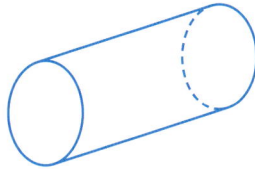

Note
Euler's Formula doesn't work for a cylinder because of the curved face.

Fill in this table with the number of faces and edges.

Number of faces	Number of edges	Number of vertices

Band 3 questions

6 Look at this cuboid.

It has a square face on each end.

a Dafydd cuts the cuboid into two pieces.

His cut is parallel to the square end.

What shape is the cut?

b Mitali has an identical cuboid.

She cuts the cuboid horizontally.

What shape is her cut?

c Prydwen says it is possible to cut the cuboid and get a triangular cut. Is she right? Explain your answer.

7 This 3D shape is called an octahedron.

It is made by joining two square-based pyramids.

Rhian says if she cuts through the octahedron she will get a cut in the shape of a diamond or rhombus.

Naga says if she cuts the octahedron she will get a square-shaped cut.

Who is right? Explain your answer.

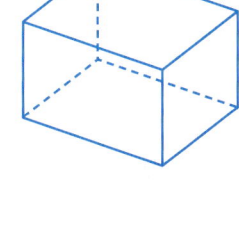

8 a This 3D shape is a square-based pyramid with the top removed.

How many faces, edges and vertices does it have?

Check your answers using Euler's Formula.

b Sketch a triangle-based pyramid with the top removed.

Find the number of faces, vertices and edges for this solid and check your answers using Euler's Formula.

This is called a **frustum** of a pyramid.

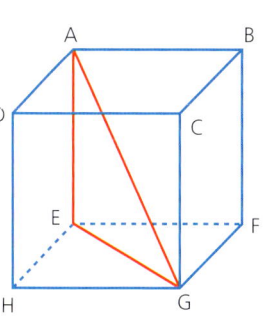

9 Different shaped triangles can be made by joining three vertices of a cube.

For example, in the diagram, AEG is a right-angled triangle.

a How many different shaped triangles can be made by joining three vertices of the cube?

b How many are there of each type?

c How many are there altogether?

d Answer questions **a–c** for a cuboid.

11.2 Nets

Skill checker

Six pupils each describe a 3D shape.

a 'My solid only has one face, which is curved.'

b 'It has five faces. The two opposite faces are equilateral triangles. The other faces are all rectangles.'

c 'It is a prism with ten faces. Its cross-section is a regular polygon.'

d 'It has six faces, twelve edges and eight vertices. All the faces are rectangular.'

e 'My solid has one face that is a 3 cm square and four identical faces that are isosceles triangles.'

f 'It is a pyramid with seven faces. Its base is a regular polygon.'

For all six, work out the names of these shapes being described.

For **e**, what can you say about the size of each triangle? Sketch the different faces (put ×4 next to the triangle, to show there are four of these).

For **d**, can you sketch the solid?

> Isometric drawing paper can be used to draw cubes and cuboids. Your teacher may have some and will show you how to use it.

▶ Nets in real life

A net is a pattern of 2D shapes that can be folded to make a 3D solid.

The same object may have many different nets.

Here is the net of a cardboard box and the box constructed.

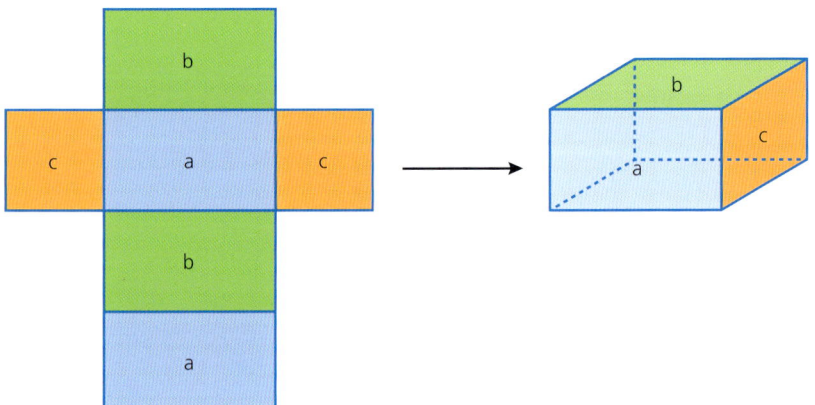

Look at this net of a triangular prism. Flaps have been added to some of the edges.

In real life, flaps are often added to the net of a cardboard box. This allows the edges to be glued together.

In this example, flap 1 will glue to edge 1, flap 2 to edge 2 and so on.

In mathematics you will not usually be asked to add flaps to your drawings of nets.

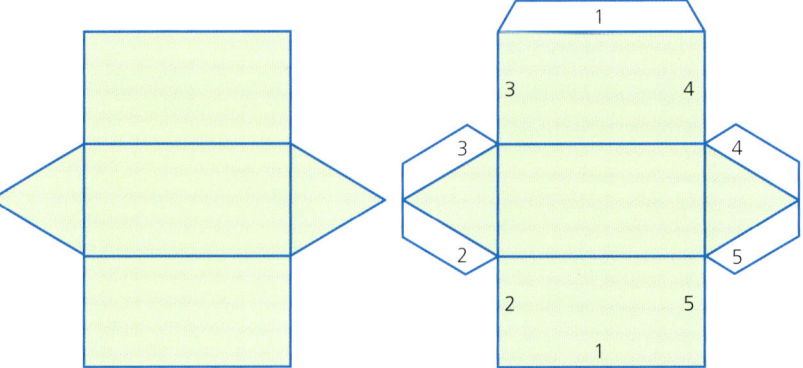

11 Properties of 3D shapes

Activity

A **hexomino** is a pattern of six connecting squares.

Some hexominoes are nets of a cube. Two of these are shown on the right.

Altogether there are 11 nets of a cube.

Can you find them?

Below are some clues to help you decide whether a hexomino is a net of a cube.

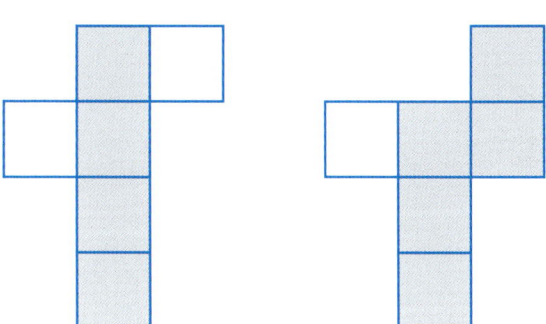

Clues

- Six of these nets have a line of four squares, like the net above on the left.
- If a hexomino has a line of four squares, the two extra squares must lie one on each side of the line of four.
- There are four nets that have a line of three, like the one above on the right.
  If there is a line of three squares, there will be another two squares in the positions shown. These four nets all have the sixth square in different positions.
  Be careful – having your sixth square in some positions will not make a net!
- There is only one net that doesn't have a line of four or a line of three.
- None of the nets of a cube include a two-by-two square.
- Watch out for rotations and reflections! For example, the three patterns below are all treated as the same net.

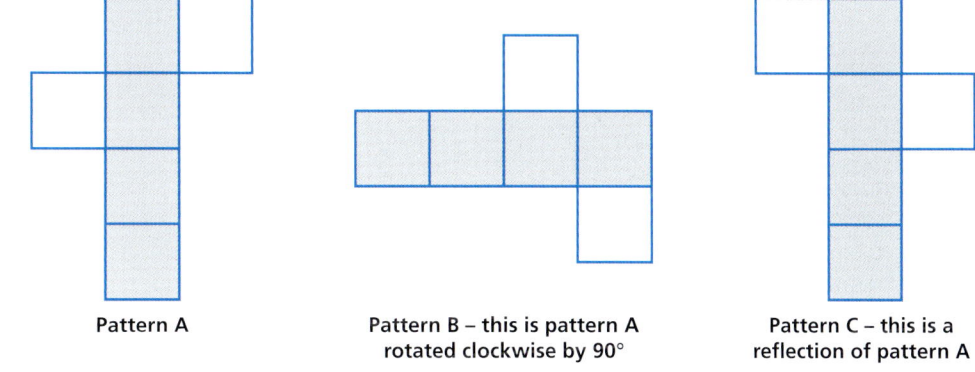

| Pattern A | Pattern B – this is pattern A rotated clockwise by 90° | Pattern C – this is a reflection of pattern A |

11.2 Now try these

Band 1 questions

1. Which two of these four patterns are nets of a cube? Copy these two nets.

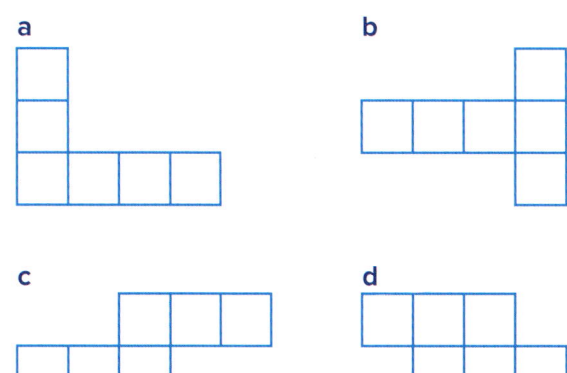

a

b

c

d

Curriculum for Wales Mastering Mathematics: Book 2

Logical reasoning

② Look at the four shapes below. Two of them are nets of a triangle-based pyramid. Copy these two nets.

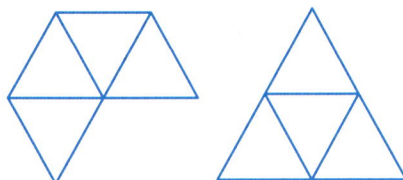

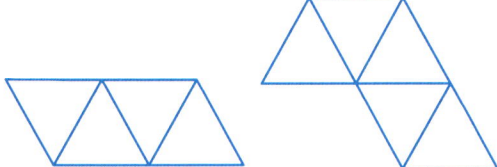

③ Look at the four shapes below. Two of them are nets of a square-based pyramid. Copy these two nets.

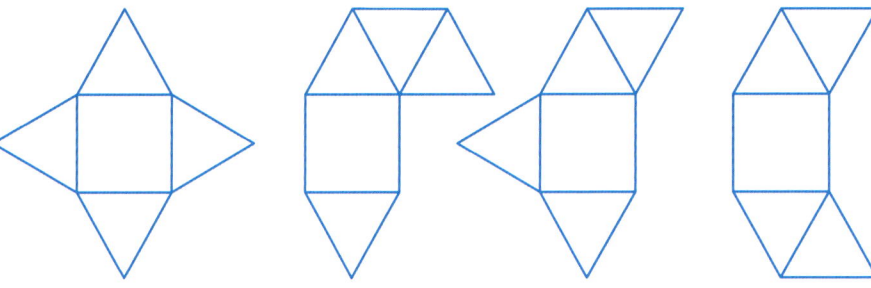

Strategic competence

④ Design a net for this cuboid

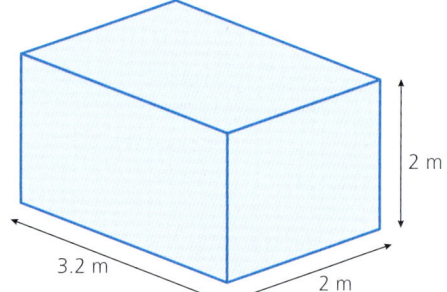

3.2 m   2 m   2 m

Band 2 questions

Logical reasoning

⑤ Two of the shapes below could be a net of a cylinder.
Copy the two nets.

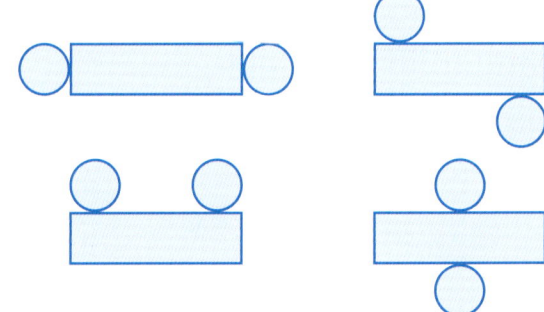

11 Properties of 3D shapes

6 The diagram shows the net of a cardboard box.

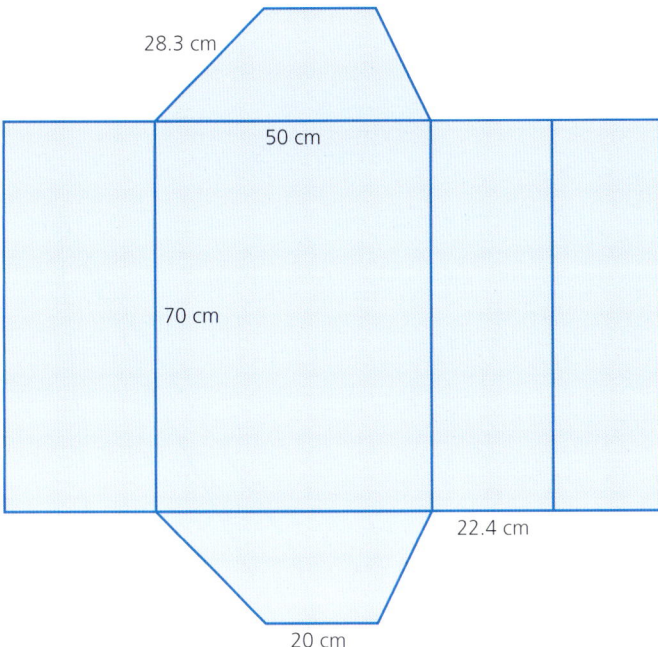

When folded, the box is the shape of a trapezoidal prism.

Copy the diagram, marking on all the missing lengths.

> A trapezoidal prism is a prism with a cross section the shape of a trapezium.

7 This net makes an **open box** with five faces and no top.

Draw a different net that will make the same box.

Band 3 questions

8 The diagram shows a cuboid and its net.

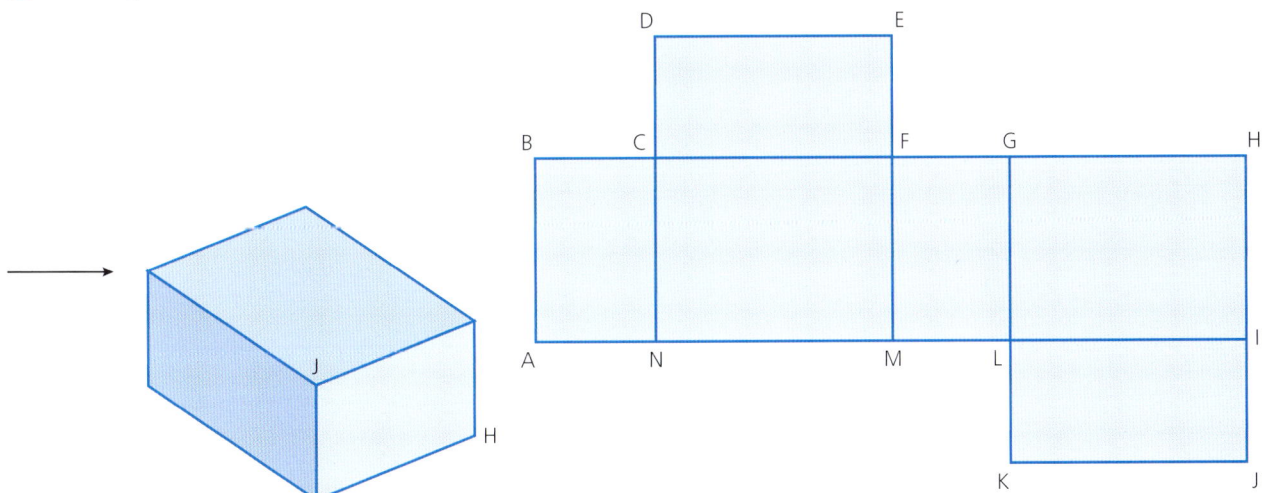

When the net is made into a cuboid:

a Which two other points meet up with H?

b Which one other point meets up with J?

c Which points meet up at the vertex shown with an arrow?

191

9 Saeed has drawn a net of a hexagonal prism, but he was not very accurate.

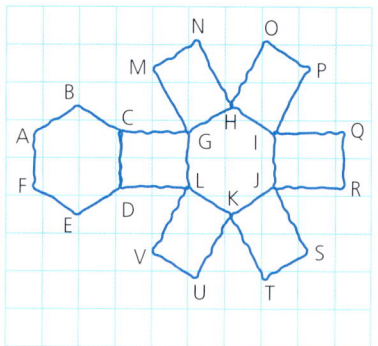

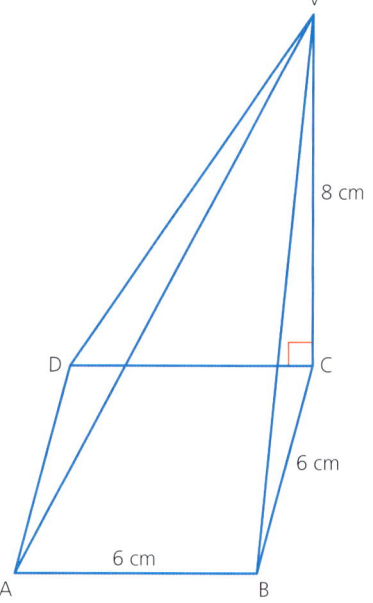

a State whether the following are true or false.
   i   ED should be the same length as UV.
   ii BC should be the same length as CG.
   iii BC should be the same length as GH.
   iv OP should be the same length as MN.
   v   Shape MNHG should be a rectangle.
b Draw the net of the hexagonal prism accurately.

10 The diagram shows a square-based pyramid VABCD.
The vertex V is directly above C.
AB = BC = CD = DA = 6 cm and VC = 8 cm.
a Draw a net for this shape.
b Cut it out and check that it works.

11.3 Surface area and volume of a cuboid

Skill checker

1 On this football pitch, the length of each rectangle is twice its width.
Find the area of:
a the 6 yard area
b the 18 yard area
c the pitch.
Give your answers in square yards.

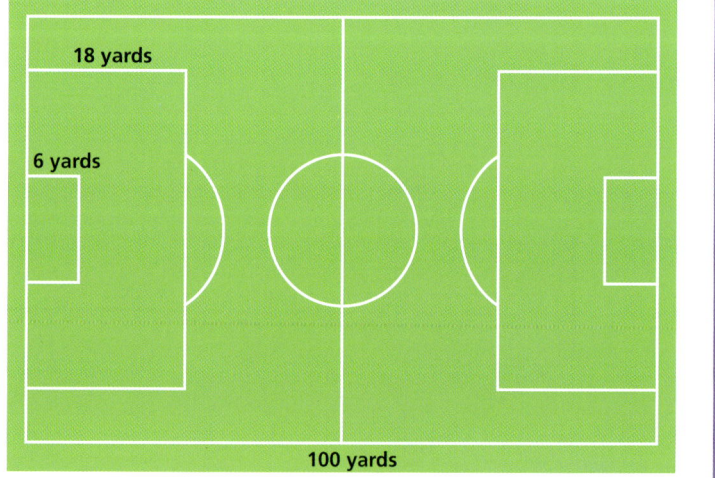

② A netball court is shown below. It is 30.5 metres long and 15.25 metres wide.
It is surrounded by a run-off zone, shown in grey.

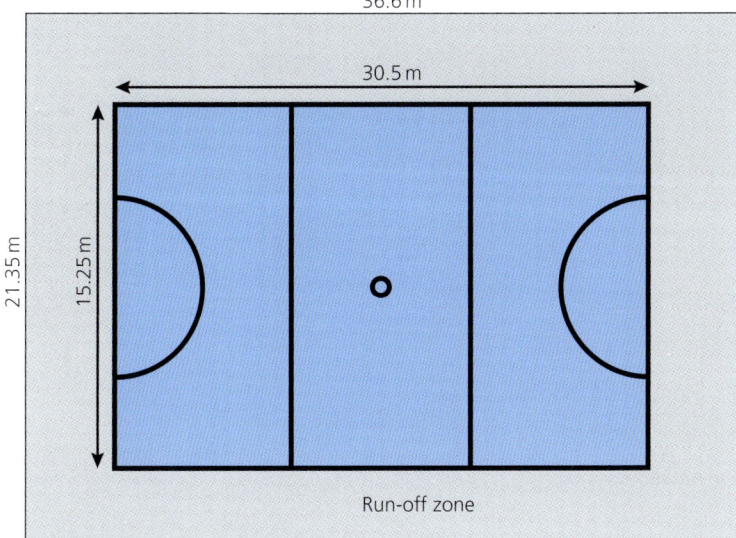

Find the area of:

a   the court area, shown in blue

b   the run-off zone.

Give your answers in square metres.

▶ Volume of a cuboid

To find the volume of a cuboid, multiply its length, width and height.

The formula for the volume of a cuboid is:

$V = lwh$

In a cube, the length, width and height are all equal.
So $V = l \times l \times l$ or

$V = l^3$

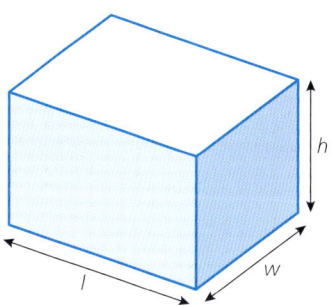

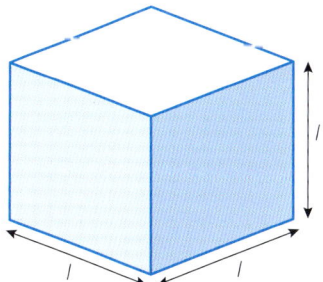

> You will need to learn and remember these formulas. In Book 3 you will also learn formulas for the volumes of other 3D shapes.

▶ Surface area of a cuboid

To find the total surface area of a cuboid, find the area of each face, then add.

The area of the base is $w \times l$. The top has the same area.

The other faces have area $l \times h$ (two of these) and $w \times h$ (two of these).

The total surface area is $2wl + 2wh + 2lh$.

Worked example

For this cuboid, find:

a the volume

b the total surface area.

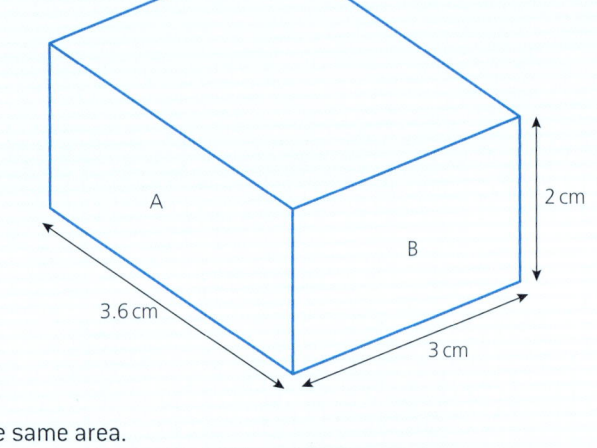

Solution

a To find the volume:

$V = lwh$
$= 3.6 \times 3 \times 2$
$= 21.6 \text{ cm}^3$

b To find the total surface area:

The area of the base is $3.6 \times 3 = 10.8 \text{ cm}^2$. The top has the same area.

The area of face A is $3.6 \times 2 = 7.2 \text{ cm}^2$. There are two faces with this area.

The area of face B is $3 \times 2 = 6 \text{ cm}^2$. There are two faces with this area.

The total surface area is $(2 \times 10.8) + (2 \times 7.2) + (2 \times 6) = 48 \text{ cm}^2$

Maths in context

Archimedes (circa 287 BC – circa 212 BC) noticed that the water in his bath rose as he got into it.

He used this method to find the volume of a crown.

Find the volume of an irregular object by dropping it into a container of water that is the shape of a cuboid.

11.3 Now try these

Band 1 questions

1 The cubes making up the cuboids in this question are centimetre cubes.

a Each layer of cuboid A is made of six cubes.

It has five layers.

What is its volume?

Copy this table and complete the row for cuboid A.

Cuboid	Number of cubes in one layer	Number of layers	Total number of cubes	Volume
A	6	5		
B				
C				
D				

11 Properties of 3D shapes

b Each layer of cuboid B is made of 10 cubes.
It has three layers.
Complete your table for cuboid B.

c Complete the table for cuboids C and D.

B

C D

② Ioan has made these cuboids with building bricks.

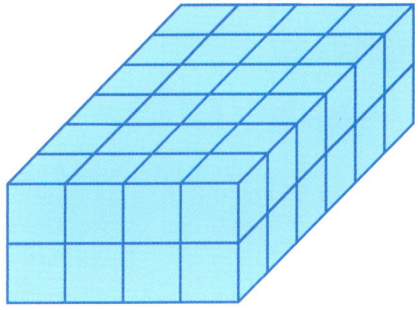

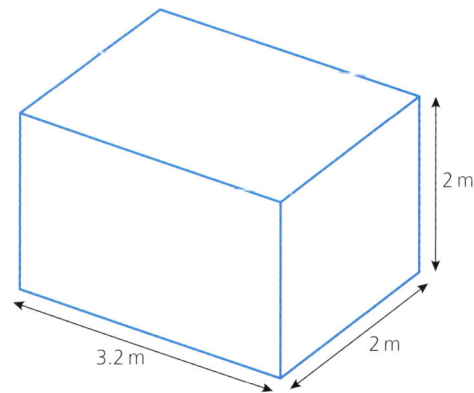

Each brick is 1 cm³.

a Find the volume of each of these cuboids.
What do you notice?

b Describe two more cuboids with the same volume.

Cross-curricular activity

In your art lesson, investigate the work of artist Rachel Whiteread. One of her famous works is 'EMBANKMENT' in which she filled the vast Turbine Hall of the Tate Modern in London with different arrangements of white blocks.

How would you arrange them as a piece of art?

③ For each of the cuboids below, find:
 i   the area of the smallest face
 ii the area of the largest face
 iii the total surface area of the cuboid
 iv the volume of the cuboid.

a

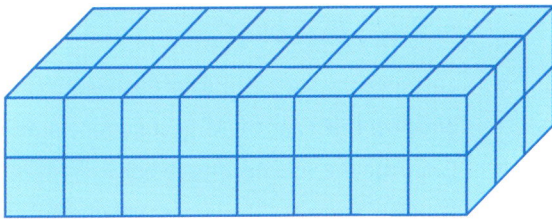

b

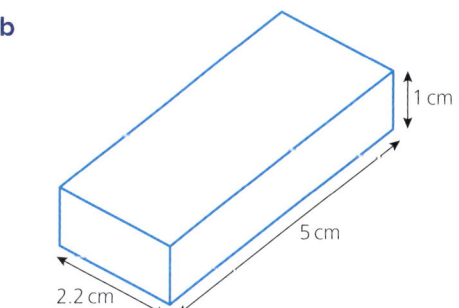

195

Band 2 questions

4 The two nets below both make small boxes in the shape of cuboids.

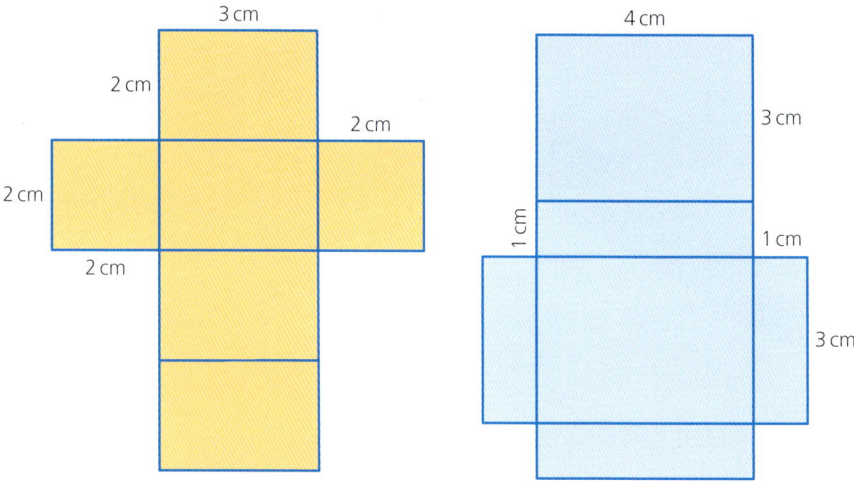

    a Calculate the surface area of each box.

       Which of them has the largest surface area?

    b Calculate the volume of each box.

       Which of them has the largest volume?

5 Look at the two cubes below.

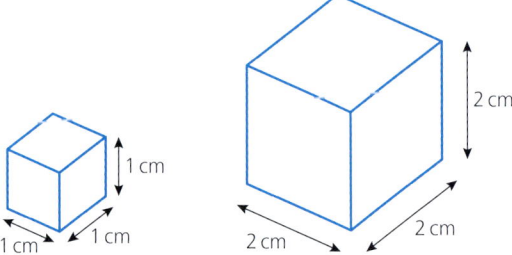

Now look at the two shapes below.

a This object is made from two of the small cubes and one big cube.

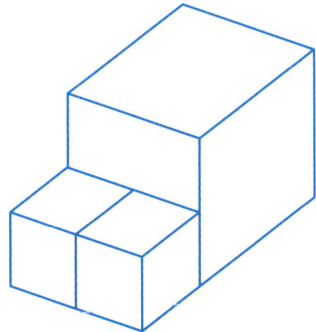

b This object is made from one big cube on top of four small cubes.

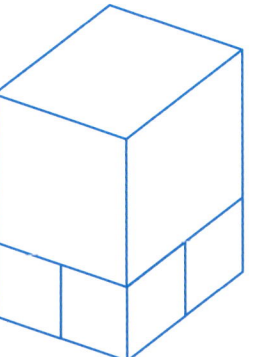

For each shape, find:

  i the total surface area

  ii the volume of the object.

6 a Look at the sugar cube and the sugar cube box.

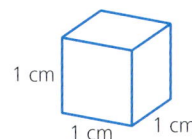

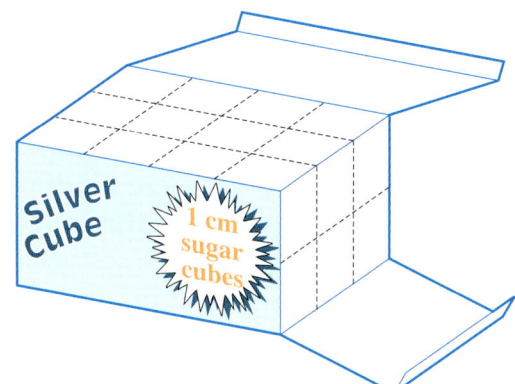

i How many sugar cubes fit in **one layer** of the box?
ii How many layers are there in the box?
iii How many cubes are there altogether?
iv Work out the volume of the box using the formula for the volume of a cuboid:

$$V = lwh$$

b i Find the volume of this box using the formula.
ii How many 1 cm sugar cubes fill the box?
iii What do you notice?

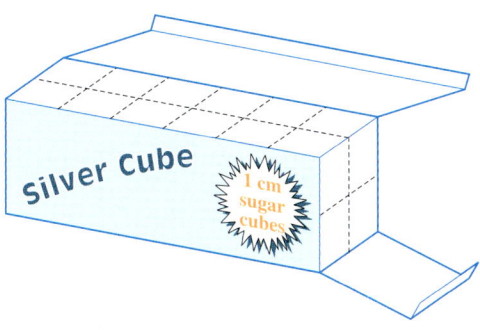

c i There are a lot of possible boxes that could hold 120 sugar cubes.

Choose a length, width and height that would be suitable.

ii Draw your box.
iii Find the surface area of your box.
iv Can you design another box (to also hold 120 sugar cubes) that has a smaller surface area?

7 What is the area of each of the rectangular faces of this regular hexagonal prism?

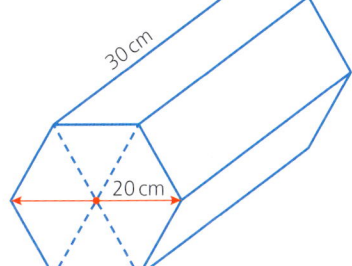

8 Three cubes are lined up on a table, from smallest to largest.

For the smallest cube, the length of each edge is 1 cm.
For the middle cube, the length of each edge is 2 cm.
For the largest cube, the length of each edge is 3 cm.

a How many times would the smallest cube fit inside the largest cube?
b What is the total volume of all three cubes?
c Ferozan remembers her earlier work on sequences.

She says, 'The three volumes make a sequence, which is the cube numbers.'

What is the nth term of this sequence?

d Ferozan then thinks about the sequence given by the three total surface areas.

She is a bit stuck coming up with a formula for this sequence.

Can you help her?

Band 3 questions

9 Bryn gets a new pair of trainers for his birthday.

The shoebox is 30 cm long, 20 cm wide and 15 cm high.

Bryn's dad wraps the shoebox in birthday wrapping paper.

   a What area of paper does he need to use?

   b When Bryn's dad unrolls the wrapping paper, he has a rectangle measuring 1 metre by 30 cm.
   Does he have enough paper to wrap the birthday present?

10 A pack of butter (menyn) weighs 250 g and has measurements 6 cm × 4 cm × 10 cm.

Packs of butter are placed in a box with measurements 18 cm × 16 cm × 50 cm.

   a How many packs of butter would fill the box?

   b What is the weight of butter in the box?

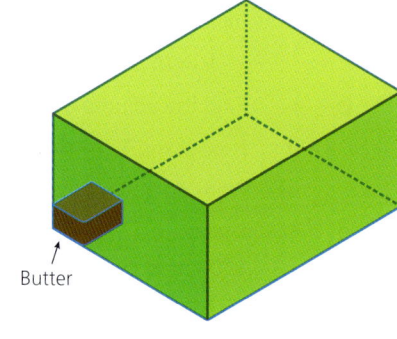
Butter

11 Three cubes are put together to make the 3D shape shown below.

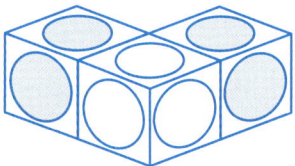

Two of the cubes have shaded circles on each face.

One cube has white circles on each face.

   a How many white circles are there on the outside of this 3D shape?

   b How many shaded circles are there on the outside of the shape?

12 A house is built using 10 000 bricks.

One of these bricks is shown in the diagram.

7000 of the bricks had just the long side face showing.

The remaining 3000 bricks were used for the corners of the building. They had both the long face and the smaller end face showing.

   a What is the total volume of all the bricks used?

   b What is the surface area of the brickwork for the house?

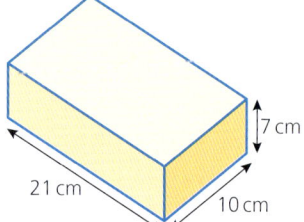

Key words

Here is a list of the key words you met in this chapter.

3D shape	Cone	Cube	Cuboid
Edge	Face	Prism	Solid
Sphere	Square-based pyramid	Vertex	

Use the glossary at the back of this book to check any you are unsure about.

11 Properties of 3D shapes

Review exercise: properties of 3D shapes

Band 1 questions

1 Match these names to the shapes below.

cuboid (ciwboid) | cylinder (silindr) | hexagonal prism (prism hecsagonol) | irregular prism (prism afreolaidd)

equilateral triangular prism (prism trionglog hafalochrog) | isosceles triangular prism (prism trionglog isosgeles)

a 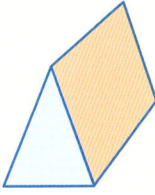   b 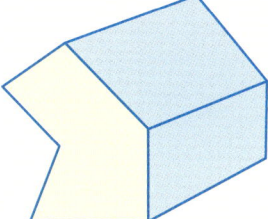   c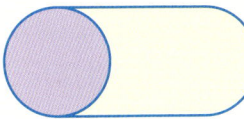

d 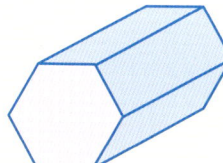   e 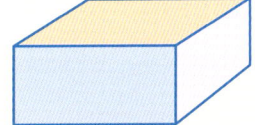   f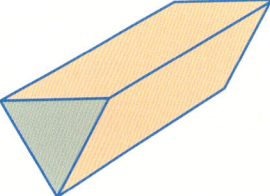

2 Two of these hexominoes are nets of a cube.

A 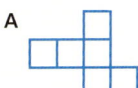   B    C    D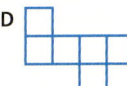

a Copy these two hexominoes.

b On a dice, the numbers on opposite faces add up to 7.

For each of the nets you have copied, put a number from 1 to 6 on each face. Make sure that when the 3D shape is constructed, opposite sides would add up to 7.

There is more than one way to do this.

3 For each of the cuboids below:

i Find the area of the smallest face.
ii Find the area of the largest face.
iii Find the total surface area of the cuboid.
iv Find the volume of the cuboid.

a 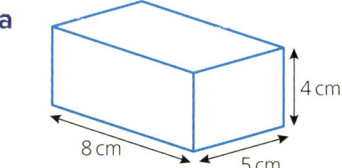   b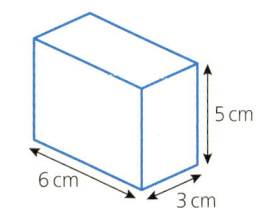

Band 2 questions

Logical reasoning

4 A solid shape is made from four identical cubes joined together.

Cubes are always joined as shown.

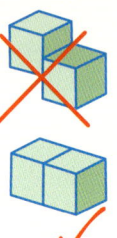

   a   Find the solid with the largest surface area.

   b   Find the solid with the smallest surface area.

   c   Sketch the solids in **a** and **b**. Your teacher may have isometric paper for this.

5 The surface area of a cube is 96 cm^2.

   a   Find the area of one face.

   b   Find the length of one edge.

   c   Find the volume of the cube.

Fluency

6 These cuboids have dimensions 9 cm × 2 cm × 2 cm and 3 cm × 3 cm × 4 cm.

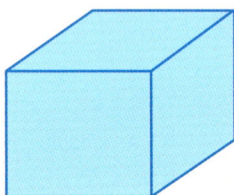 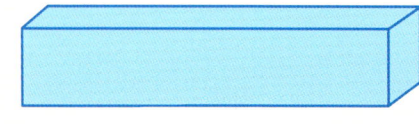

   a   Draw a net for each cuboid.

   b   Both cuboids are to be made from cardboard nets. Which cuboid needs less cardboard?

Band 3 questions

Strategic competence

7 Vijay is giving clues to his friend Huw about the dimensions of a cuboid.

The length, width and height are all whole numbers in centimetres.

Clue 1: Vijay says, 'It has a volume of 96 cm^3.'

Huw says, 'I think it has a length of 48 cm, a width of 2 cm and a height of 1 cm.'

Vijay tells Huw he is wrong.

Clue 2: Vijay says, 'The width and height are the same.'

Huw says, 'I think it has a length of 24 cm, a width of 2 cm and a height of 2 cm.'

Vijay tells Huw he is wrong again.

Clue 3: Vijay says, 'The length, width and height are all more than 1 cm.'

What should Huw's next guess be? He should not need any more clues.

8 Alwena wants to make a cardboard box with a volume of 64 cm^3.

The cardboard box will be a cube or cuboid, with 6 faces, including a lid.

There are many ways to do this, for example:

	Length	Width	Height
Method 1	8 cm	8 cm	1 cm
Method 2	2 cm	2 cm	16 cm

   a   Alwena works out that, if she uses Method 1, the surface area will be 160 cm^2.

   Find the surface area of the box if she uses Method 2.

   b   What is the smallest amount of cardboard Alwena could use to construct her box?

   What length, width and height would this box have?

9 Hywel is a painter and has been asked to paint an office room.
The room is the shape of a cuboid.
Hywel must paint all four walls and the ceiling.
The length of the office is 6 metres and the width is 4.5 metres. Its height is 2.5 metres.
a What is the total surface area Hywel will need to paint?
b One litre of paint covers approximately 12 m² of wall.
One tin of paint contains 2.5 litres.
How many tins of paint will Hywel need to buy?

10 Look at this sequence of 3D shapes made from 1 cm cubes.

      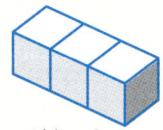

Object 1   Object 2   Object 3

a Copy and complete this table for the sequence of shapes.
Include object 4, which would be made from four 1 cm cubes.

Object number, n	Length	Surface area, S	Volume
1	1 cm	6 cm²	1 cm³
2	2 cm		
3			
4			

b Look at the sequence of numbers in the surface area column of your table.
Find a formula for the nth term of this sequence.
$S = $ _____

12 Percentages

Coming up...

- Working with percentages
- Working out percentage increase and decrease
- Calculating percentage change

Percentage snap

This is a game for two players.

Start with a deck of playing cards with all the picture cards removed. The aces have a value of 1.

Deal all forty cards out (20 to each player).

Each player turns over one card at the same time.

The players must try to work out one card as a percentage of the other.

Call 'SNAP' when the percentage matches any of these:

| 20% | 25% | $33\frac{1}{3}$% | 50% | 75% | $66\frac{2}{3}$% | 75% | 120% | 150% |

The player calling 'SNAP' must also say what percentage they have spotted.

For example, if the two cards are the 6 of hearts and the 4 of spades, a player could call 'SNAP $66\frac{2}{3}$%' or 'SNAP 150%'. If correct, the player keeps the pair of cards. If not, their opponent takes the pair.

After all the cards have been turned over, the player who has collected the most pairs wins the game.

12.1 Working with percentages

Skill checker

Copy and complete this bar model diagram, filling in all the rectangles with the correct percentage, fraction and decimal values.

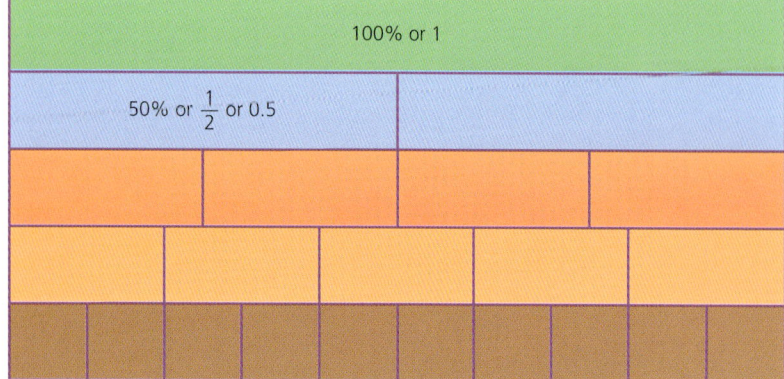

12 Percentages

▶ Percentages revision

Remember the important skills you have learnt relating to percentages:

- Percentages add up to 100%.
- Finding a percentage of a quantity; for example, 20% of £300.
- Writing one number as a percentage of another; for example, writing 6 cm as a percentage of 10 cm.
- Converting between fractions, percentages and decimals.

Worked example

There are 50 seats on a bus. It is full when it leaves Cardiff bus station.

At the first stop 20% of the passengers get off the bus.

a What percentage of the passengers remain on the bus?
b How many passengers get off at the first stop?
c How many passengers remain on the bus?
d At the second stop, 5 more people get on. How full is the bus now? Give your answer as a percentage.
e Write your answer to part **d** as a fraction and a decimal.

Solution

a If 20% get off the bus, 80% remain on the bus.

b 20% of 50 is $\frac{1}{5}$ of 50, which is 10 passengers who get off the bus.

c 80% of 50 is $\frac{4}{5}$ of 50, which is 40 passengers who stay on the bus.

d 40 + 5 = 45

The bus is now $\frac{45}{50}$ full.

$\frac{45}{50} = \frac{90}{100} = 90\%$

So the bus is now 90% full.

e As a fraction the bus is $\frac{45}{50}$ full, which simplifies to $\frac{9}{10}$ full. As a decimal this is 0.9.

▶ Percentages greater than 100%

In some contexts, it is possible to have percentages greater than 100%. For example:

- A company's profits are 120% of last year's profits.
- The number of people in a workplace is 150% of the number last year.
- The number of people with an illness in the UK has increased by 250% over the last 10 years.
- The blood sugar level of a patient is 130% of the normal level.

In other contexts, percentages greater than 100% do not make sense. For example:

- You cannot score more than 100% in an exam.
- It is not possible for a footballer to put in 110% effort.
- You cannot arrive at school on time 150% of the time.

To convert any percentage to a decimal, divide by 100.

So, for example, 152% = 1.52.

To convert to a fraction:

$152\% = \frac{152}{100} = \frac{76}{50} = \frac{38}{25} = 1\frac{13}{25}$

When converted, a percentage greater than 100% becomes:
- a decimal greater than 1; or
- a **top-heavy** or **improper** fraction, which can be converted to a **mixed number**.

Worked example

In 2010, 3 million people in the UK had a disease called diabetes.

By 2018 that number had increased to 3.6 million.

a What percentage of the 2010 figure is the 2018 figure?
(You should not need a calculator for this.)

b What percentage of the 2018 figure is the 2010 figure?
(You may use a calculator for this.)

Solution

a First, write the numbers as a fraction and simplify:

$$\frac{3.6 \text{ million}}{3 \text{ million}} = \frac{3.6}{3} = \frac{1.2}{1} = 1.2$$

As a percentage, $1.2 = 120\%$.

In 2018, the number of people with diabetes was 120% of the number in 2010.

b Again, write the numbers as a fraction and change to a decimal.

$$\frac{3 \text{ million}}{3.6 \text{ million}} = \frac{3}{3.6} = \frac{30}{36} = \frac{5}{6}$$

$$\frac{5}{6} = 0.8333\ldots$$

As a percentage, $0.8333\ldots$ is 83.3% (correct to one decimal place).

In 2010, the number of people with diabetes was 83.3% of the number in 2018 (correct to one decimal place).

Activity

Find out the same information about diabetes in Wales.

Does the data mirror the UK percentages?

12.1 Now try these

Band 1 questions

1. Find:
   a 20% of 40
   b 40% of 250 kg
   c 120% of 6 litres
   d 6% of 50
   e 8% of 25 cm.

2. Copy and complete the following.
   a 15 is ___% of 100.
   b 6 is ___% of 30.
   c 20 is ___% of 80.
   d 28 is ___% of 50.
   e 84 is ___% of 112.
   f 15 seconds is ___% of 45 seconds.
   g 70 cm is ___% of 200 cm.
   h £2 is ___% of £5.
   i 32 g is ___% of 80 g.
   j 2.5 km is ___% of 500 m.

3
a What percentage of 60 is 15?
b What percentage of 40 is 28?
c What percentage is 60 of 90?
d What percentage is 105 cm of 140 cm?
e 184 miles is what percentage of 230 miles?
f 1200 litres is what percentage of 1600 litres?

4
a Convert these decimals to percentages.
   i 0.27   ii 0.04   iii 1.9   iv 0.075   v 0.1825
b Convert these fractions to percentages.
   i $\frac{1}{4}$ ii $\frac{7}{10}$ iii $\frac{3}{5}$ iv $\frac{9}{5}$ v $\frac{31}{25}$
c Convert these percentages to decimals.
   i 82%   ii $5\frac{1}{2}\%$ iii 120%   iv 231%   v $54\frac{1}{5}\%$
d Convert these percentages to fractions in their simplest forms.
   i $33\frac{1}{3}\%$ ii $12\frac{1}{2}\%$ iii 150%   iv $37\frac{1}{2}\%$ v $6\frac{1}{4}\%$

5 In these situations, do you think it is possible for the blank space to be a percentage greater than 100% or not?
a Meirion got the best score in his year in his RE exam. He scored _____ %.
b A bus leaving the town centre was _____ % full.
c The class is now _____ % of its size last year.
d A jar of coffee now contains _____ % of the previous amount.
e Mr Evans was a very rich man. When he died, he left _____ % of his wealth to charity.
f Elin was a great sailor, but even she had accidents. On about _____ % of her expeditions the boat capsized.

Band 2 questions

6 In a chapel choir, there are 40 people.
25% are children under 18.
a What percentage of the choir are adults?
22 members of the choir have stated that they are female.
b What percentage have stated that they are female?
c What percentage have not stated that they are female?

7 Sioned is running a marathon. A marathon is 26 miles or 40 km long.
Sioned passes a sign saying:

16 km

a What percentage of the race has Sioned completed?
b What percentage does she still have to complete?
c Sioned's friend Dai is 3 km behind her.
   What percentage of the race has he completed?

8 Ashok has a 25 acre field. He plants $15\frac{1}{2}$ acres with wheat.
What percentage of his field does he plant with wheat?

9 Classes 8E and 8F voted for a pupil who would sit on the school council.
There were three pupils to choose from.
Ioan Bevan received 42% of the votes.
28% voted for Carys Jones.
a What percentage of pupils voted for Llwyd Llewellyn?
b 50 pupils voted altogether.
   How many voted for the winning pupil?

10 28% of a juice drink is pure apple juice.

How much pure apple juice is in these?

   a   A 500 ml carton   **b**   A 1 litre bottle   **c**   A 200 ml glass

11 On the first day of the new school year, the headteacher of Ysgol Glanuchaf, Mrs Hughes, says, 'At Ysgol Glanuchaf we have increased our student population to 105% of last year's number!'

Last year there were 800 pupils in the school. How many pupils are there now?

Band 3 questions

12 Sixty Year 8 pupils in Class 8AC and Class 8GH were asked about their favourite school subject. The results are summarised in this table.

Unfortunately, the pupils who conducted the survey have lost some of the numbers.

   a   Copy the table below and fill in the blank spaces.

	Maths (Mathemateg)	English (Saesneg)	Science (Gwyddoniaeth)	PE (Addysg Gorfforol)	History (Hanes)	Music (Cerddoriaeth)	Art (Celf)	Total (Cyfanswm)
Class 8AC (Dosbarth 8AC)		6	5	5	1		?	35
Class 8GH (Dosbarth 8GH)	?			?	0	1		
Total (Cyfanswm)	13	?	11		1	6	10	

   b   What percentage of Class 8AC said their favourite subject was Art?

   c   Of those pupils who said Art was their favourite subject, what percentage were in Class 8AC and what percentage were in Class 8GH?

   d   What percentage of Class 8GH said that Maths was their favourite subject?

   e   What percentage of all pupils chose PE as their favourite subject?

You can do this activity with your class!

13 In 2021, Simi's annual insurance premium was 2% of the value of her house contents. Her house contents were valued at £25 000.

   a   Find how much Simi paid for her house insurance in 2021.

   b   The insurance company wrote Simi a letter telling her that, in 2022, the premium would be 104% of the 2021 premium. How much must Simi pay for her 2022 house insurance?

12.2 Percentage increase and decrease

Skill checker

Dafydd did a survey of the people taking packed lunches to school.

He asked each person whether they had a favourite sandwich.

The results are shown in the pie chart.

Favourite sandwiches (Hoff frechdanau)

Ham (Ham) (30%)
Cheese (Caws) (20%)
Cheese and pickle (Caws a phicl)
Cucumber (Ciwcymbr) (15%)
Cheese and ham (Caws a ham) (25%)

① Unfortunately, Dafydd lost the percentage figure for the cheese and pickle category.

What percentage should appear in this section of the pie chart?

② Twenty pupils took part in the survey.

How many pupils said they preferred:

   a   cheese and ham sandwiches

   b   ham sandwiches

   c   cucumber sandwiches?

12 Percentages

▶ Percentage increase and decrease

There are several ways to work out percentage increases and decreases.

Worked example

Find the amount when £120 is:

a increased by 15%

b decreased by 15%.

Solution

There are various methods for finding 15% of £120.

Method 1: Finding 1%

1% of £120 is $\frac{1}{100} \times £120 = £1.20$.

15% is $15 \times £1.20 = £18$.

Method 2: Finding 10%, then 5%

10% of £120 is $\frac{1}{10} \times £120 = £12$.

5% is £6.

15% is £12 + £6 = £18.

a An increase of 15% gives £120 + £18 = £138.

b A decrease of 15% gives £120 − £18 = £102.

> Remember
>
> You can often replace the word 'of' with a times sign.

> Sometimes finding 1% will be the only method. Other times you can build up a percentage using combinations of 5% and 10%.

> In Book 3 you will learn a calculator method called **multipliers**.

▶ Simple interest

If you put money into a bank or building society account, you will be given **interest** every year.

Using **simple interest**, the amount of interest given will remain the same each year.

> The other type of interest calculation is **compound interest**. Using this, the interest paid changes each year if the balance increases each year.

Worked example

Govinda puts £200 into a bank savings account.

The bank account pays 6% simple interest per annum.

Find out how much Govinda has in his savings account at the end of:

a 1 year   b 2 years   c 3 years.

Solution

1% of £200 is $\frac{1}{100} \times £200 = £2$.

6% of £200 is $6 \times £2 = £12$.

Using simple interest, the interest is £12 per year.

a After 1 year Govinda has £212 in the account.

b After 2 years the balance is £224.

c After 3 years it is £236.

> 'Per annum' means 'per year'.

12.2 Now try these

Please note: You cannot use a calculator for questions 1 and 2, but calculators are allowed for the remainder of the questions in this section.

Band 1 questions

1 a Copy and complete this table.

Number	10%	5%	20%
60	6	3	12
25	2.5		
240			
520			
2600			

b Using the answers in your table to help you, answer this.

   10% of 60 is _____.

   60 increased by 10% is 60 + _____ = _____.

c Now answer these.
   i   Increase 25 by 10%.
   ii Find 520 increased by 10%.
   iii Increase 240 by 15%.
   iv Find 60 decreased by 15%.
   v   Find 2600 increased by 20%.
   vi Decrease 2600 by 15%.

2 A coat costs £80.

a Copy and complete this table.

Percentage	100%	10%	5%	15%
Amount	£80			

Using the table, find the price of the coat if its price:

b increases by 15%
c decreases by 15%.

3 Niall invests £360 in the Gwlad Building Society for one year. The building society pays simple interest of 4.5% per year. How much interest will Niall receive each year?

Band 2 questions

4 The cost of a subscription on a website was £8 per year in 2021.

a In 2022, the cost of the subscription increased by 10%.
   How much is the subscription in 2022?

b The company announced that for 2023, it would reduce the cost of the subscription by 10% from the 2022 price.
   How much is the subscription in 2023?

5 The length of a yacht in a marina is 16 metres.
The super-yacht next to it is 250% longer. How long is the super-yacht?

12 Percentages

6 The ingredients in a pasta sauce are shown in the pie chart.
  a What weight of tomatoes is in a 400 gram jar of this sauce?
  b How much sugar, in grams, is there in the same jar?
  c The ingredients for this type of pasta sauce are changed slightly.
     The amount of sugar is reduced by 25%.
     How much sugar is in the sauce now?
  d What percentage of the sauce is now sugar?

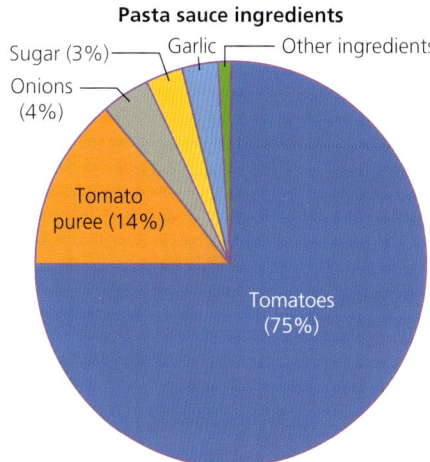

Band 3 questions

7 A standard tin of baked beans contains 0.375 g of salt per 100 g.
The standard tin weighs 400 g.
  a How much salt is in a standard tin? Give your answer in grams.
In the standard tin, 5% of the weight is sugar.
  b How much sugar is in the standard tin?
The tin shown has reduced levels of salt and sugar.
  c How much salt is in this tin of beans?
  d How much sugar is in this 400 g tin?
A third variety of beans has no added salt or sugar.
It has 97% less salt and 99% less sugar than the standard tin.
  e How much salt and sugar is in a 400 g tin of this variety?

8 Copy the arithmagon below.

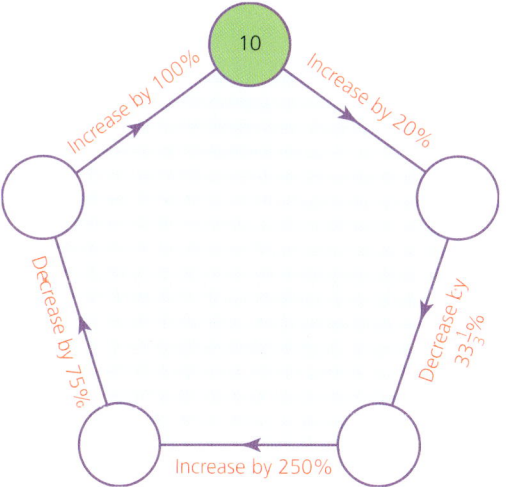

  a Start at the green number 10.
     Follow the arrows to complete the numbers in the circles.
     Do you get back to 10?
  b Draw the arithmagon again, this time with a different multiple of 10 in the green circle.
     Follow the same instructions.
     Do you still get back to your starting number?

12.3 Percentage change

Skill checker

1. Convert these fractions to percentages.

   a $\frac{4}{5}$ b $\frac{7}{20}$ c $\frac{3}{50}$ d $\frac{4}{25}$ e $\frac{9}{20}$

   f $\frac{11}{50}$ g $\frac{6}{5}$ h $\frac{5}{2}$ i $\frac{7}{4}$ j $\frac{39}{60}$

2. What fraction of each triangle is shaded?

   a   b    c    d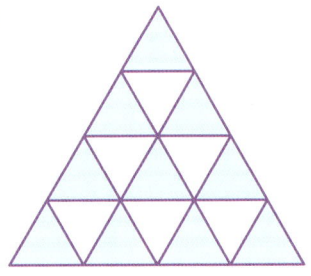

▶ Calculating percentage change

Worked example

The population of an island increases from 500 to 650 people. What is the percentage increase?

Solution

The increase is 650 − 500 = 150 people.

As a fraction this is $\frac{150}{500}$

- $\frac{150}{500}$ ← The increase.
- 500 ← The original population.

To convert to a percentage increase, either:

- find an equivalent fraction with 100 as the denominator:

  $\frac{150}{500} = \frac{30}{100} = 30\%$

or

- write the fraction as a decimal:

  $\frac{150}{500} = 0.3 = 30\%$.

Cross-curricular activity

In a series of PE lessons with your friends, do an activity in which you can record your times or measurements.

This could be, for example, running around a track or doing the long jump. It is up to you!

Then work out the percentage increase or decrease of your improvement over the sessions.

What conclusions can you draw from the information?

▶ Percentage error

Worked example

Tomos measures an oil tank as 1600 mm wide.
Its actual width is 1580 mm.
What is the **percentage error** in Tomos's measurement?

Solution

The error in the measurement is 1600 − 1580 = 20 mm.
To find the percentage error, divide the error by the actual width:

Percentage error = $\frac{20}{1580}$ = 0.013 = 1.3%

▶ Percentage profit and loss

Worked example

Maya's job is restoring musical instruments.

a   She buys a guitar for £75 and restores it. She sells it for £120.
    Find her percentage profit.
b   Maya bought a piano for £1200, but only managed to sell it for £1020.
    Find her percentage loss.

Solution

a   Profit = £120 − £75 = £45

    As a fraction this is $\frac{45}{75}$.   *(The profit over the original price.)*

    $\frac{45}{75} = \frac{90}{150} = \frac{30}{50} = \frac{60}{100}$ = 60%

    So Maya's percentage profit is 60%.

b   Loss = £1200 − £1020 = £180

    As a fraction this is $\frac{180}{1200}$.

    $\frac{180}{1200} = \frac{18}{120} = \frac{9}{60} = \frac{3}{20} = \frac{15}{100}$ = 15%

    So Maya's percentage loss is 15%.

Worked example

Zorro was an overweight dog. He was 14.0 kg.
His owner put him on a programme of long daily walks.
After six months his weight is 9.8 kg.
What percentage weight has Zorro lost?

Solution

Weight lost = 14.0 − 9.8 = 4.2 kg
As a fraction:
$\frac{4.2}{14.0} = \frac{42}{140} = \frac{6}{20} = \frac{30}{100}$ = 30%
The percentage decrease is 30%.
So Zorro has lost 30% of his weight.

Remember

When you work out a percentage increase or decrease, always start by writing a fraction with the increase or decrease as the numerator and the **original quantity** as the denominator.

Activity

① The green square shown has a side length of 10 cm.

It is enlarged to become the purple square with a side length of 20 cm.

a What is the percentage increase in the side length?
b Find the area of the green square.
c Find the area of the enlarged square.
d What is the percentage increase in the area?

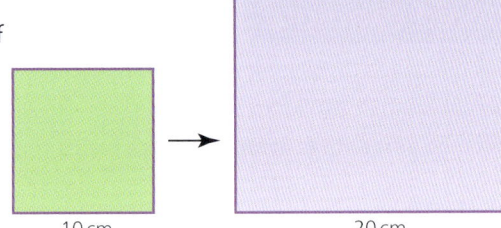

② For the shapes below, sketch an enlarged shape with all the side lengths doubled.

Your sketch does not need to be accurate or to scale, but remember that every side will double in length.

Then repeat steps **a** to **d** above for each shape.

Shape	i 10 cm, 10 cm (triangle)	ii 10 cm, 5 cm (rectangle)	iii $a = 8$ cm, $b = 12$ cm, 5 cm (trapezium)
Formula for area	Area = $\frac{1}{2}$ × base × height	Area = base × height	Area = $\frac{1}{2}$ × $(a + b)$ × height

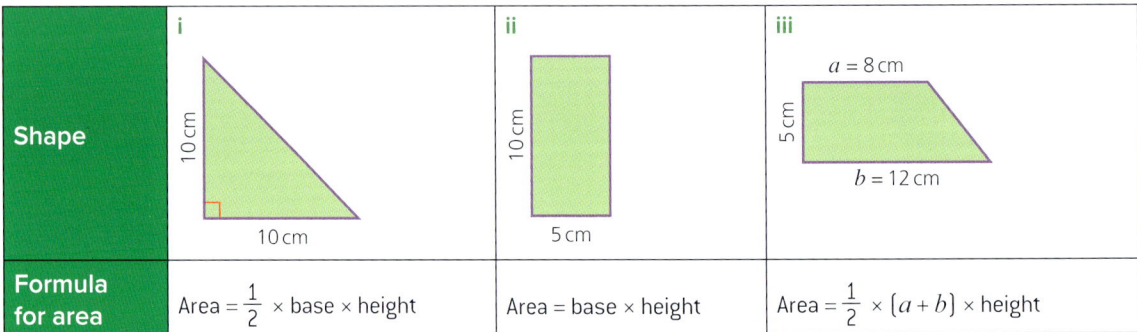

Summarise your results in the table below.

Shape	Percentage increase in side length	Percentage increase in area
(square)		
(triangle)		
(rectangle)		
(trapezium)		

12 Percentages

12.3 Now try these

Band 1 questions

1. The cost price of a mobile phone is £49. It is sold for £70.
   What is the percentage profit?

2. The cost price of a poster is £4.50. It is sold for £3.60.
   What is the percentage loss?

3. For each situation, calculate the percentage profit.
   Give your answers to the nearest whole number.
   a Cost price 20p, selling price 45p
   b Cost price 75p, selling price 85p
   c Cost price £100, selling price £124.90
   d Cost price £1500, selling price £2300

4. For each situation below, calculate the percentage loss.
   Give your answers to the nearest whole number.
   a Cost £45, selling price £30
   b Cost 50p, selling price 18p
   c Cost £1200, selling price £900

5. A bottle of a brand of soft drink contains 30 grams of sugar.
   a To make it a healthier drink, the factory reduces the sugar to 18 grams. What percentage reduction is this?
   b Sales of the drink go down and the company increases the sugar content back to 30 grams.
   What percentage increase is this?

Band 2 questions

6. A motor race is over 20 laps of a track that is 2.5 km long.
   a What distance is the race?
   b For the new season, the course organisers increase the length of the track by 10%.
   What is the new length of the track?
   c The organisers also increase the number of laps by 10%.
   How many laps do the cars have to travel now?
   d What is the new distance for the entire race?
   e What is the percentage increase in the length of the race?

7. Find the error and the percentage error in each of these measurements.
   a Measurement = 4.6 cm, actual length = 4.5 cm
   b Measurement = 4.5 cm, actual length = 4.6 cm
   c Measurement = 360 g, actual mass = 350 g
   d Measurement = 290 g, actual mass = 280 g
   e Measurement = 580 ml, actual volume = 640 ml

8. Find the percentage change that takes you from A to B in each case.

A	70	120	88	40	75	140	110	125
B	84	174	110	52	66	63	88	80

Look at your answers for going from 88 to 110 and from 110 to 88. What do you notice?
Would you expect them to be the same? Explain why they are not.

9 This table gives the prices of seven foods in six consecutive years.

Food	1972	1973	1974	1975	1976	1977
1 lb sausages (selsig)	21p	24p	29p	32p	37p	44p
4 oz coffee (coffi)	29p	30p	32p	40p	41p	72p
1 lb potatoes (tatws)	2p	2p	2p	3p	7p	12p
12 eggs (wyau)	20p	20p	47p	31p	39p	48p
2 lb sugar (siwgr)	10p	9p	10p	29p	23p	21p
1 pint milk (llefrith/llaeth)	6p	6p	6p	5p	9p	10p
1 lb carrots (moron)	3p	4p	5p	7p	7p	14p

a Copy and complete this table to show the percentage change for each food.

Food	1972 price	1977 price	Price increase	Increase / Original	As a percentage
1 lb sausages	21p	44p	23p	$\frac{23}{21} = 1.095$	109.5%
4 oz coffee	29p	72p			
1 lb potatoes	2p	12p			
12 eggs	20p	48p			
2 lb sugar	10p	21p	11p		
1 pint milk	6p	10p			
1 lb carrots	3p	14p			

b Mrs Brown's shopping list is shown here.

What was the percentage increase in the cost of Mrs Brown's shopping between 1972 and 1977?

> 1 lb sausages
> 8 oz coffee
> 5 lb potatoes
> 6 eggs
> 3 pts milk
> 1 lb carrots

10 The table below shows the company profits at Nikobar Software for the years 2017 to 2022.

Year	2017	2018	2019	2020	2021	2022
Profit (£)	50 100		62 500	65 000	110 000	105 000
Percentage increase/decrease (compared with previous year)		15%		8.5%		

a Copy and complete the table, finding:

   i   The company profit in 2018.

   ii   The percentage increase in profit in 2020, when compared with the 2019 figure.

   iii The percentage increase in profit in 2021, when compared with the 2020 figure. Give your answer to one decimal place.

   iv   The percentage decrease in profit in 2022, when compared with the 2021 figure.

b In which year was the biggest increase in profits?
   What might explain this?

Cross-curricular activity

Spreadsheets are extremely useful for analysing tabulated data and working out different types of percentage changes.
Learn how to use spreadsheets to work out information like that in questions 9 and 10.

Band 3 questions

11 Jacqueline always felt it was unfair that she had such a long name.

a When she reached the age of 8, she decided to shorten her name to make it quicker to write. She started calling herself Jackie.

What is the percentage decrease in the number of letters she now had to write?

b Two years later, she shortened her name again to Jak.

What is the percentage decrease in the number of letters from Jackie to Jak?

c What is the overall percentage decrease in the number of letters?

12 For a school project, Imran records the amount of sunshine each day during a school week. The results are shown on the bar chart below.

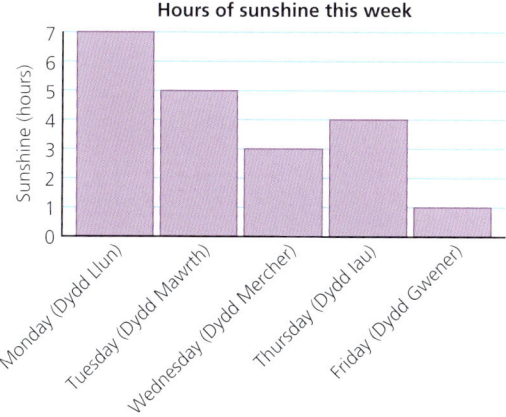

a Monday was the sunniest day this week. What percentage of the total hours of sunshine was recorded on Monday?

b What percentage of the sunshine was recorded on Friday?

c What was the percentage decrease in the sunshine recorded between Tuesday and Wednesday?

Imran continued his sunshine measurements for the entire month of June, including weekends.

His measurements are shown in the bar chart below.

This bar chart shows the number of times he recorded each amount of sunshine.

For example, on three days there was no sunshine.

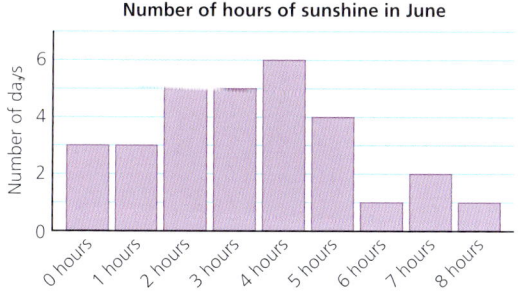

d What percentage of days were completely cloudy?

e On what percentage of days were there 4 hours of sunshine?

f Nerys had collected similar sunshine measurements during May.

Nerys recorded 80 hours of sunshine during May. Imran recorded 100 hours of sunshine in June.

Imran says, 'That's a 20% increase from May to June!'

Is he correct? Explain your answer.

Strategic competence

⑬ This information is taken from government statistics on transport and travel in Great Britain since 1961.

Transport and travel	1961	1971	1981	1991	2001	2011
Road vehicles (millions)						
Licensed road vehicles	9.0	14.0	19.3	24.5	28.9	34.5
Motor vehicles registered for the first time	1.3	1.7	2.0	1.9	2.9	2.4
Length of network (thousand km)						
All public roads	314.0	325.0	342.0	360.0	392.0	393.0
Motorways	0.2	1.3	2.6	3.1	3.5	3.6

> Research the same information for Wales.

Give your answers to the following questions to one decimal place.

a   How many more licensed vehicles were on the road in 2001 compared to 1961?

b   Calculate this increase as a percentage of the number of vehicles in 1961.

c   Which 10-year period had the greatest percentage increase in the number of road vehicles?

d   Calculate the percentage increase in the length of motorways over each 10-year period.

Key words

Here is a list of the key words you met in this chapter.

Decimal	Fraction	Interest	Percentage
Percentage change	Percentage decrease	Percentage error	Percentage increase
Percentage loss	Percentage profit		

Use the glossary at the back of this book to check any you are unsure about.

12 Percentages

Review exercise: percentages

Please note: You cannot use a calculator for questions 1 to 5, but calculators are allowed for the remainder of the questions in this section.

Band 1 questions

1. Each week Harri saves £3 of the £12 he earns. What percentage of his earnings does he spend?

2. Nia takes 56 letters to the post office for posting. Fourteen are first class and the remainder are second class. What percentage of the letters go second class?

3. Catrin's gross wage is £240 per week. Her take-home pay is £156.
   What percentage is this of her gross wage?

   Gross wage means her pay before tax is deducted.

4. Work out these.
   - a Increase 30 by 10%.
   - b Decrease 600 by 10%.
   - c Increase 20 by 25%.
   - d Decrease 40 by 10%.
   - e Increase 36 by $33\frac{1}{3}$%.
   - f Decrease 72 by $66\frac{2}{3}$%.
   - g Decrease 45 by 10%.
   - h Increase 48 by 100%.
   - i Decrease 27 by 100%.
   - j Increase 30 by 150%.

5. The value of a classic car increased by 10% during the past year.
   If it was worth £16 400 a year ago, what is its value now?

6. A shop bought printers at £88 each and sold them for £123.20.
   What was the percentage profit?

7. Tomos buys a new car for £8720 and sells it for £5789 three years later.
   What is his percentage loss?

Band 2 questions

8. a A house was bought for £160 000 and sold for £210 000.
      Calculate the percentage profit.
   b A flat was bought for £90 000 and sold for £83 000.
      Calculate the percentage loss.

9. One week in August, 120 flights operated by Cabin Fever Airlines left London Heathrow airport.
   Of these, 15% flew to the USA.
   How many of these flights:
   - a flew to the USA?
   - b flew elsewhere?

   The following week, Cabin Fever Airlines announced that they would increase their flights to the USA to 150% of the previous number.
   - c How many flights does the airline fly to the USA per week now?

10. In 2018, 89.2 million passengers used King's Cross underground station in London.
    In 2017, the figure was 86.5 million.
    What percentage increase in passengers was there between 2017 and 2018?

Band 3 questions

11 There are 25 pupils in class 8M and 20% of them have their birthday in April.

   a   How many pupils in class 8M have their birthday in April?

   b   Of those pupils with a birthday in April, 40% have their birthday on 1st April.
       How many pupils have their birthday on that date?

12 Cwmfawr United's average attendance for the 2019 season was 8% higher than it was during the 2018 season.

   a   If the average attendance during 2018 was 3000, what was the 2019 season's average attendance?

   b   The 2017 season was Cwmfawr United's best year. They had an average attendance of 3600.
       Find the percentage change in the 2018 season when compared with the 2017 season.

   c   Work out the same percentage comparisons with your favourite sports team or local team.
       Is there a reason for these percentage changes?

13 Multiplicative reasoning

Coming up...
- Reviewing ratio and proportion
- Using conversion graphs
- Understanding best buys

Table tennis

Dewi and Ceri are playing table tennis. The first person to reach 10 points wins the game.

Dewi gets the first point and Ceri gets the second one.

The score is now 1 – 1.

The ratio of their scores is 1 : 1.

They continue with the game.

a   Copy the table below and fill in the blanks.

   Remember to simplify the ratios where you can!

	Who wins the point?	Dewi's score	Ceri's score	Ratio Dewi's score : Ceri's score
2nd point	Ceri	1	1	1 : 1
3rd point	Ceri	1	2	
4th point	Ceri	1		1 : 3
5th point	Dewi		3	2 : 3
6th point		2	4	1 : 2
7th point	Ceri	2		2 : 5
8th point	Ceri	2	6	
9th point	Dewi	3		1 : 2
10th point	Dewi			2 : 3
11th point	Ceri	4		4 : 7
12th point	Ceri	4	8	
13th point		5		
14th point	Dewi			3 : 4
15th point	Ceri	6		
16th (final) point				3 : 5

b   Who wins the game and what is the final score?

13.1 Ratio and proportion review

Skill checker

Speed test: simplify these ratios.

Can you get all 20 right in under 2 minutes?

8 : 2	4 : 16	10 : 6	25 : 15	300 : 200
7 : 35	9 : 21	99 : 77	17 : 51	16 : 14
70 : 75	650 : 850	98 : 49	36 : 63	108 : 144
2 : 4 : 6	100 : 50 : 25	108 : 144 : 24	2 : 6 : 20	15 : 33 : 60

▶ Splitting a quantity in a given ratio

Worked example

On a Christmas tree farm, 240 trees must be delivered to two different shops, Festive Fun and Trees R Us, in the ratio 3 : 5.

There are two delivery people.

Bryn delivers to Festive Fun.

Daf delivers to Trees R Us.

a How many trees does Bryn deliver?
b How many trees does Daf deliver?
c What fraction of all the trees does Bryn deliver?
d What fraction of all the trees does Daf deliver?

Solution

The total number of parts is 3 + 5 = 8.

To find one part, divide the total number of trees by 8:

240 ÷ 8 = 30

One part is 30 trees.

a Since Bryn is delivering three parts, he delivers:
   3 × 30 = 90 trees

b Daf is delivering five parts. He delivers:

5 × 30 = 150 trees

Bryn delivers 90 trees and Daf delivers 150.

> As a check, add the two amounts:
> 90 + 150 = 240
> This is the total number of trees.

This solution can be illustrated using a bar model.

240 Christmas trees							
30 trees	30 trees	30 trees	30 trees	30 trees	30 trees	30 trees	30 trees
Bryn delivers 3 × 30 = 90 trees			Daf delivers 5 × 30 = 150 trees				

c Bryn delivers $\frac{90}{240}$ trees.

$$\frac{90}{240} = \frac{9}{24} = \frac{3}{8}$$

d Daf delivers $\frac{150}{240}$ trees.

$$\frac{150}{240} = \frac{15}{24} = \frac{5}{8}$$

Bryn delivers $\frac{3}{8}$ of the trees and Daf delivers $\frac{5}{8}$ of them.

▶ Proportion

Worked example

A dung beetle is making dinner for his family. He follows this recipe:

> Feeds 15 dung beetles
> 20 grams cow dung
> 18 grams horse manure
> 30 sheep droppings

The dung beetle only has 12 family members.
How much of each ingredient should he use?

Solution

Find how much of each ingredient the dung beetle would need for 1 beetle, then multiply by 12:

> Divide by 15 to find how much he needs for 1 beetle in this column.

Ingredients	For 15 beetles	For 1 beetle	For 12 beetles
Cow dung	20 g	$20 \div 15 = 1\frac{1}{3}$ g	$1\frac{1}{3} \times 12 = 16$ g
Horse manure	18 g	$18 \div 15 = 1.2$ g	14.4 g
Sheep droppings	30	2	24

> Multiply by 12 to find how much he needs for 12 beetles.

13.1 Now try these

Band 1 questions

1. Simplify these ratios.
   - a $14:2$
   - b $5:20$
   - c $3:15$
   - d $12:21$
   - e $42:30$
   - f $90:63$
   - g $38:57$
   - h $18:27:6$
   - i $40:2:10$
   - j $18:4:10$

2. Find the number that goes in the box to make an equivalent ratio.
   - a $3:5=\square:60$
   - b $3:4=6:\square$
   - c $1:11=\square:33$
   - d $5:2=15:\square$
   - e $7:3=\square:15$
   - f $8:7=24:\square$
   - g $16:9=48:\square$
   - h $1:7:4=\square:\square:28$
   - i $5:11:4=\square:77:\square$
   - j $27:7:2=\square:28:8$

3. Write these ratios in their simplest form.
   - a 1 metre : 25 cm
   - b 6 kg : 500 g
   - c 250 ml : 2 litres
   - d 15 cm : 1.65 metres

4. Write $2.8:2.1$ in its simplest form, using whole numbers.

5. Divide 45 in the ratio $2:7$.

6. Three pieces of timber have lengths 1.2 m, 2 m and 2.2 m.
   Find the ratio of these lengths in its simplest form.

Band 2 questions

7. In a school, the ratio of the number of students having school dinners to those taking packed lunches is $4:7$.
   - a What fraction of students have school dinners?
   - b If there are 330 students in the school, how many have packed lunches?

8. Vladimir gives Aaron £14 million.
   Aaron splits the money in the ratio $5:2$ between himself and his friend Naginder.
   - a How much money does Aaron keep?
   - b How much money does Naginder receive?

9. Carwyn is making a mosaic using a ratio of three blue tiles to five red tiles to four green tiles.
   Blue mosaic tiles cost 30p each; red ones cost 25p each; green tiles cost 16p each.
   If 66 blue tiles are used, how much money will Carwyn spend:
   - a on red tiles
   - b on green tiles
   - c altogether?

10. Carys receives a present of £200 from her nain.
    She spends 20% of the money on a magazine subscription.
    The rest of the money she splits between clothes, food and savings in the ratio $5:2:1$.
    How much does she save?

11. a Meringue is made by mixing egg whites and cups of sugar in the ratio $5:1$.
       How many cups of sugar are needed if 12 egg whites are used in the mixture?
    b Apple crumble topping is made using flour, butter and sugar in the ratio $5:2:1$.
       How much flour is needed if 80 grams of butter is used?

Band 3 questions

12 Jo, Kevleen and Lakar are friends. They kept a count of the number of books they read last year.
Jo read twice as many books as Kevleen. Lakar reads four times as many books as Jo.
  a Write down the ratio of the number of books read by Jo to Kevleen to Lakar.
  b In total, the three friends read 110 books.
     How many books did each of the friends read?

13 The side lengths in shapes X and Y below are in the ratio 1 : 2.
The side lengths of trapezium Y are double the side lengths of the trapezium X.

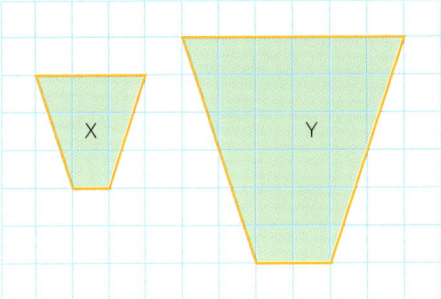

  a For each of these pairs of shapes, what is the ratio of lengths in shape X to shape Y?

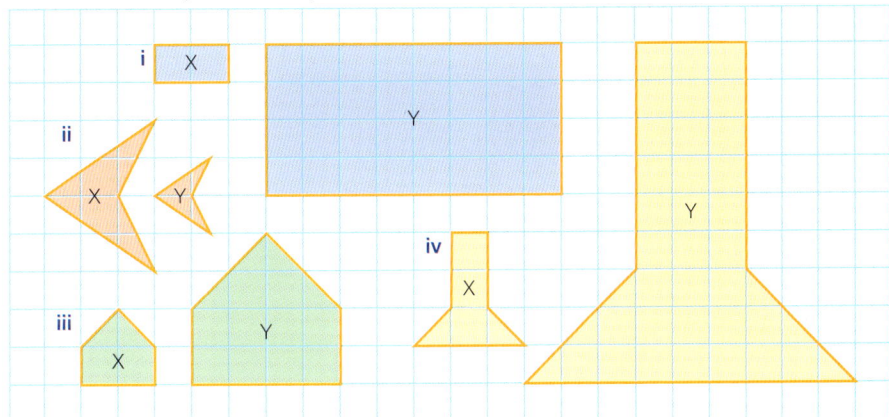

  b For the two pentagons in part **iii**, work out the area of both shapes X and Y by counting the small squares.
     What is the simplified ratio of the area of X to the area of Y?
  c Work out the ratio of the areas of shape X to shape Y in part **iv**.

14 Here are the ingredients for five pancakes.

```
150 g flour
1 egg
200 ml milk
```

  a Write the ratio of flour to milk in its simplest form.
  b How much milk is needed for seven pancakes?

15 There are 24 people working in a hotel restaurant:
  • four men under 18
  • four women under 18
  • six men over 18
  • ten women over 18.
  a Write the number of men to the number of women as a ratio in its simplest form.
  b What fraction of the people working in the kitchen are men?
  c Find the ratio of the number of women to the total number of workers in its simplest form.

13.2 Conversion graphs

Skill checker

Plot the points below to crack the code and answer this riddle:

Did you hear about the two men who stole a calendar?

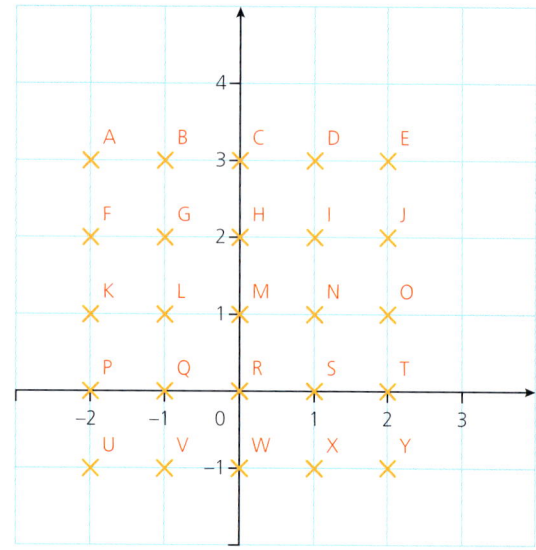

(2,0), (0,2), (2,3), (2,−1) / (−1,3), (2,1), (2,0), (0,2) / (−1,2), (2,1), (2, 0) / (1,0), (1,2), (1,−1) / (0,1), (2,1), (1,1), (2,0), (0,2), (1,0) !

▶ Converting units

Worked example

This is a conversion graph between ounces and grams.

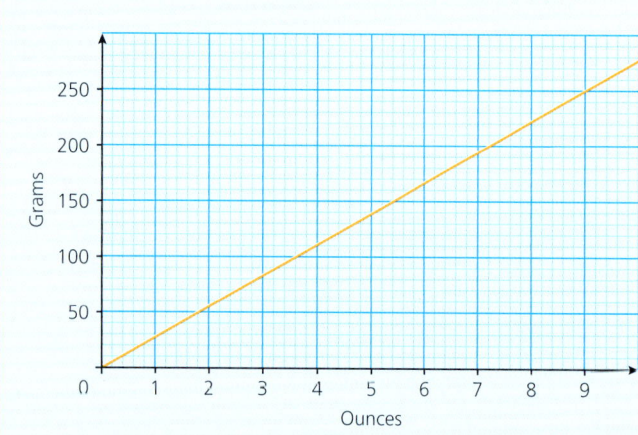

a A packet of butter is labelled 250 g.
   What is this in ounces?
b Pawl buys 8 ounces of cheese.
   What is this in grams?
c How many ounces are 500 grams?
d Convert 80 ounces to grams.

Solution

a The blue line on the graph shows that 250 g is about 9 ounces.

b The green line on the graph shows that 8 ounces is about 225 grams.

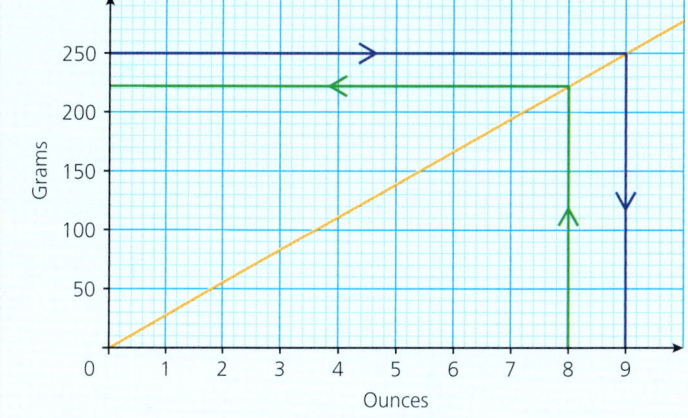

c Using the answer from **a**:
   250 g is roughly 9 ounces
   so 500 g is roughly 18 ounces.   *Multiply by 2.*

d Using the answer from **b**:
   8 ounces is roughly 225 g
   so 80 ounces is roughly 2250 g (or 2.25 kg).   *Multiply by 10.*

Worked example

Penny is going on holiday to Turkey, where the currency is lira.

The graph shows the conversion rate between British pounds and Turkish lira.

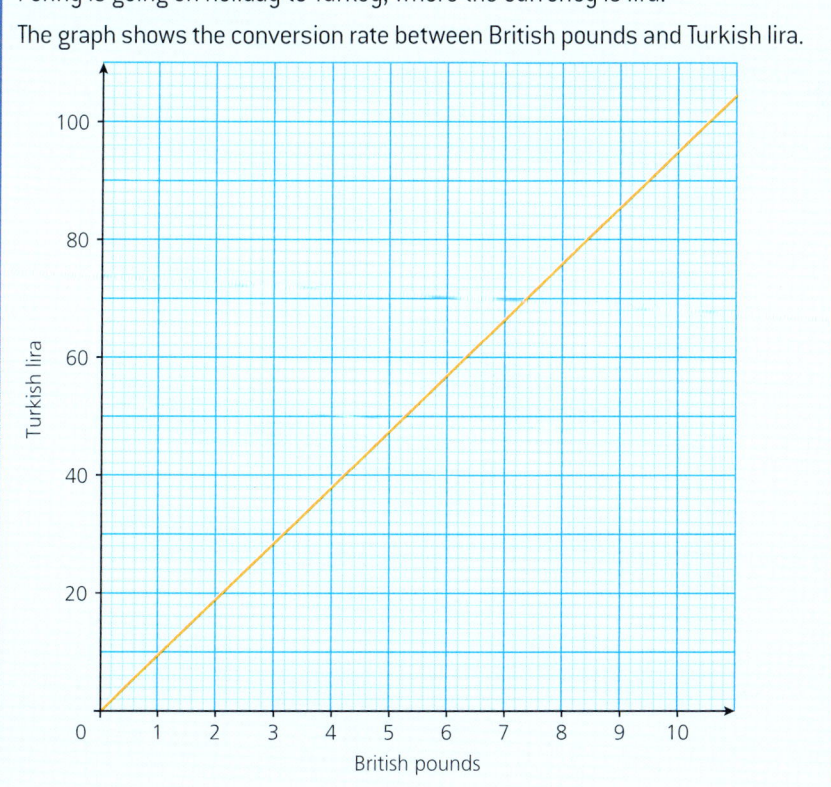

a Copy and complete the following, using the conversion graph above to help you.
  i £10 = ___ lira
  ii £100 = ___ lira
  iii ___ is roughly 90 lira.
  iv ___ is roughly 180 lira.

b Sunhats cost £7.50 in Penny's local shop in the UK.

Penny forgets to buy a sunhat before she leaves and buys one in a Turkish market for 80 lira.

Did Penny spend more getting her sunhat in Turkey, instead of buying it before leaving?

Solution

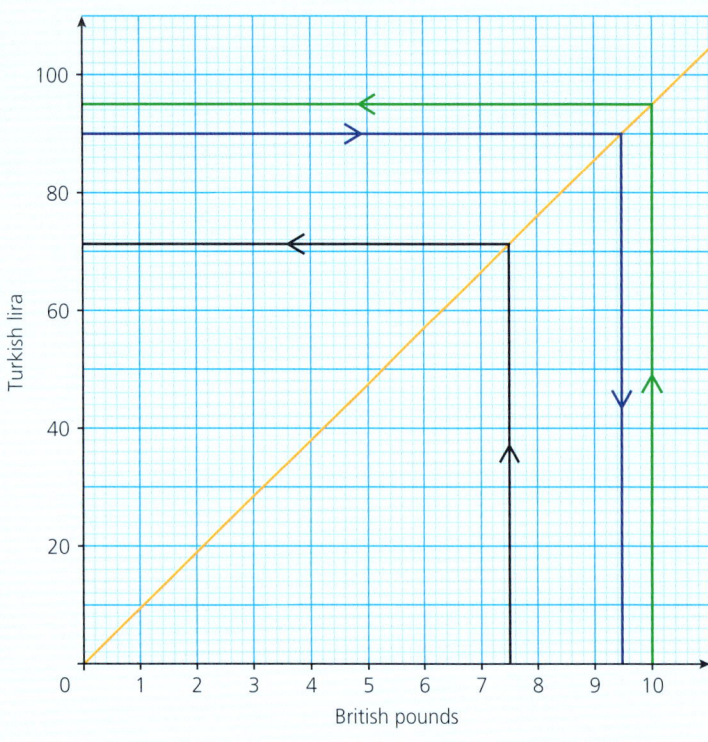

a i Using the green line on the graph, £10 = 95 lira.
  ii £10 = 95 lira
   so £100 = 950 lira. *Multiply by 10.*
  iii Using the blue line on the graph, 90 lira is roughly equivalent to £9.50.
  iv 90 lira is roughly £9.50
   so 180 lira is roughly £19. *Multiply by 2.*
b Using the black line on the graph, £7.50 is roughly 71 lira.
Penny paid 80 lira, so she paid more for her sunhat in Turkey.

13 Multiplicative reasoning

13.2 Now try these

Band 1 questions

Fluency

1 5 gallons is roughly 22.5 litres.

    a Copy and complete the following.
        i ___ gallons = 45 litres
        ii ___ gallons = 90 litres
        iii 15 gallons = ___ litres
        iv 1 gallon = ___ litres

    b Plot the points above on a copy of the conversion graph below.
        Join the points with a line.
        One point has been plotted for you.

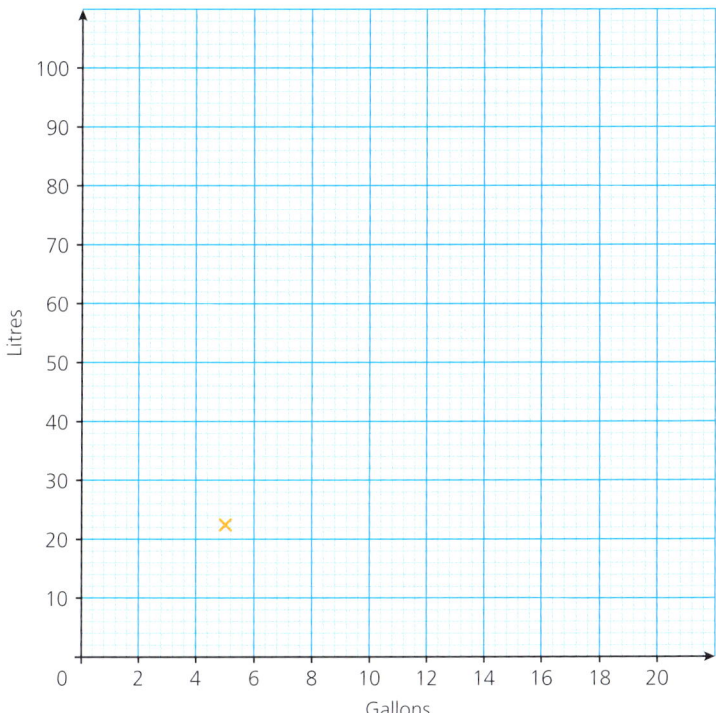

Logical reasoning

2 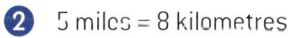 5 miles = 8 kilometres

    a Copy and complete the following, using the conversion above to help you.
        i 10 miles = ___ km
        ii 20 miles = ___ km
        iii ___ miles = 40 km

    b Plot the points above on a copy of the conversion graph on the right.
        Join the points with a line.
        You should have a single straight line passing through (0, 0).

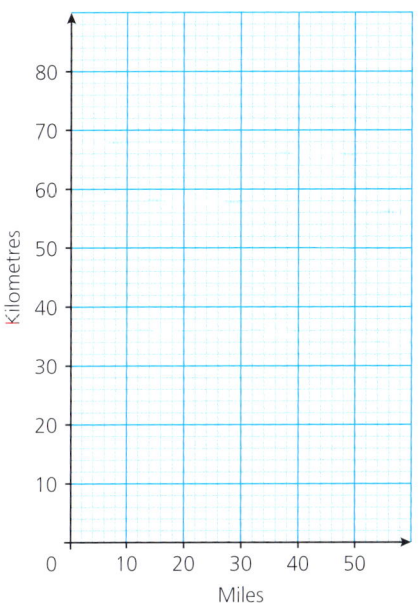

c Jac is on holiday in France.

Convert these distances to miles.

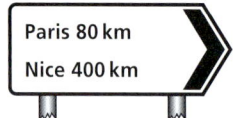

i 80 km

ii 800 km

iii 400 km

You can do this using the conversion in part a or using your graph from part b.

Band 2 questions

3 Myfanwy plots some points for a conversion graph to convert inches to centimetres, but it doesn't look right.

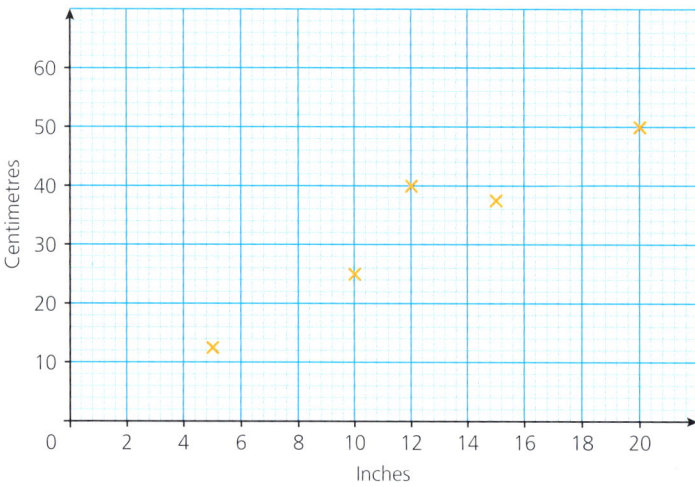

a Which one of the points appears to be in the wrong place?

b Copy Myfanwy's graph, missing out the incorrect point.

Draw a straight line through the other four points.

c Use your graph to correct the point that was plotted wrongly.

d Use your graph to estimate how many centimetres are equivalent to 18 inches.

e The world's largest butterfly is the Queen Alexandra Birdwing.

It has a wingspan of 28 cm.

Write the wingspan in inches.

4 This conversion graph is for changing between pounds and kilograms.

Use the graph to answer the following questions.

a A newborn baby weighs 3 kg.

What is this in pounds?

b A turkey is labelled 5 kg.

What is this in pounds?

c Lowri wants to buy 8 pounds of potatoes.

What is this in kilograms?

d A cat basket is suitable for weights of up to 10 pounds.

What is this in kilograms?

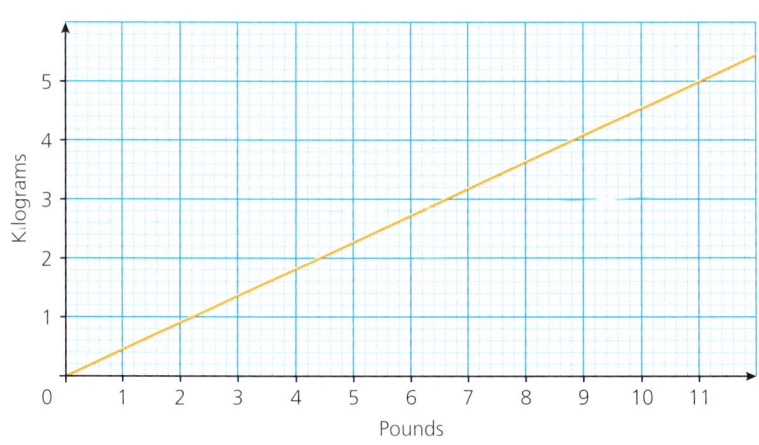

5 £70 is roughly the same as €80.

   a Draw a conversion graph between British pounds and euros.

   b A hire car costs €70 for one day.

   How much is the price in British pounds?

   c André buys a meal for £25.

   How much does this cost in euros?

   d Meinir buys a present for €53.

   Convert the price into British pounds.

6 Amelia visits her sister Kay, who lives in the USA.

The graph on the right shows the exchange rate from British pounds to US dollars.

   a How much in dollars is £10 worth?

   b Using the information from part **a**, complete the following:

   £1 = $___

   c Amelia plans to change £500 to US dollars for her holiday.

   How many US dollars does she get?

   d The following summer Kay visits Amelia in the UK.

   Kay changes $520 into British pounds for her trip.

   How many British pounds does Kay bring?

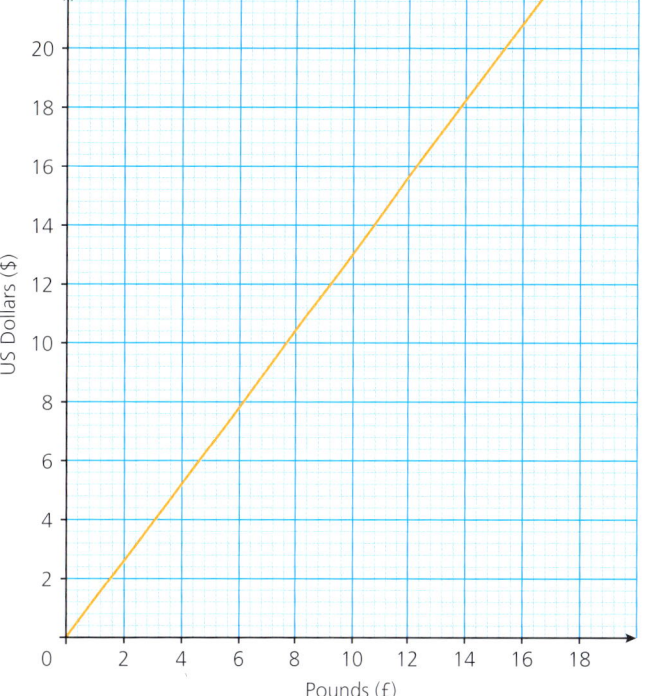

7 The Japanese Bullet Train can reach a top speed of 320 km per hour.

However, over long journeys its average speed is reduced to 225 km per hour due to stopping at stations.

   a Using an average speed of 225 km per hour, how far can the Bullet Train travel in:

   i 60 minutes **ii** 90 minutes **iii** 120 minutes?

   b Draw a conversion graph with time (in minutes) on the x-axis and distance (in kilometres) on the y-axis.

   Plot the three points from part **a**.

   c How long would it take the Bullet Train to travel from Tokyo to Shin-Aomori, a distance of 675 km?

   d Roughly how far would the Bullet Train travel in $2\frac{1}{2}$ hours?

Band 3 questions

8 The graph shown is a conversion graph for calculating the cost of a hire car, based on the number of miles driven.

   a Pedr and Chris hire a car and go camping for the weekend.

   They drive a total of 400 miles.

   What is the cost of their hire car?

   b Carys and Leri also hire a car.

   They have a budget of £110.

   How far can they travel?

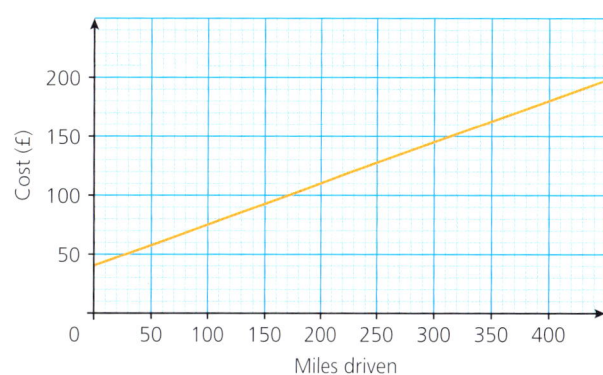

9 In the UK and the rest of Europe, temperatures are measured in degrees Celsius.

In the USA, temperatures are measured in degrees Fahrenheit.

You can use the formula below to convert a temperature measured in degrees Celsius to degrees Fahrenheit.

$F = 1.8C + 32$

For example, to convert a temperature of 40°C:

$F = 1.8 \times 40 + 32$

= 104 degrees Fahrenheit

a Copy and complete this table.

Convert the temperatures given to degrees Fahrenheit using the formula.

Temperature in degrees Celsius	20	30	40	60	80
Temperature in degrees Fahrenheit			104		

b Draw a conversion graph. Use the x-axis for degrees Celsius with values from 0 to 100.

Use the y-axis for degrees Fahrenheit with values from 0 to 220.

Plot the points from your table, for example (40, 104).

Join the points with a straight line.

Note that your line **will not** pass through (0, 0).

Cross-curricular activity

How could you create this table using the conversion formula in a spreadsheet?

c The freezing point of water is 0 degrees Celsius.

Use your graph to find the freezing point of water in degrees Fahrenheit.

d The boiling point of water is 100 degrees Celsius.

Use your graph to find the boiling point of water in degrees Fahrenheit.

10 At the end of the summer term in Ysgol Glanuchaf, Year 8 pupils take exams in all their subjects.

Mrs Abacus, the Maths teacher, sets an exam with a maximum score of 40 marks.

Mr Bunsen, the Science teacher, sets an exam with a maximum score of 60 marks.

Here are the scores of three pupils in the two subjects.

	Daniel	Priya	Lucinda
Maths score out of 40	32	30	24
Science score out of 60	30	54	36

The headteacher Mr Power asks all teachers to convert the pupils' scores to percentages for the school reports.

a Mrs Abacus converts all her pupils' scores to percentages correctly.

Find the percentage scores in the Maths exam for Daniel, Priya and Lucinda.

Copy and complete this table.

	Daniel	Priya	Lucinda
Maths score out of 40	32	30	24
Maths percentage			

b Plot a conversion graph to convert from Maths exam score to percentage.

Use the x-axis for exam score, using values from 0 to 40.

Use the y-axis for percentage, using values from 0 to 100.

Plot the points in your table in part **a** and join the points with a straight line.

Note

Again, this type of work could be done on a spreadsheet, especially if the whole class results were being recorded.

c Mr Bunsen makes a mistake converting the Science scores to percentages.

He adds 30 to each pupil's exam score.

Copy and complete this table with the incorrect percentage scores Mr Bunsen gives.

	Daniel	Priya	Lucinda
Science score out of 60	30	54	36
Science incorrect percentage score	60		

d Plot a conversion graph to convert from Science exam score to Mr Bunsen's incorrect percentage score.

Use the x-axis for exam score, with values from 0 to 60.

Use the y-axis for percentage score, with values from 0 to 100.

Plot the points in your table in part **c** and join the points with a straight line.

Label the line 'Science % incorrect'.

e What is the minimum possible percentage, using Mr Bunsen's incorrect percentage method?

f What is the maximum possible incorrect percentage score?

g Correct Mr Bunsen's percentage scores.

Find the three pupils' Science percentage scores correctly.

Copy and complete this table.

	Daniel	Priya	Lucinda
Science score out of 60	30	54	36
Science percentage (corrected)			

h Plot these points on your graph from part **d**.

Join the points with a second straight line and label the line 'Science % correct'.

i Alun is a pupil in Mr Bunsen's science class.

He says, 'It doesn't make any difference to me! I got the same percentage score using both methods!'

What score out of 60 did Alun get in the exam?

What was his percentage score?

13.3 Best buys

Skill checker

Here are the ingredients needed in a recipe for chocolate chip cookies.

How much of each of these ingredients would you need to make 25 cookies?

a Flour
b Butter
c Soft brown sugar
d White sugar

125 g	Unsalted butter
115 g	Soft brown sugar
110 g	White caster sugar
1	Medium egg
1 tsp	Vanilla extract
220 g	Self-raising flour
$\frac{1}{2}$ tsp	Salt
200 g	Chocolate chips
Makes 20 cookies	

▶ Comparing unit costs

You can find out the cost of one item, one litre or one kilogram, and so on.

This is called the **unit price**.

To find the unit price, divide the price by the number of items, the number of litres or the number of kilograms.

Worked example

Orange juice can be bought in differently sized cartons.

Holly needs to buy several cartons of orange juice for a party.

Which size carton should Holly buy?

Solution

Larger carton

2 litres cost £3.00.

1 litre costs £3.00 ÷ 2 = £1.50.

The unit cost is £1.50.

Smaller carton

1.5 litres cost £2.40.

1 litre costs £2.40 ÷ 1.5 = £1.60.

The unit cost is £1.60.

The larger carton has a lower unit cost.

Holly should buy the larger carton.

▶ The unitary method

After finding the unit cost, you can find the cost of any number of items by multiplying.

Worked example

Copy and complete these tables to compare the cost of buying ten fish fingers for each of these brands.

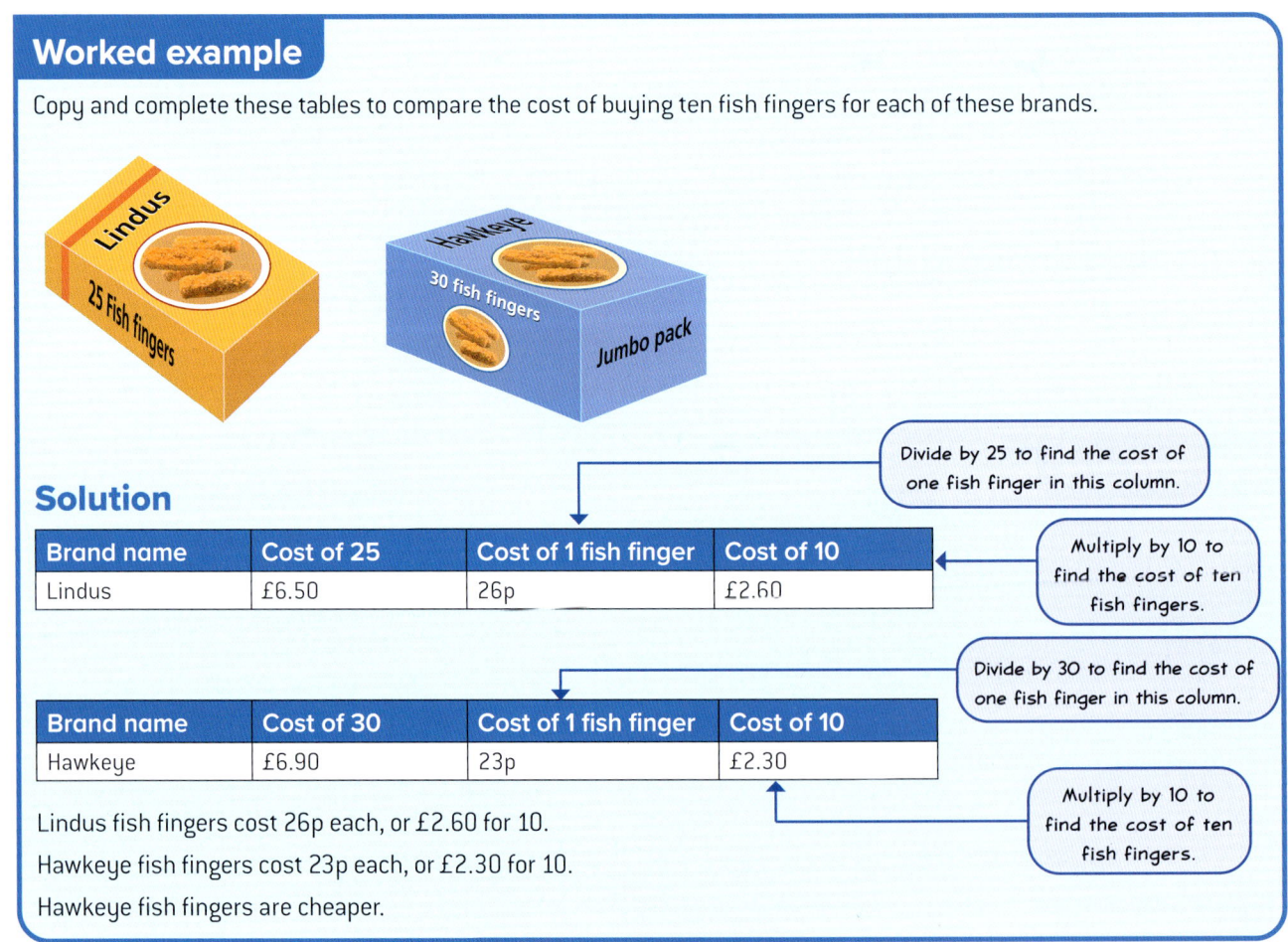

Solution

Brand name	Cost of 25	Cost of 1 fish finger	Cost of 10
Lindus	£6.50	26p	£2.60

Divide by 25 to find the cost of one fish finger in this column.

Multiply by 10 to find the cost of ten fish fingers.

Brand name	Cost of 30	Cost of 1 fish finger	Cost of 10
Hawkeye	£6.90	23p	£2.30

Divide by 30 to find the cost of one fish finger in this column.

Multiply by 10 to find the cost of ten fish fingers.

Lindus fish fingers cost 26p each, or £2.60 for 10.

Hawkeye fish fingers cost 23p each, or £2.30 for 10.

Hawkeye fish fingers are cheaper.

13 Multiplicative reasoning

Activity

Can you spot the problem here?

① What do you think this was supposed to say?

> **Irresistible Chocolate Cakes**
>
> **£2.25**
>
> Equivalent to £2.25 per item.

② Anything wrong here?

> **Cookie mix 1.5 kg**
>
> **£1.90**
> £1,266.67 per kg

③ Finally, what do you think of this sign?

> **Orange and mango smoothie**
>
> Sale: Only **£2** each   Was £2

13.3 Now try these

Band 1 questions

1. Orange juice is sold in two different cartons, with volumes of 700 ml and 500 ml.
   a. Find the cost for 100 ml of juice for the 700 ml carton.
   b. Find the cost for 100 ml of juice for the 500 ml carton.
   c. Which carton of juice gives the best value for money?

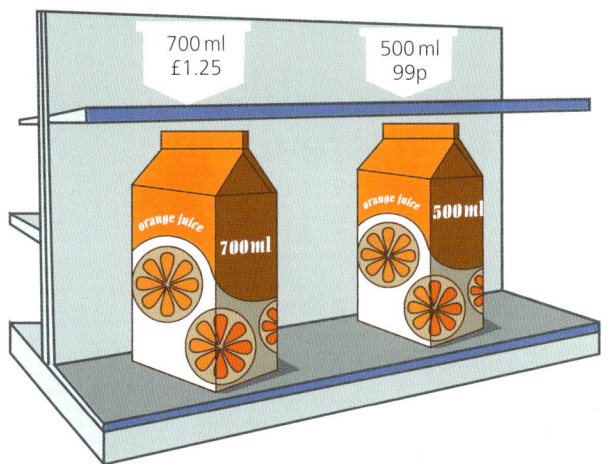

2. In Superstuff supermarket, barbecue charcoal is sold at £7.50 for 10 kg.

   In Stainberry's supermarket, similar charcoal is on sale at £4.98 for 6 kg.
   a. What is the cost of charcoal per kilogram in each supermarket?
   b. Which supermarket offers the best value for money?

3. Jam is sold in two sizes of jar: 454 g and 340 g.
   Which size jar gives the better value for money?

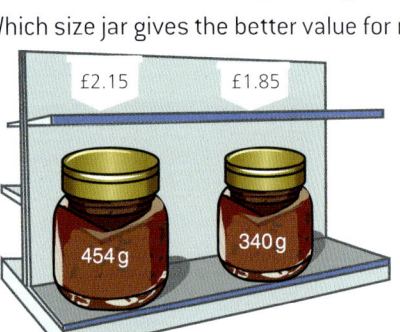

4. Cornflakes are sold in two differently sized boxes.
   a. What is the cost of 100 grams of cornflakes for the 750 g box?
   b. What is the cost of 100 grams of cornflakes for the 1.4 kg box?
   c. Which box is the better buy?

5 In Superstuff supermarket, three different brands of bread flour are available: Meredith's, McDermott's and Superstuff's own brand.

Each brand of flour comes in a differently sized bag. The prices are shown below.

Meredith's Bread Flour	McDermott's Bread Flour	Superstuff Bread Flour
500 g	1 kg	1.5 kg
88p	£1.28	£1.40

a Copy and complete this table.

Brand name	Cost of 100 g	Cost of 1 kg
Meredith's		
McDermott's		
Superstuff		

b Which brand of bread flour is the cheapest?

Band 2 questions

6 Dai Owens is a politician standing in an election.

If he is elected, he promises to spend £350 million pounds on the National Health Service **every week**.

Nerys Williams is standing in the same election.

If she is elected, she promises to spend £1800 million pounds on the National Health Service **every month**.

a Assuming there are 30 days in a month, find who is promising the most money for the health service per day.

b How much money is each politician promising the health service every year?

7 a Work out the price of one chocolate in each of these packets of chocolates.

b Which of the packets offers the best value for money?

89p

£1.99

£2.45

c Apart from the price of one chocolate, what other factors might affect your decision about which packet of chocolates to buy?

Band 3 questions

8 In Superstuff supermarket there are three differently sized bags of frozen peas.

Superstuff Frozen Peas	Superstuff Frozen Peas	Superstuff Frozen Peas
375 g	800 g	1 kg
£1.30	£2.15	£2.50

a Copy and complete this table to compare these prices.
Remember to fill in the empty spaces in the header row.
There are many ways to do this.

Bag weight	Price	Cost of _____ g	Cost of _____
375 g			
800 g			
1 kg			

b Which size bag is the best value for money?

9 In a clothing shop, green, white and blue T-shirts are on sale.

A pack of 3 green T-shirts costs £16.50.

A pack of 10 white T-shirts costs £34.50.

A pack of 2 blue ones is £9.90.

a Find the unit cost of all three types of T-shirt.

b Which colour T-shirt has the cheapest unit cost?

c Pagat is hoping to buy some new T-shirts.

Is the unit cost the only factor Pagat considers when deciding which T-shirts to buy?

10 These two differently sized boxes of crisps are the same price.

If you buy the smaller box, however, you get a second box half price.

Which deal is better value for money? Explain your answer.

Key words

Here is a list of the key words you met in this chapter.

Best buy	Conversion graph	Currency conversion	Proportion
Ratio	Unitary method		

Use the glossary at the back of this book to check any you are unsure about.

Review exercise: multiplicative reasoning

Band 1 questions

1 Cari has some apples and bananas in her fruit bowl. The ratio of bananas to fruit is 2 : 9.
  a What fraction of Cari's fruit are bananas?
  b What is the ratio of bananas to apples?

2 Write these ratios in their simplest form.
  a 260 g : 2 kg
  b 25 cm : 4.25 metres
  c 5 litres : 750 ml
  d 4 metres : 60 cm

3 Work out the unit cost in each case and decide which of these gives the best value for money.
  a i   A box of 26 chocolate biscuits for £3; or
     ii A packet of 20 chocolate biscuits costing £2.50.
  b i   A bag of 5 oranges marked at £2.90; or
     ii A punnet of 12 oranges costing £5.00.
  c i   A packet of yeast weighing 30 grams and costing £1.90; or
     ii A box containing six sachets of yeast, each weighing 7 grams. The whole box costs £2.20.

Band 2 questions

4 Ralia is going back to school after the summer holidays.
  She needs some plastic folders for her schoolwork.
  She sees these different packs of folders on an internet shopping site.

Clear Plastic Folders	**Useful Plastic Folders**	**Large Plastic Folders**
Made from tough plastic	Quite large	Made from strong, clear plastic
Big enough for homework	£5.90 for 8	Large enough for all schoolwork
£4.80 for 6		£7.50 for 10

  a Work out the unit price for all three different types of plastic folder.
  b Which folders are cheapest?
  c Which ones do you think Ralia should buy? Explain your answer.

5 David does an experiment to see how electrical resistance varies with the length of wire used.
  Here are his results.

Length of wire (cm)	10	20	30	40	50
Resistance (ohms)	2	4	6	8	10

  a Draw a pair of axes.
     Use values from 0 to 50 centimetres on the horizontal axis and from 0 to 10 ohms on the vertical axis.
     Plot the points and join them with a straight line.
  b The resistance of a piece of wire is 7.4 ohms.
     What is the length of the wire?
  c David says, 'If you double the length of the wire, it doubles the resistance.' Is he right?

6. Mr Rafferty is thinking about buying a house to rent out to other people.

   He sees these three adverts.

   House 1
   FOR SALE

   2 bedroom house in quiet road of busy town
   £220 000
   Plenty of shops and workplaces nearby
   Near to train station

   House 2
   HOUSE FOR SALE
   £360 000

   4 bedroom house in great condition
   Excellent location in busy and friendly town
   Close to the motorway

   House 3

   House for sale
   5 bedrooms
   £300 000
   In need of repair
   Location: middle of nowhere
   Nearest school 20 miles away
   No shops anywhere nearby
   Transport links: rubbish

   a Mr Rafferty doesn't know which house to buy, so he tries to work out which one is the best value for money.
      He decides to work out the cost per bedroom.
      Copy and complete this table.

	House 1	House 2	House 3
Cost per bedroom	£220 000 ÷ 2 = £110 000		

   b Using cost per bedroom as the unit cost, which house gives the best value for money?

   c Do you think Mr Rafferty should buy the house that gives the best value for money?
      What other factors should he consider?

7. Look at the two angles on a straight line.

   a What is the ratio of angle A : angle B in its simplest form?
   b What fraction of the total does each angle make?

   Angle A is made 10° larger and angle B is made 10° smaller.

   c What is the ratio of their sizes now in its simplest form?
   d What fraction of the total does each angle make now?

   Angles A and B are returned to their original sizes.

   e By how many degrees would you need to decrease A and increase B so that the ratio of angle A : angle B is 1 : 9?

8 One British pound is worth roughly 140 Japanese yen.

a Copy and complete this table.

British pounds	1	3	5
Japanese yen			

b Use your table to draw a graph to convert British pounds to Japanese yen.

Use the x-axis for pounds with values from 0 to 10.

Use the y-axis for yen with values from 0 to 1500.

c Use your graph to answer the following.

  i How many Japanese yen would you get for £10?

  ii How many British pounds would you get for 500 yen?

d i A British company orders laptop computers for all its staff from a Japanese electronics company.

  The British company pays £10 000.

  How much is this in Japanese yen?

  ii A Japanese tourist visits London.

  She changes 100 000 yen to British pounds.

  Roughly how many British pounds would she get?

Band 3 questions

9 The organisers of a marathon race expect 2000 people to take part.

a They place an order for 1500 oranges. Each orange is cut into 8 segments.

  How many orange segments are available for the runners?

The following year the number of runners increases to 2300.

b How many orange segments should there be?

c How many oranges should be bought?

10 This graph is used to convert between pints and litres.

a Explain how to use the graph to convert:

  i 4 pints into litres

  ii 12 pints into litres.

b Convert:

  i 7 pints into litres

  ii 2 litres into pints

  iii 2.8 pints into litres

  iv 1.5 litres into pints.

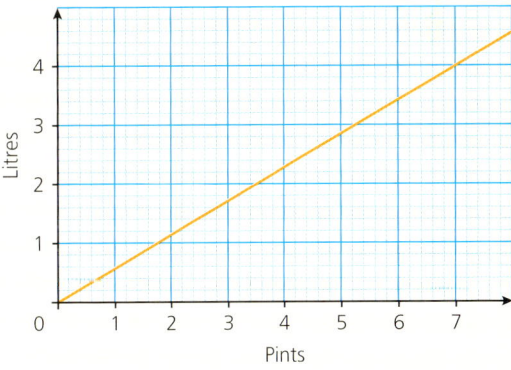

Consolidation 5: Chapters 11–13

Band 1 questions

1. Simplify these ratios.
   - a 12 : 2
   - b 4 : 10
   - c 3 : 27
   - d 15 : 10
   - e 18 : 3
   - f 30 : 18
   - g 16 : 10
   - h 42 : 40
   - i 90 : 36
   - j 46 : 69

2. Find the number that goes in the box to make an equivalent ratio.
   - a 3 : 5 = ☐ : 40
   - b 2 : 5 = 6 : ☐
   - c 2 : 11 = ☐ : 33
   - d 7 : 2 = 21 : ☐
   - e 7 : 3 = ☐ : 15
   - f 8 : 9 = 16 : ☐
   - g 17 : 9 = 51 : ☐
   - h 11 : 5 : 1 = ☐ : 30 : 6
   - i 7 : 9 : 2 = ☐ : 99 : ☐
   - j 22 : ☐ : 2 = ☐ : 12 : 4

3. Find:
   - a 30% of 50
   - b 20% of 350 litres
   - c 125% of 16 kg
   - d 12% of 800
   - e 85% of 40 cm.

4. Copy and complete the following.
   - a 94 is ___% of 100.
   - b 54 is ___% of 60.
   - c 36 is ___% of 120.
   - d 37 is ___% of 50.
   - e 124 is ___% of 155.
   - f 45 seconds is ___% of 60 seconds.
   - g 180 g is ___% of 300 g.
   - h £300 is ___% of £500.
   - i 32 km is ___% of 80 km.
   - j 3.5 m is ___% of 50 cm.

Band 2 questions

5. a Which two of these patterns are nets of a cube?
      Copy the two nets.
   b Imagine folding the two nets you have drawn into cubes.
      Answer these questions by copying and completing the table below.
      - i Which face will appear opposite the red face?
      - ii Which face will appear opposite the orange face?
      - iii Which face will appear opposite the blue face?

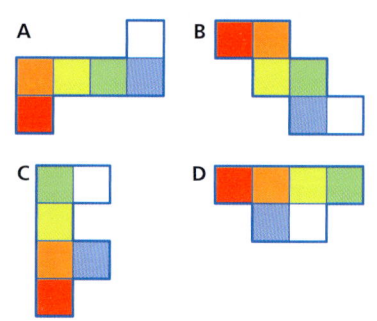

Net (A, B, C or D)	Face opposite red	Face opposite orange	Face opposite blue

Consolidation 5

6 a Ffion has a pedometer, which counts the number of steps she takes each day. On Monday she took 5500 steps. On Tuesday her number of steps increased by 20%.

How many steps did Ffion take on Tuesday?

b Wyn the ice sculptor is making an ice sculpture of a horse. As he chips away at the block of ice, it loses 15% of its weight.

The original block weighed 72 kg. What will the weight of the sculpture be when Wyn has finished?

c Emlyn has a voucher for 15% off a meal at Pizza YumYum.

His bill came to £30. How much does Emlyn actually pay?

d VAT is added on to the cost of work done by a heating engineer at a rate of 20%.

What is the total bill for a job that cost £40 before VAT was added?

e An airline has just increased its prices by 12.5%.

A flight to Croatia cost £340 before the price rise.

What is the new price of the flight?

7 Nargis saves £300 in a savings account that offers 5% simple interest per annum.

She keeps the money in the savings account for 6 years.

a How much interest does Nargis earn in one year?

b How much interest does she earn in six years?

c What is the final balance on Nargis' account after six years?

8 For each situation below, calculate the percentage profit or percentage loss.

a Cost £5, selling price £3.50

b Cost £1.50, selling price £1.30

c Cost £6, selling price £18

9 Here are the ingredients for a spaghetti carbonara recipe.

> **Spaghetti carbonara**
> 100 g ham
> 50 g parmesan cheese
> 30 g pecorino cheese
> 3 large eggs
> 360 g spaghetti
> 2 cloves of garlic
> 50 g unsalted butter
> sea salt and freshly ground black pepper
> *Serves 4 people*

How much of each of the following ingredients would you need for five people?

a Ham

b Spaghetti

c Butter

d Total quantity of cheese

10 A bottle of perfume contains 30 ml of scent and 120 ml of water.

a What percentage of the total liquid is scent?

b To reduce costs, the factory reduces the volume of scent to 12 ml. What is the percentage decrease in the volume of scent?

c After complaints from customers, the company increases the volume of scent from 12 ml back to 30 ml.

What is percentage increase in the volume of scent?

11 At a bingo last year, there were 25 prizes worth £5 each.
  a What was the total value of all the prizes last year?
  b This year, the number of prizes has been reduced by 20%.
    How many prizes are given out this year?
  c The organisers also decrease the value of the prizes by 10%.
    What is the value of each prize this year?
  d What is the **total value** of all the prizes for both years?
  e What is the percentage decrease in the total value of the prizes?

12 Find the error and the percentage error in each of these measurements.
  a Measurement = 303 ml, actual volume = 300 ml
  b Measurement = 2.6 cm, actual length = 2.5 cm
  c Measurement = 2.5 cm actual length = 2.6 cm
  d Measurement = 36 g, actual mass = 40 g
  e Measurement = 40 g, actual mass = 36 g

Band 3 questions

13 Petrol is sold in litres. It used to be sold in gallons.
This conversion graph can be used to convert between gallons and litres.

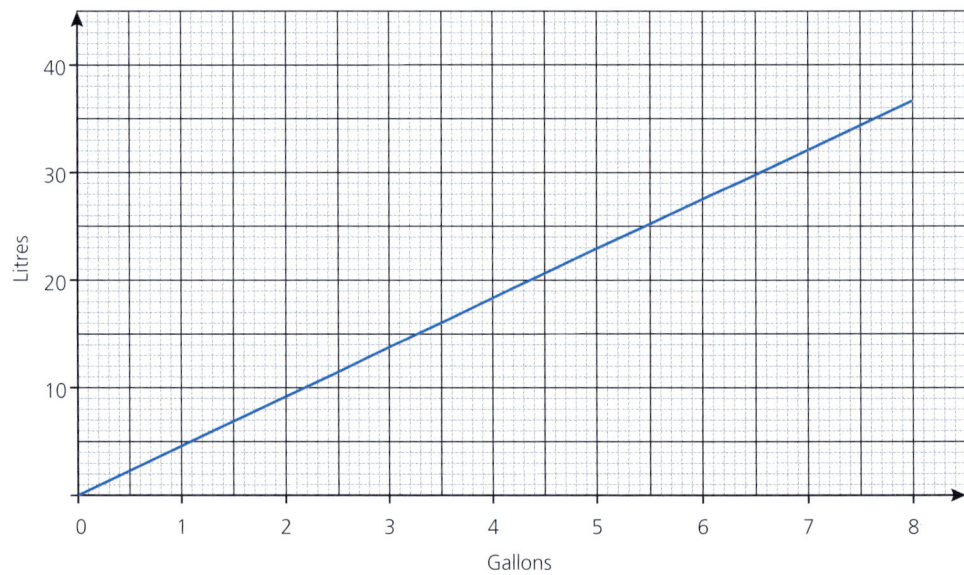

  a Manon buys 35 litres of petrol.
    How many gallons is this?
  b Emyr's car takes 7.5 gallons of petrol.
    Use the graph to convert this to litres.
  c 1 gallon is equal to 8 pints.
    i How many gallons are there in 10 litres?
    ii How many pints are there in 10 litres?
    iii How many pints are there in 1 litre?

Consolidation 5

14 A turret at one corner of a castle is in the shape of a hexagonal prism with a hexagon-based pyramid attached to one end.

a Copy and complete this table for this 3D shape.

Number of faces	Number of vertices	Number of edges

b Use Euler's Formula to check your answers. Copy and complete:

$F + V =$ _____

$E + 2 =$ _____

c Three different 2D shapes make up the surface of the turret.

What are the names of these shapes and how many of each are there?

Copy and complete this table.

Name of 2D shape	Number

15 A 3D model of a letter T is shown.

Answer the following questions using the clues below.

a Find the total surface area of the model.

b Find the volume of the model.

c Find the ratio of the total height to the total width of the model.

Clue 1: The model is made from two identical blocks: one vertical and one horizontal.	Clue 2: The base of the vertical block is a square, with side length 2 cm.	Clue 3: The height of the vertical block is three times its width.

16 Tolu and Pedro both have the Coolphone 20.

Tolu can use 2 GB of data each month and she can make 500 minutes of calls.

Her plan costs £20 per month.

Pedro can use 1.5 GB of data each month and he can make 400 minutes of calls.

His plan costs £18 per month.

a Copy and complete the following.

Tolu's plan

She pays £20   for 2 GB of data.
She pays £☐   for 1 GB of data.   ÷ 2

She pays £20   for 500 minutes of calls.
She pays £☐   for 100 minutes of calls.   ÷ ☐

Pedro's plan

He pays £☐   for ☐ GB of data.
He pays £☐   for 1 GB of data.   ÷ ☐

He pays £☐   for ☐ minutes of calls.
He pays £☐   for 100 minutes of calls.   ÷ ☐

b Whose plan gives the better value for money?

14 Working with data

Coming up...
▶ Finding averages from frequency tables
▶ Drawing and interpreting pie charts

An average spider

Each leg describes the four numbers in the spider's eyes along with the missing number in the corresponding foot.

Find a whole number to go in each foot of the spider so that the statement on each leg is true.

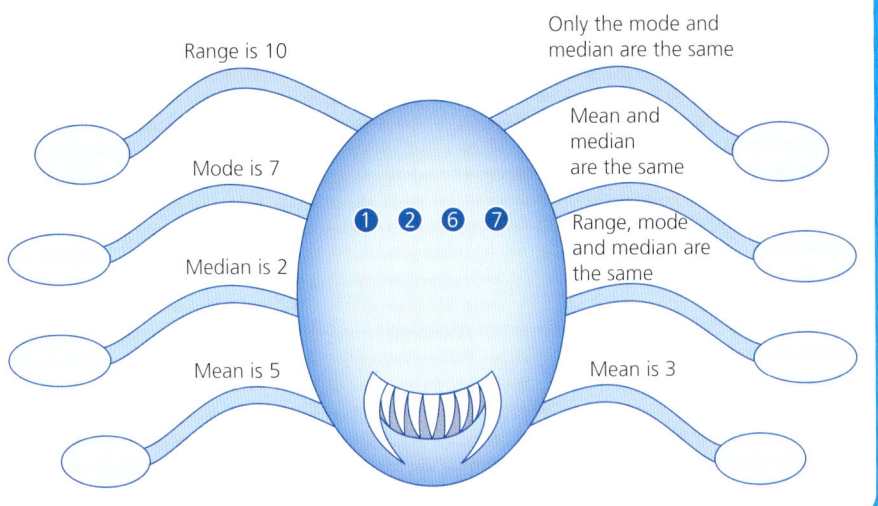

- Range is 10
- Mode is 7
- Median is 2
- Mean is 5
- Only the mode and median are the same
- Mean and median are the same
- Range, mode and median are the same
- Mean is 3

Eyes: 1, 2, 6, 7

14.1 Frequency tables

Skill checker

① Solve these puzzles.

| If the week begins on Monday, what is the middle day of the week? | What is the date and time of the middle moment in November? |

| What is the date of the middle day in January? | What is the date and time of the middle moment in a non-leap year? |

| What is the date and time of the middle moment in January? | What is the date and time of the middle moment in a leap year? |

> There are 365 days in a year. A leap year has an extra day.

244

▶ Review of averages and range

In Book 1 you learnt that an **average** is a value that is representative or typical of all the values in the data set.

There are three averages that you need to know how to use.

① **Mode:** This is the **most common** item of data.
The mode doesn't have to be a number.

② **Median:** Put all the numbers in **order of size** and find the **middle value**.

When you have n items of data the median is the $\left(\dfrac{n+1}{2}\right)$th value.

> If you have 9 numbers, the middle number is the $\left(\dfrac{9+1}{2}\right)$th = 5th number.
> 3, 5, 6, 7, ⑧, 11, 14, 16, 19

Watch out! When there is an even number of data values, there are two values in the middle.

Find the number halfway between these two values by:
- adding them together
- dividing your answer by 2.

③ **Mean:** Add up all the items of data and divide by the total number of items.
This is often written as:

$$\text{mean} = \dfrac{\text{sum of all data values}}{\text{number of data values}}$$

> **Take care!**
> The mean and median can only be used for numerical data.
> When there is an even number of values, then the median might not be one of your actual data values.
> The mean might not be a number in the data set, but it will never be larger than the largest value or smaller than the smallest value.

The **range** is used to show how spread out the data is.
The range is the **difference** between the highest and the lowest values.

> A higher range means the data is very spread out (varied). A small range means the data is more consistent.

Activity

Five number cards are drawn from a pack of cards.

① Find the:
  a mean   b median   c mode   d range.

② A sixth card is drawn from the deck.
  The median and mode are the same as **part 1**.
  The mean of all the cards is 6.
  What number is on the card?

③ A seventh card is drawn from the deck.
  The mean, median and mode are the same as **part 2**.
  What number is on the card?

Cross-curricular activity

Find out the length of the ten longest rivers in Wales and another country of your choice.

Work out the mean, mode, median and range of the length of the rivers of both Wales and your chosen country.

What do you notice? What conclusions can you make from the averages you have found? Do they depend on the size of the country, its landscape, its climate, etc.?

Discuss your results with your teacher.

Larger sets of data can be organised into a **frequency table**.
You can work out the averages and range from a frequency table without having to see the original data.

> In the real world, it isn't that useful to find the average of a very small data set, because you can simply look at what the data is!

Worked example

Manon is doing a project on Welsh castles.

She asks some students at her school how many castles they have visited in Wales.

No one that Manon asked had visited more than six castles.

Here are the results.

> **Number of castles visited:**
> 0 0 0 0 0 0 0 0 1 1 1 1 1 1 1 1 1 1 1 1 1 2 2 2 2 2 2 2 2 3 3 3 3 3 3 3 4 4 4 4 4 4 5 5 5 5 6 6

← This is called the raw data.

Number of castles visited	Frequency
0	8
1	13
2	10
3	7
4	6
5	4
6	2

Remember

Frequency means 'how many'.

← So seven people had visited three castles.

← Manon has organised the data into a **frequency table**.

a How many people did Manon ask?
b Work out the range.
c What is the modal number of castles visited? ← This is another way of asking for the mode.
d Work out the median number of castles visited.
e Work out the mean number of castles visited.

Solution

a Add up all the frequencies.

$8 + 13 + 10 + 7 + 6 + 4 + 2 = 50$

So Manon asked 50 people altogether.

b The most castles visited is 6. The least castles visited is 0.

So the range is $6 - 0 = 6$. ←

Range is highest value – lowest value.

Watch out! Make sure you work out the answer to $6 - 0$.

c The mode is 1. ←

Look for the group with the **highest frequency**. More people made one castle visit than any of the other responses.

d The median is the middle value when the data is in order of size.

Manon asked 50 people, so the middle two people are the 25th and 26th people.

$\frac{20 + 1}{2} = 25.5$ so the median is halfway between the 25th and 26th values.

Imagine Manon wrote out all the data values like this:

0, 0, 0, 0, 0, 0, 0, 0
She writes 0 eight times as eight people have not visited a castle.

1,1,...1,1
The next fourteen numbers are 1. She has written down 22 numbers so far.

2,2,...2
The next ten numbers are 2, so the 25th and the 26th numbers are 2.

Can you see how to work this out without writing down any of the numbers?

The 25th and 26th numbers are 2 so the median is 2.

e To find the mean, you have to add up all the items of data and divide by the total number of items.

You could go back to Manon's raw data and add up all 50 items of data, but it is quicker to use the frequency table. Add another column to the table to help you.

Number of castles visited	Frequency	Number of castles visited × frequency
0	8	0 × 8 = 0
1	13	1 × 13 = 13
2	10	2 × 10 = 20
3	7	3 × 7 = 21
4	6	4 × 6 = 24
5	4	5 × 4 = 20
6	2	6 × 2 = 12
Total	50	110

13 people have visited one castle.

10 people have visited two castles.

The quickest way to work out 3 + 3 + 3 + 3 + 3 + 3 + 3 is 7 × 3 or 3 × 7.

Altogether there are 0 + 13 + 20 + 21 + 24 + 20 + 12 = 110 visits to castles.

There are 110 castle visits by 50 people.

So the mean is $\frac{110}{50}$ = 2.2 castles visited.

The mean doesn't have to be a whole number!

Manon wants to show her school as Welsh castle enthusiasts, so which average should she use? Why?

Cross-curricular activity

Find out how to use a spreadsheet to work out the mean from a frequency table.

14.1 Now try these

Band 1 questions

1. Write each of these as a multiplication.
   - **a** 3 + 3 + 3 + 3 + 3 + 3
   - **b** 7 + 7 + 7 + 7 + 7 + 7 + 7 + 7 + 7
   - **c** 4 + 4 + 4 + 4 + 4

2. Complete each of these statements.
   - **a** 4 + 4 + 4 + 5 + 5 + 5 + 5 + 5 + 6 + 6 = ☐ × 4 + ☐ × 5 + ☐ × 6
   - **b** 7 + 7 + 7 + 8 + 8 + 8 + 9 + 9 + 9 + 9 + 9 + 9 + 9 = ☐ × 7 + ☐ × 8 + ☐ × 9

3. Evaluate $\frac{n+1}{2}$ when:
   - **a** $n = 7$
   - **b** $n = 35$
   - **c** $n = 73$
   - **d** $n = 12$
   - **e** $n = 50$
   - **f** $n = 200$.

Curriculum for Wales Mastering Mathematics: Book 2

4 A group of runners are numbered from 1 to 99.

The runners line up in number order.

Which runner is standing in the middle?

5 Find the **i** mean and **ii** range of each of these data sets.

a 4, 9, 11, 5, 6

b 1, 1, 2, 2, 2, 3, 3, 4

c 5, 0, 3, 10, 12, 6, 5, 1, 13, 8

Band 2 questions

6 Find the median of each of these data sets.

a 4, 5, 7, 8

b 10, 12, 13, 15, 16, 18

c 12, 6, 20, 12, 4, 15, 13, 9

d 6, 4, 10, 3, 9, 12, 11, 7

7 The table shows the temperature in two seaside resorts each day for two weeks in June.

	Mon (Llun)	Tues (Maw)	Wed (Mer)	Thurs (Iau)	Fri (Gwe)	Sat (Sad)	Sun (Sul)
Llandudno	21 °C	24 °C	20 °C	19 °C	19 °C	20 °C	20 °C
	22 °C	23 °C	24 °C	23 °C	20 °C	19 °C	20 °C
Tenby (Dinbych y Pysgod)	24 °C	24 °C	23 °C	21 °C	21 °C	20 °C	21 °C
	22 °C	21 °C	23 °C	24 °C	22 °C	20 °C	21 °C

a Construct a frequency table for each resort.

b Find the mode for each resort.

c Find the range for each resort.

8 Aneirin asks everyone in his class how much pocket money they get each week.

Here are his results.

£8	£6	£10	£5	£6	£10	£8	£6	£10	£6
£7	£6	£9	£7	£9	£8	£10	£8	£9	£7
£8	£8	£6	£8	£5	£6	£6	£7	£9	£7

a Complete the frequency table of Aneirin's results.

Amount	Tally	Frequency
£5	II	2
£6		
£7		
£8		
£9		
£10		

b Find the modal amount of pocket money.

c How many people did Aneirin survey?

Aneirin's class decide to put all one weeks' pocket money together to raise money for a party.

d How much money do they raise?

e What is the mean amount of money that each person contributes?

Cross-curricular activity

Find out the monthly average temperatures of different parts of the world. Choose locations which are very far North (or South), near the Equator, inland and near the sea.

Compare their temperature ranges (and possibly compare them with where you live in Wales).

Are the results what you expected?

Are the temperature ranges explained by their geographical locations?

9 Gwawr is captain of her school hockey team.

She keeps a record of the number of goals the team scores in each match.

Here is Gwawr's record for the first twenty matches.

```
2 0 1 3 2 2 0 1 4 3
2 3 0 1 4 6 2 2 2 4
```

a Construct a frequency table.
b Find the mode.
c Find the range.
   In the next five matches, the team scores 30 goals.
d Is Gwawr's team improving? Explain your answer.

Cross-curricular activity

Do the same investigation in your school (or for a team you play for outside of school).

Ask your PE teacher for the scores of the school's hockey team (or any other sport).

Do the same analysis as in question 9.

10 Four groups of students each line up in age order from youngest to oldest.

Work out the age of the middle student in each group.

What average have you found?

| Group 1 ||
Age (years)	Number of students
11	6
12	3

| Group 2 ||
Age (years)	Number of students
11	7
12	4
13	6

| Group 3 ||
Age (years)	Number of students
11	6
12	5
13	7
14	11

| Group 4 ||
Age (years)	Number of students
12	7
13	6
14	4
15	13
16	5

11 Josh has a biased dice.

A 'biased dice' means that the scores 1, 2, 3, 4, 5 and 6 are not equally likely.

He rolls the dice and records the score in a frequency table.

Score	Frequency	Score × frequency
1	8	
2	3	6
3	4	
4	2	
5	5	
6	3	
Total		

a What is the modal score?
b How many times does Josh roll the dice?
c What is the median score?
d Copy and complete the table.
e What is the total of all Josh's rolls?
f What is the mean score?

Band 3 questions

Fluency

12 Caitlin is a tour operator and works for the Welsh Tourist Board.

She gathers some data on the number of times people have visited the Pembrokeshire Coast National Park.

Number of visits	Frequency
1	10
2	22
3	43
4	13
5	9
6	3

a How many people has she asked in the survey?
b What is the range of Caitlin's data?
c What is the modal number of visits?
d Work out the median number of visits.
e Work out the mean number of visits.

Logical reasoning

13 Zoltan records the ages of students who visit the Cardiff Central Library one Saturday lunchtime.

Age	Frequency
11	8
12	17
13	10
14	7
15	16
16	12

a Calculate the:
   i range   ii mode   iii median   iv mean.

b Another student, who is 12 years old, visited the Cardiff Central Library on this Saturday lunchtime.
   How will your answers to part **a** change if this student is included in Zoltan's data?
   State whether each answer will increase, decrease or stay the same.

14 Ioan and Llinos are training for their school swimming team.

They record their practice times, in seconds, for the 50 m breaststroke.

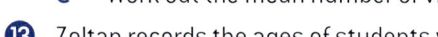

| Ioan | 47 | 52 | 49 | 40 | 52 | 47 | 48 | 47 | 51 | 52 | 49 | 50 | 50 | 50 | 51 | 48 | 50 | 51 | 46 | 49 |
| Llinos | 52 | 53 | 54 | 49 | 47 | 53 | 51 | 49 | 48 | 52 | 52 | 50 | 53 | 54 | 51 | 49 | 52 | 51 | 50 | 49 |

a Construct frequency tables for Ioan and Llinos.
b Calculate the mean time for:
   i Ioan   ii Llinos.
c Find the range for:
   i Ioan   ii Llinos.

There is only one place left on the swimming team.

d Who would you choose and why?

Cross-curricular activity

This is an exercise you could do in your PE lessons, especially if you are recording times or measurements on your school's sports day.

Discuss with your PE teacher how you could record this information.

Also, you could take this data and use a spreadsheet to work out the averages as in question 14.

15 Karl has some tomato plants.

Some are treated with Grow-well fertiliser, the rest are not.

Karl counts the number of ripe tomatoes he picks from each plant.

With Grow-well	5	4	4	2	7	8	10	2	3	5	4	6	6	10	7	9	8	8	8	2
Without Grow-well	3	5	3	5	6	9	3	3	5	4	6	5	6	5	3	6	5	6	6	3

a Construct a frequency table for each set of data.

b For each set of data, find:

  i the mean

  ii the median number of tomatoes picked per day for each plant.

c Does the data support the claim that Grow-well increases the average number of ripe tomatoes each plant produces?

  Explain your answer.

Cross-curricular activity

This could be an experiment in a Biology lesson to see how fertilisers improve plant growth.

16 Danya carried out a survey to find out how many cups of coffee some students in his school had drunk yesterday.

Here are his results.

Number of cups of coffee	Frequency
0	y
1	14
2	9
3	6
4	x
5	1

Danya surveyed 45 students.

The mean number of cups of coffee drunk is 1.4.

Calculate the value of x and of y.

14.2 Pie charts

Skill checker

① Measure each of these angles.

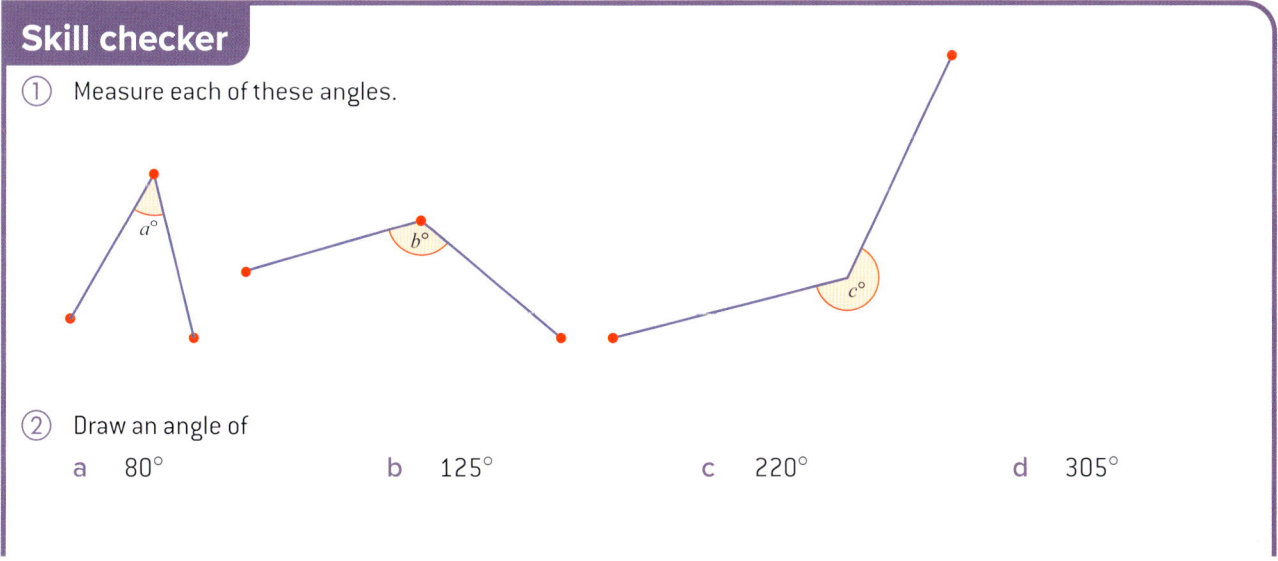

② Draw an angle of

  a 80° b 125° c 220° d 305°

③ Work out the following.
   a $\frac{2}{3}$ of 60
   b $\frac{3}{4}$ of 80
   c $\frac{5}{12} \times 36$
   d $\frac{25}{40} \times 200$

④ Write each of these fractions as a percentage.
   a $\frac{3}{4}$
   b $\frac{9}{20}$
   c $\frac{10}{25}$
   d $\frac{5}{8}$

▶ Using pie charts

Data that can be put into different groups like colour, types of sports, favourite pets, and so on, is called **categorical** or **qualitative** data. A **pie chart** is a useful way to display categorical data.

A pie chart is a circle divided into sectors. Each sector represents a different group.

Slice of pie.

A pie chart shows the different proportions of each group – the group with the largest 'slice of pie' is the modal group.

Worked example

Seren asked 40 students from her school which Welsh artist they like best.
The pie chart shows her results.

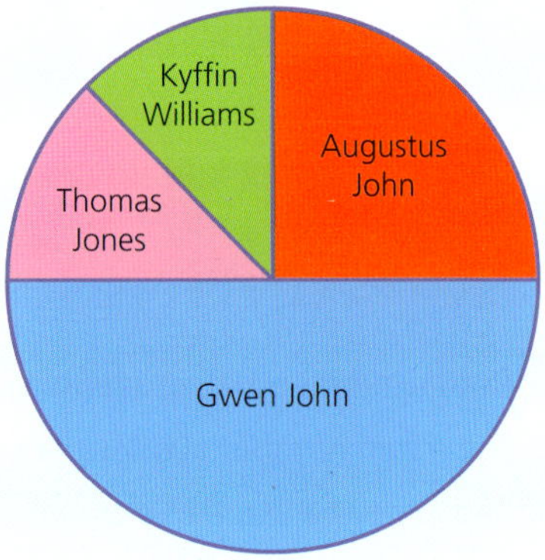

a What percentage of the students liked Augustus John best?
b How many students like Kyffin Williams best?
c What is the mode?

Solution

a The sector for Augustus John is $\frac{1}{4}$ of the pie chart.
$\frac{1}{4} = 1 \div 4$
$= 0.25$
$= 25\%$

Write $\frac{1}{4}$ as a decimal. Then multiply by 100 to change it to a percentage. So 25% of the students like Augustus John best.

Note: Sometimes you need to measure the angle with a protractor.
Augustus John has an angle of 90°, so 90° out of the whole 360° is Augustus John.

A whole turn is 360°.
'Out of' means divide.

As a fraction this is $\frac{90}{360} = \frac{1}{4}$ (÷ 90)

b The Kyffin Williams sector has an angle of 45°.

Use your protractor to measure the angle.

$\frac{45}{360} \times 40 = 5$

Find $\frac{1}{360}$ of 40 and then multiply by 45.

So five students like Kyffin Williams best.

c The mode is Gwen John.

The mode is represented by the largest sector.

Cross-curricular activity

1 In your Art lesson, find out more about the works of these (and other) Welsh artists.
2 In your History lesson, find something historically significant about when these artists were alive.

14 Working with data

To draw a pie chart, you need to work out the angle for each sector.

Worked example

The table shows which streaming service 30 people were watching on Monday evening.

Streaming service	Frequency
Netflix	11
iPlayer	6
Amazon Prime	5
Sky	8

a Draw a pie chart to represent the data.
b What percentage of people watched iPlayer?

Solution

a Start by working out the angle needed for each person in the pie chart.

÷ 30 30 people are represented by 360°. ÷ 30
One person is represented by 12°.

Service	Frequency	Angle of sector
Netflix	11	$11 \times 12° = 132°$
iPlayer	6	$6 \times 12° = 72°$
Amazon Prime	5	$5 \times 12° = 60°$
Sky	8	$8 \times 12° = 96°$
Total	30	360°

← Check your angles total 360°.

Now draw a pie chart. Make sure you include a key.

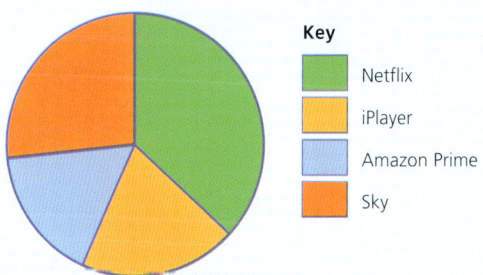

Key
- Netflix
- iPlayer
- Amazon Prime
- Sky

b Six out of 30 people watched iPlayer.

$\frac{6}{30} = 0.2$ ← Turn the fraction into a decimal by finding numerator ÷ denominator.

$0.2 = 20\%$ ← Turn the decimal into a percentage by multiplying by 100.

So 20% of people watched iPlayer.

Sometimes the angles don't work out to be whole numbers. When this is the case, you should round each angle to the nearest degree before drawing the pie chart.

Cross-curricular activity

Find out how you can use spreadsheets to create pie charts from the data you have collected.

Activity

Baan surveys some students at his school to see if they have climbed Yr Wyddfa (Snowdon), Pen y Fan or neither.

Yr Wyddfa	14
Pen y Fan	8
Neither	25

Baan wants to draw a pie chart of these results.

a   What angle should Baan use to represent one person?

Store this number in your calculator to use in question 2.

b   Copy and complete this table.

	Frequency	Angle of sector (to nearest degree)
Yr Wyddfa	14	
Pen y Fan	8	
Neither	25	
Total		

c   What is the total of the angles? Why does this happen?

d   Draw a pie chart for Baan's data. How should you deal with the missing 1°?

14.2 Now try these

Band 1 questions

1 Work out these.

a $\frac{30}{360} \times 60$
b $\frac{18}{360} \times 80$
c $\frac{120}{360} \times 90$
d $\frac{100}{360} \times 180$

2 Measure these angles.

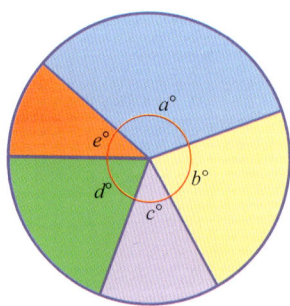

3 Rhys draws a pie chart to show the favourite pets of people in his class.

a   What is the mode?

b   What fraction of the class like cats best?

c   Rhys surveyed 28 people.
How many like rabbits best?

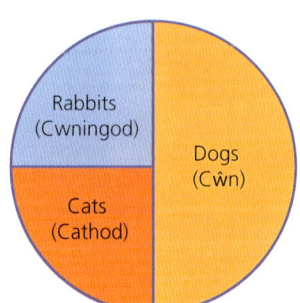

④ The pie chart shows how Sioned spent one day.

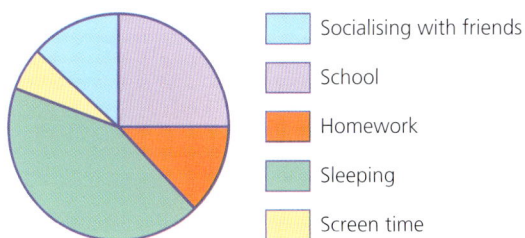

a What does she do for longer than anything else?
b Which activity takes up $\frac{1}{4}$ of her day?
c On which two activities does she spend the same amount of time?

Band 2 questions

⑤ Nerys surveys some groups of people to find out about their hobbies.

She draws some pie charts of her results.

Work out the angle she should use to represent one person when she has surveyed:
a 180 people
b 90 people
c 45 people
d 30 people.

⑥ This pie chart shows which language 120 students at a school are studying.

- $\frac{1}{4}$ of the students study German.
- $\frac{1}{8}$ of the students study Spanish.

a What fraction of the students study French?
b What angle is this represented by on the pie chart?
c How many students study French?

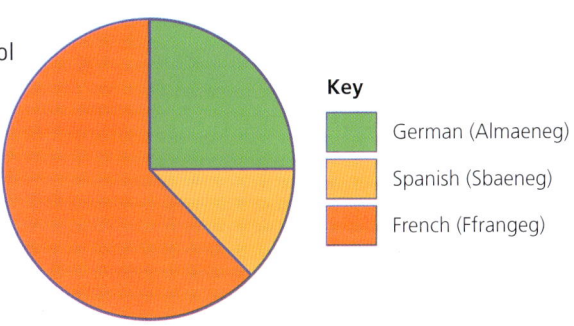

⑦ This table shows the pets of Megan's friends.

Pet	Frequency
Dog (Ci)	3
Cat (Cath)	6
Pony (Merlen)	2
Hamster (Bochdew)	1

a Draw a bar chart to show the data.
b Megan wants to draw a pie chart of her results.
   What angle should she use to represent one person?
c Show the data as a pie chart.
d Explain the advantages and disadvantages of both diagrams.

Note

Bar charts can also be created in spreadsheets.

8 Isobel surveys 60 people to find out their favourite sport.

Here are her results.

Isobel wants to draw a pie chart of her data.

Sport	Frequency
Rugby (Rygbi)	5
Football (Pêl-droed)	20
Tennis (Tenis)	6
Hockey (Hoci)	15
Netball (Pêl-rwyd)	5
Swimming (Nofio)	9

a Work out the angle that Isobel should use to represent one person.

b Work out the angle that Isobel should use for each sport.

c Draw a pie chart of Isobel's data.

9 Speedy the Snail has won 87 medals in the World Championship Snail Races.

Speedy's proud owner wants to draw a pie chart of his results.

	Frequency	Angle of sector (to nearest degree)
Gold (Aur)	42	
Silver (Arian)	33	
Bronze (Efydd)	12	
Total		

a What angle should the owner use to represent one medal? *Store this number in your calculator to use in part b.*

b Copy and complete the table.

c What is the total of the angles? Why does this happen?

How should you adjust your answers when you draw the pie chart?

d Draw a pie chart to show the medals won by Speedy.

Band 3 questions

10 The table shows the UK holiday destinations of the customers at a travel agency.

England (Lloegr)	35%
N Ireland (Gogledd Iwerddon)	15%
Scotland (Yr Alban)	30%
Wales (Cymru)	20%

Draw a pie chart to illustrate this data.

Hint Work out 35% of 360° to find the angle for England.

11 The bar chart shows the number of sweets of each colour in a packet of sweets.

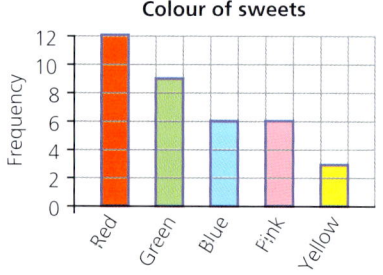

Draw a pie chart to show this data.

14 Working with data

12 Eifion asks his friends what their favourite meal is.

He records his results in a table and plans to draw a pie chart of his results.

a Copy and complete Eifion's table.

Meal	Frequency	Angle of sector
Pizza	36	90°
Curry	12	
Pasta		45°
Burger and chips	48	
Cawl		60°

b Draw a pie chart for Eifion's data.

13 The pie charts show some information about sales at two different coffee shops.

Tina's Teashop   Carlo's Coffee

a Tina's Teashop sold 50 cups of tea.

How many cakes did she sell?

b Aled says, 'The pie charts show that Tina's Teashop sold more cups of tea than Carlo's Coffee.'

Is Aled right? Explain your answer fully.

Cross-curricular activity

1 In a Geography lesson, do an investigation which compares the sizes of the populations of the countries of the United Kingdom. Show the results in a pie chart.
   ▶ What difficulties did you encounter when drawing the pie chart?
   ▶ Were you surprised by the relative size of the sectors?
   ▶ Is there any other data about the United Kingdom that would benefit from being displayed as a pie chart?

2 Compare the populations of the continents of the world using a pie chart.
   ▶ Were you surprised by the results?
   ▶ Compare your pie charts with the ones drawn on a spreadsheet.
   ▶ Do they look similar?

Key words

Here is a list of the key words you met in this chapter.

Average	Categorical data	Frequency	Mean
Median	Mode	Pie chart	Range

Use the glossary at the back of this book to check any you are unsure about.

Review exercise: working with data

Band 1 questions

1 Pablo counts the number of sweets in each of 15 packets. Here are his results:

9, 12, 10, 10, 14, 9, 8, 11, 11, 12, 10, 13, 10, 12, 11

Find the median number of sweets in a packet.

2 Ten students sit three examinations. Their marks, out of 40, are shown below.

Mathematics (Mathemateg)	20	37	30	10	22	27	29	15	16	19
English (Saesneg)	12	18	22	27	33	25	35	15	31	39
Science (Gwyddoniaeth)	15	19	21	23	25	29	32	34	5	16

a Work out the range of marks for each exam.

b Work out the median for each exam.

3 Tomos records how many people live at each house on his street.

Number of people	Frequency
1	4
2	6
3	3
4	7
5	5
6	1

a How many houses are on Tomos's street?

b What is the modal number of people in a house?

c What is the range?

d How many people live on Tomos's street?

4 A teacher records the number of late students to her school every morning for a week.

Here is a pie chart of her results.

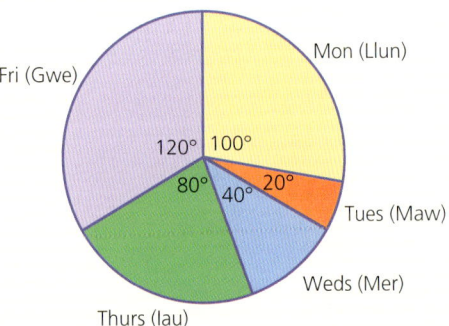

Six students were late on Friday.

a What fraction of students were late on Friday?

b How many students were late that week altogether?

c What angle represents one student?

d Work out how many students were late on each of the other days.

Band 2 questions

5 The mean distance a jogger runs for 14 days is 5 km.

She then runs a mean distance of 13 km each day for the next seven days.

Find the total distance she has run.

6 The manager of a bookshop collects some data on the number of sales one Monday.

Type of book	Number of sales	Angle of sector
Fiction	16	
Children's books	20	
Biography	18	
Science fiction	4	
Crime fiction	14	
Other	8	

> **Cross-curricular activity**
>
> This is an exercise you could do in an English lesson.
>
> Find out everyone's favourite genre of book and draw a pie chart to show this.

The manager wants to draw a pie chart for her data.

a   What angle should she use to represent one book?

b   Copy and complete the table.

c   Draw a pie chart to show this information.

7 Hanna asks everyone in her class how much pocket money they get each week.

Her results are shown in the table.

Hanna says that the average pocket money is £5.

a   Why has Hanna chosen £5?

b   Do you agree with Hanna? Give a reason for your answer.

Amount	Frequency
£5	9
£6	7
£7	5
£8	6
£9	2
£10	1

8 Ami records the shoe size of everybody in her class.

Shoe size	$2\frac{1}{2}$	3	$3\frac{1}{2}$	4	$4\frac{1}{2}$	$5\frac{1}{2}$	5	6	7
Number of people	3	6	4	4	3	4	2	1	1

a   Find the median shoe size.

b   Find the mean shoe size.

9 The table shows the results of the last 18 games played by the football team Cwmfelin United.

Result	Number of games
Lost (Colli)	4
Won (Ennill)	12
Drawn (Cyfartal)	2

a   Draw a pie chart to show this information.

b   Dai says the modal result is 'Won'. Is Dai correct? Explain your answer.

c   Efa says the median result is 'Won'. Is Efa correct? Explain your answer.

Band 3 questions

10 The pie chart shows the items borrowed from a library one day.

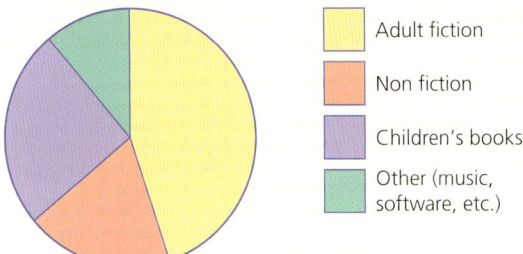

- a What is the largest category of items borrowed?
- b 200 items are borrowed in total.

  What percentage of the items borrowed are children's books?
- c How many children's books are borrowed?
- d 45% of the items borrowed are adult fiction.

  How many adult fiction books are borrowed?
- e Draw a different type of chart to represent this information.

  What are the advantages and disadvantage of each chart?

11 Eliza is a zoologist. She records the number of baby meerkats born to some female meerkats.

Number of babies	Frequency
1	3
2	4
3	7
4	x
5	8
6	3

The total number of meerkat babies is 130.

- a Calculate the value of x.
- b Work out the mean number of meerkat babies.

  Give your answer correct to two decimal places.
- c Work out the median number of meerkat babies.

12 Sukith records the number of emails he receives each day for y days.

Number of emails	Frequency
4	5
5	4
6	6
7	x
8	4
9	8

The mean number of emails Sukith receives each day is 6.7.

- a Calculate the value of:

  i x ii y.
- b Sukith says the median number of emails he received was 6.

  Is Sukith correct? Explain your reasoning fully.

15 Circles

Coming up...

▶ Finding the circumference of a circle

▶ Finding the area of a circle

▶ Solving problems involving circumference and area of a circle

Circle patterns

You can make interesting patterns using only circles.

All you need is a pair of compasses and a sharp pencil.

Follow these steps and look at the diagrams below.

Make sure that your compasses are not too loose!

Step 1 Draw a circle in the middle of your page.

Step 2 Put your compass point on a point (B) on the circumference (edge) of your circle.

Don't alter the setting of your compasses for steps 2 to 8.

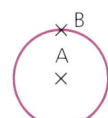
Steps 1 and 2

Step 3 Draw another circle with its centre at B.

Step 4 Put your compass point on the point (C) where your two circles meet.

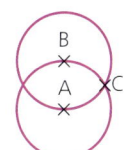
Steps 3 and 4

Step 5 Draw another circle with its centre at C.

Step 6 Put your compass point on the point (D) where your last circle meets your first circle.

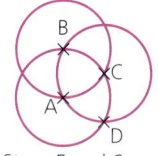
Steps 5 and 6

Step 7 Draw another circle with its centre at D.

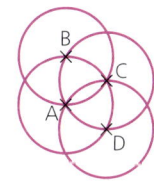
Step 7

Step 8 Keep drawing more circles in the same way until you complete the circle pattern.

Step 9 Make your compasses twice as wide as they were to start with. Draw a final circle with centre at A to complete the pattern.

Step 10 Colour your pattern in!

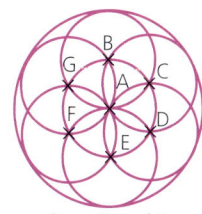
Steps 8 and 9

> **Cross-curricular activity**
>
> Look on the internet for examples of circle spin paintings by the British artist Damien Hirst.
>
> In an Art lesson, look at more of his artwork. Some of them are famous and you may recognise them.
>
> Also, look for other pieces of art that involve using circles.

15.1 Circumference

Skill checker

1. Work out the value of $2r$ when:
   - a $r = 5$
   - b $r = 2.4$
   - c $r = 5.8$
2. Round each of these numbers to one decimal place.
   - a 19.513
   - b 13.491
   - c 3.95
3. Solve these equations.
   - a $5r = 12$
   - b $3.2r = 10$
   - c $6.28r = 39.5$
4. Convert these measurements to centimetres.
   - a 14 mm
   - b 234 mm
   - c 1230 mm
   - d 2.4 m
   - e 0.8 m
   - f 0.05 m
5. Convert these measurements to millimetres.
   - a 3 cm
   - b 3.2 cm
   - c 0.7 cm

▶ Finding the circumference of a circle

You met circles in Book 1. In this chapter you will learn how to work out the area and circumference of a circle.

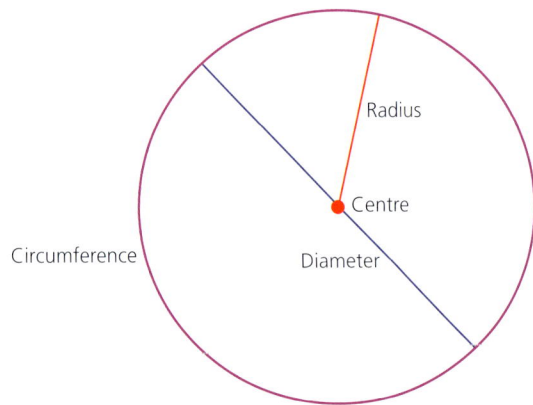

Remember:
- The **circumference** is the distance around the edge of a circle.
- The **radius** of a circle is the distance from the centre to the circumference.
- The **diameter** is the distance across the circle through the centre.

Make sure you learn these words.

15 Circles

Activity

Draw five circles on a piece of paper.

Measure the diameter of each circle.

Use a piece of string to help you measure around the circumference of your circles.

Make sure you use the same units when you measure your diameter and circumference.

Copy and complete this table.

Circumference (cm)	Diameter (cm)	Circumference / Diameter

What do you notice?

In the activity you found that when you divide the circumference of a circle by its diameter, you get a number a little bit bigger than 3.1.

$$\frac{C}{d} \approx 3.1$$

The decimal part of π carries on forever and never repeats – mathematicians have so far calculated over 31 000 000 000 000 digits of pi.

In fact, if you could measure them perfectly you would get the number 3.141 592 654...

The number 3.141 592 654... is called 'pi' and has the symbol π (a Greek letter).

$$\frac{C}{d} = \pi$$

π is used as it is the Greek letter 'p' for 'perimeter'.

So the circumference, C, of a circle of diameter d is:

$$C = \pi d$$

Since the diameter is twice the radius, you can replace d with $2r$ and say $C = 2\pi r$.

You can use 3.14 instead of π, but it is more accurate to use the π button on your calculator.

Sometimes you are told the circumference and you need to work out the radius.

Worked example

Find the circumference of this circle.

Give your answer correct to one decimal place.

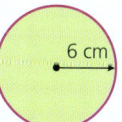
6 cm

Solution

The radius is 6 cm so the diameter is 2 × 6 cm = 12 cm.

You know the radius, so the first step is to work out the diameter.

Substitute $d = 12$ into $C = \pi d$

$= \pi \times 12$

Use the π button on your calculator.

$= 37.699...$

The three dots show that the decimal part carries on.

So the circumference is 37.7 cm to one decimal place.

Note

Your calculator display may show the answer as 10π.

You have a button on your calculator to change the display to a decimal.

It may look like this $\boxed{S \Leftrightarrow D}$ or $\boxed{F \Leftrightarrow D}$.

Make sure you learn how to use this button.

Sometimes you need to solve problems involving part of a circle.

> **Worked example**
>
> The circumference of a circle is 45.6 cm.
>
> Work out the diameter of the circle.
>
> Give your answer correct to one decimal place.
>
> ### Solution
>
> You can use $C = \pi d$ to work out the diameter.
>
> $45.6 = \pi d$ — *Replace C with 45.6.*
>
> $\div \pi$ ... $\div \pi$ — *Use the π button on your calculator.*
>
> $\dfrac{45.6}{\pi} = d$
>
> $14.51... = d$
>
> So the diameter is 14.5 cm to one decimal place.

> **Worked example**
>
> Erin pushes her bicycle 1.85 metres along the road.
>
> Her front wheel makes one complete turn.
>
> Calculate:
>
> **a** the radius
>
> **b** the diameter of her bicycle wheel.
>
> Give your answers correct to the nearest centimetre.
>
> ### Solution
>
> **a** The circumference of the front wheel is 1.85 m.
>
> *Erin's wheel makes one complete turn so the bicycle will move forward the same distance as the circumference of the wheel.*
>
> You can use $C = \pi d$ to work out the diameter.
>
> $1.85 = \pi d$ — *Replace C with 1.85.*
>
> $\div \pi$ ... $\div \pi$
>
> $\dfrac{1.85}{\pi} = d$
>
> $0.588... = d$
>
> So the diameter of the wheel is 0.59 m or 59 cm to the nearest cm.
>
> **b** The radius is half the diameter so $r = \dfrac{d}{2}$.
>
> $r = \dfrac{0.588...}{2}$
>
> $= 0.294...$
>
> So the radius is 0.29 m or 29 cm to the nearest cm.

Use the unrounded value of d from your calculator to avoid rounding errors. What answer would you have got if you had used 0.59 m instead?

15 Circles

Half of a circle is called a **semicircle** and a quarter of a circle is called a **quadrant**.

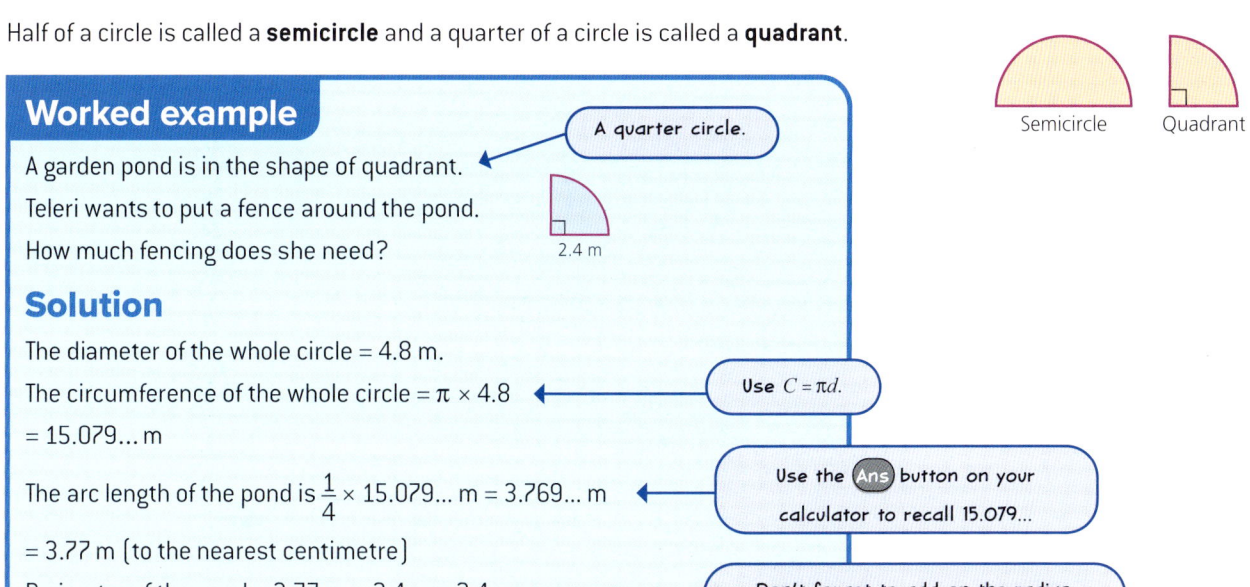

Semicircle   Quadrant

Worked example

A garden pond is in the shape of quadrant.
Teleri wants to put a fence around the pond.
How much fencing does she need?

A quarter circle.

2.4 m

Solution

The diameter of the whole circle = 4.8 m.
The circumference of the whole circle = $\pi \times 4.8$

Use $C = \pi d$.

= 15.079... m

The arc length of the pond is $\frac{1}{4} \times 15.079... \text{ m} = 3.769... \text{ m}$

Use the Ans button on your calculator to recall 15.079...

= 3.77 m (to the nearest centimetre)
Perimeter of the pond = 3.77 m + 2.4 m + 2.4 m
= 8.57 m (to the nearest centimetre)

Don't forget to add on the radius twice when you find the perimeter!

Cross-curricular activity

The Greek symbol for pi (π) was first used by the Welsh mathematician William Jones (1675–1749).
In your History lesson, find out more about William Jones.
Were there any other important mathematicians alive at that time?
What was Wales and the world like then? Were there any events of historical significance?

15.1 Now try these

Round your answers to one decimal place, unless the question says otherwise.

Band 1 questions

1 Work out the diameter of a circle with a radius of:
   a 12 m   **b** 3.2 cm   **c** 16 mm   **d** 14.7 cm.

2 Work out the radius of a circle with a diameter of:
   a 7 cm   **b** 46 mm   **c** 2.8 m   **d** 12.6 cm.

3 Work out the value of each of these. Round your answers to two decimal places.
   a $\pi \times 2$ **b** $\pi \times 3$ **c** $\pi \times 4$ **d** $\pi \times 5$

4 Which of these lines shows the diameter of the circle?
Explain how you know.

Use the π button on your calculator.

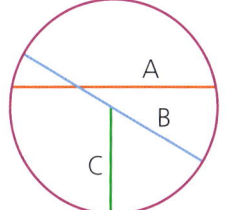

Curriculum for Wales Mastering Mathematics: Book 2

Fluency

5 Use the formula $C = \pi d$ to calculate the circumference of each of these circles.

Write down the correct unit for each of your answers.

a 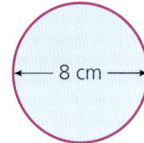   b 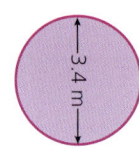   c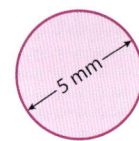

6 Find the circumference of these circles. Remember to find the diameter first.

Write down the correct unit for each of your answers.

a 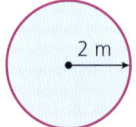   b 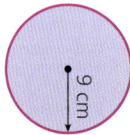   c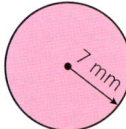

Band 2 questions

Strategic competence

7 Work out the circumference of a circle with:

    a   diameter = 9 cm   b   radius = 2.5 m.

8 Work out the circumference of this circle.

> You'll need your ruler for this question!

9 A circle has a circumference of 6.28 m.

Calculate:

    a   the diameter of the circle   b   the radius of the circle.

10 Copy and complete this table.

Radius	Diameter	Circumference (to 2 d.p.)
4 cm	8 cm	
	12 cm	
		21.98 cm
	15 cm	
12.5 cm		
21.6 cm		
		1.26 m
		31.40 km

11 A lawn is laid in a garden.

The lawn is a circle of radius 3.5 m.

    a   What is the diameter of the lawn?

    b   What is the circumference of the lawn?

        Give your answer correct to the nearest centimetre.

Edging is fitted round the circumference of the lawn to keep it neat.

The edging comes in flexible 1.5 m lengths and costs £5.99 per length.

    c   How many lengths of edging are needed?

    d   How much does the edging cost?

15 Circles

12 The large wheel on Huw's wheelchair has a diameter of 60 cm.

Huw pushes the wheel round exactly once.

a How far has Huw moved?

Huw crosses a busy road that is 20 m wide.

b How many times does he have to rotate the large wheel to do this?

> **Remember**
> Perimeter is the distance all the way around a shape.

13 Work out the perimeter of each of these shapes.

a

32 mm

b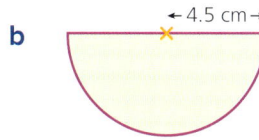
← 4.5 cm →

c
← 6.5 m →

d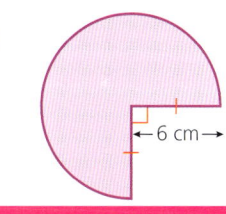
← 6 cm →

Band 3 questions

14 The perimeter of a swimming pool consists of a semicircle and a square.

A lifeguard walks all the way around the edge of the pool.

a How far does she walk?

b By the end of the lifeguard's shift, she has walked 50 times around the pool and her fitness tracker shows that she has walked 1250 steps.

What is the lifeguard's stride length?

> Your stride length is the length of one of your steps.

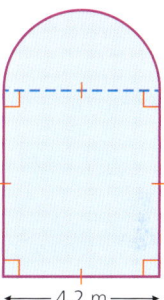
← 4.2 m →

15 Work out the perimeter of these shapes.

a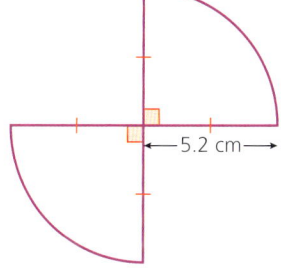
← 5.2 cm →

b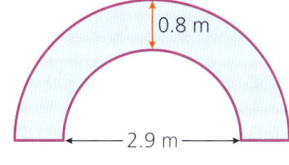
0.8 m
← 2.9 m →

16 On Big Ben's famous clock tower in London, the hour hand is 2.7 m long and the minute hand is 4.5 m long.

a How far does the tip of the minute hand travel in one hour?

b How far does the tip of the minute hand travel in 12 hours?

c How far does the tip of the hour hand travel in 12 hours?

d How far does the tip of the hour hand travel in one hour?

e A bird perches exactly in the centre of the hour hand for half an hour.

How far does the bird travel? Give your answer correct to the nearest centimetre.

17 A 400 m racetrack consists of two 100 m straights and two 100 m semicircles, one at each end, as shown in the diagram.

Find the distance marked x.

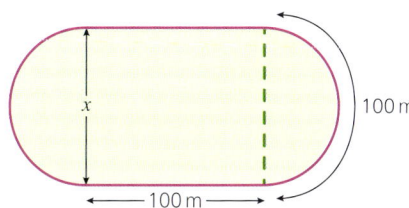
100 m
← 100 m →

Curriculum for Wales Mastering Mathematics: Book 2

15.2 Area of a circle

Skill checker

1. Work out the shaded area of these shapes.

   a 12 m, 3 m, 1.5 m (square hole)

   b 8.1 cm, 7.2 cm, 10.3 cm (triangle)

   c 5 cm, 3.6 cm, 4.1 cm (parallelogram)

   d 18 mm, 10.6 mm, 12 mm (trapezium)

2. Work out the circumference of these circles.

   a 4 m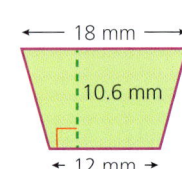

   b 11 cm

▶ Finding the area of a circle

Activity

You will need a piece of 1-centimetre-squared paper, some scissors and glue for this activity.

Draw a circle on a piece of centimetre-squared paper and cut it out.

1. Estimate the area of your circle by counting squares.

2. Fold your circle in half four times.

   Cut your circle up into 16 small sectors and arrange them like this.

   A sector looks like a slice of cake.

   Cut this sector in half ...

   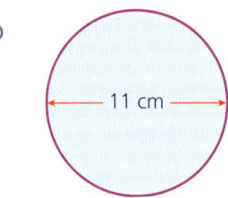

   ... add one half here, and stick the other half at the other end.

   Cut one of your sectors in half and add it at either end of your arrangement.

3. a Copy and complete this sentence.

   I have made a shape that is quite close to a _____.

   Use a ruler to help you.

   b Estimate the area of the circle.

4. How can you make your estimate even closer to the actual area of the circle?

   What shape would the sectors make if they were as thin as possible?

5. Dilys draws a circle with radius 8 cm.

   She cuts it into very thin sectors and arranges them to make a rectangle.

   a Work out the length and width of Dilys's rectangle.

   b Work out the area of Dilys's circle.

15 Circles

In the activity you found that you can cut a circle up and arrange the sectors into a shape that is fairly close to a rectangle. If you could cut the circle up into the thinnest sectors possible, you would make a perfect rectangle.

- The width of the rectangle is the same as the radius of the circle r.
- The length of the rectangle is the same as half the circumference of the circle.

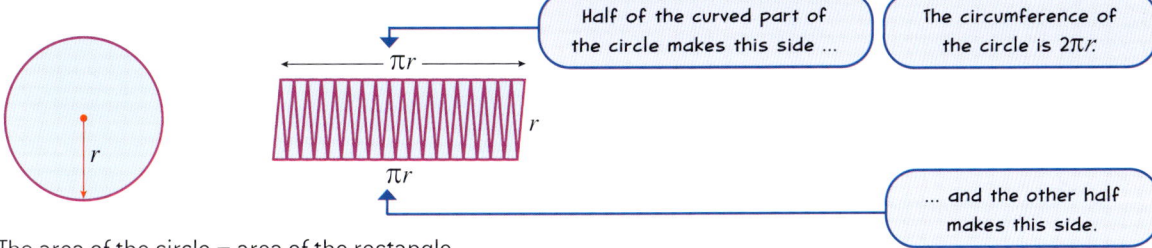

Half of the curved part of the circle makes this side ...

The circumference of the circle is $2\pi r$.

... and the other half makes this side.

The area of the circle = area of the rectangle
$$= \pi r \times r$$

You write this as:

Area of circle = πr^2

Worked example

Find the area of this circle.

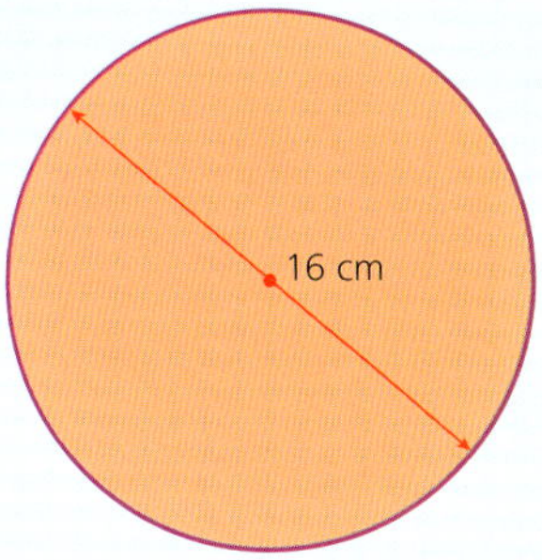

[Circle with diameter labelled 16 cm]

Solution

You need to work out the radius first.

Radius = 16 cm ÷ 2 = 8 cm

$A = \pi r^2$

$= \pi \times 8^2$ ← Substitute $r = 8$ into the formula.

$= \pi \times 64$ ← Remember BIDMAS — square first and then multiply.

$= 201.06...$

So the area is 201.1 cm² ← Don't forget the units!
(to one decimal place).

Remember
The radius is half the diameter.

You may need to work out the area of a more complicated looking shape. Use the same methods that you have already met for finding areas:

- Work out the area of each part of the whole shape and add the areas together.
- Remember to subtract the area of any holes.
- To find the area of a semicircle, work out the area of the full circle and then divide by 2.

Worked example

A path is made around a circular pond. The path is 1.2 m wide.

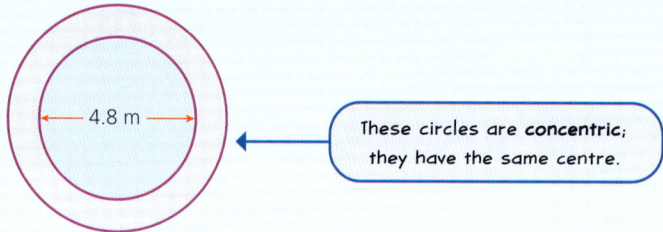

These circles are **concentric**; they have the same centre.

Work out the area of the path.

Solution

Radius of pond = 2.4 m

Radius of pond and path together (whole circle)
= 2.4 m + 1.2 m = 3.6 m

Area of path and pond = $\pi \times 3.6^2 = 40.71...$ m²

Area of pond = $\pi \times 2.4^2 = 18.09...$ m²

Area of path = area of path and pond − area of pond
$= 40.71...$ m² $- 18.09...$ m²
$= 22.61...$ m²

So the area of the path is 22.6 m² (to one decimal place).

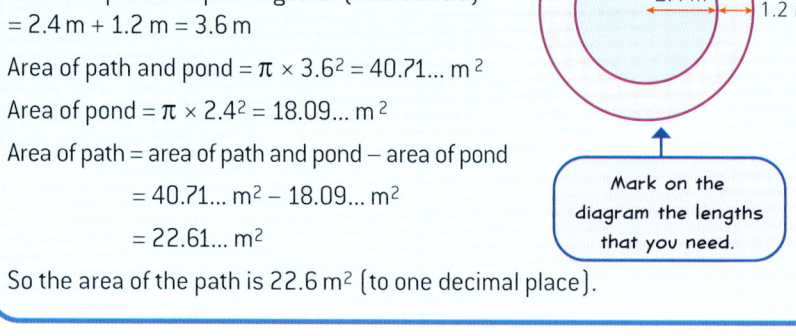

Mark on the diagram the lengths that you need.

Sometimes you will be given the area and you will need to work back to find the radius or diameter.

Worked example

The area of this circle, is 50 cm².
Calculate the radius of the circle.
Give your answer to the nearest centimetre.

Area = 50 cm²

Solution

Method 1: Using a number machine

To find the area of a circle, you can input the radius into this number machine.

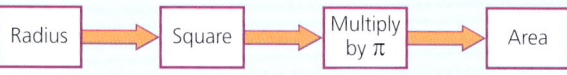

Radius → Square → Multiply by π → Area

You can reverse the number machine to find the radius.

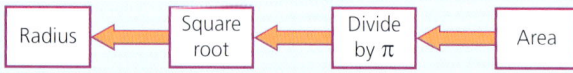

Radius ← Square root ← Divide by π ← Area

Input area = 50 cm² into the reversed number machine.

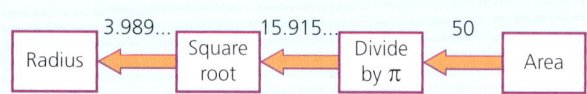

3.989... ← 15.915... ← 50
Radius ← Square root ← Divide by π ← Area

Note

- The reverse of multiply is divide.
- The reverse of square is square root.
- Do not round until you reach your final answer.

So the radius = 3.989… cm
= 4 cm (to the nearest cm).

Method 2: Solving an equation

$\pi r^2 = A$

$\pi r^2 = 50$

$\div \pi$ ⟲ ⟳ $\div \pi$

$r^2 = 15.915…$ ⟵ $\dfrac{50}{\pi} = 15.915…$

$r = \sqrt{15.915…}$
$= 3.989…$

So the radius is 4 cm (to the nearest cm).

15.2 Now try these

Round your answers to one decimal place, unless the question says otherwise.

Band 1 questions

1 Look at this function machine.

Radius → Square → Multiply by π → Area

Use the function machine to work out the area when:

   a radius = 10 cm   **b** radius = 7 cm   **c** radius = 9 cm.

2 Work out the value of each of these.   *Use the π button on your calculator.*

Round your answers to two decimal places.

   a $\pi \times 3^2$ **b** $\pi \times 4^2$ **c** $\pi \times 5^2$ **d** $\pi \times 6^2$

3 Work out the area of each of these circles.

   a 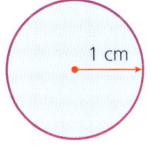 1 cm

   b  3.2 cm

   c 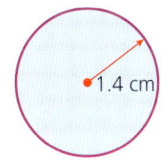 1.4 cm

4 For each circle:

   i work out the radius

   ii use your answer to work out the area.

   a 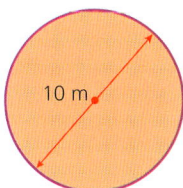 10 m

   b 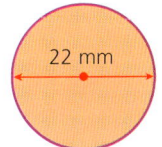 22 mm

   c 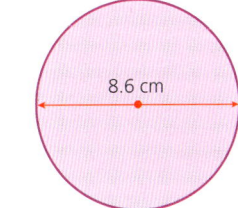 8.6 cm

5 Jamal and Anwen both work out the area of a circle with radius 2 cm.

Here are their workings.

Anwen
Area = π × 2²
= π × 4
= 12.56…
Area = 12.6 cm²

Jamal
Area = π × 2²
= 6.28…²
= 39.47…
Area = 39.5 cm²

Who has got the right answer?

What mistake has the other person made?

Band 2 questions

6 Calculate the areas of the circles with these dimensions.

a Radius = 15 cm   b Diameter = 2.4 km   c Diameter = 52 mm   d Radius = 2.6 m

7 A sumo wrestling ring has a diameter of 24 m.

Find the area of the ring.

8 This stained-glass window is made from three large red circles and two small blue circles.

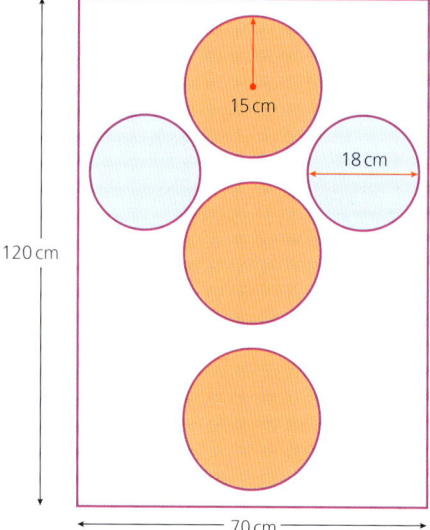

The surrounding glass is clear.

a Calculate the area of one of the red circles.

b Calculate the area of one of the blue circles.

c Calculate the area of the clear glass.

9 The top of this birthday cake is a circle of radius 12 cm.

a What is the area of the top of the cake?

The cake is displayed on a circular board of radius 14 cm.

b What is the area of the cake board?

c What is the area of board that is visible when the cake is placed on it?

The design on the cake includes a fish of area 80 cm² and three circular bubbles radius 1 cm, 1.5 cm and 2 cm.
- **d** What is the total area of the design?
- **e** What area of icing is not covered by the design on the top of the cake?

10 Work out the area of each of these semicircles.

a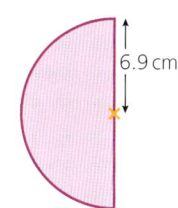

b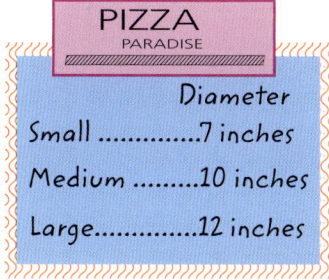

> **Hint**
> Work out the area of the whole circle first.

11 The small pizza below serves two people.

```
PIZZA PARADISE
Diameter
Small ............. 7 inches
Medium ......... 10 inches
Large ............. 12 inches
```

- **a** How many people would a medium pizza serve?
- **b** What about a large pizza?

A village in Italy is having a pizza festival.

They want to make one giant pizza for 100 people.
- **c** What diameter does the pizza need to have?

> **Hint**
> Compare the areas of a small and a medium pizza.

12 Use the number machine below to calculate the radius of a circle of area:
- **a** 80 cm²
- **b** 120 cm²
- **c** 200 cm².

Give your answers correct to two decimal places.

Band 3 questions

13 Copy and complete this table.

Radius	Diameter	Area (to 2 d.p.)
4 cm	8 cm	
	12 cm	
		12.56 m²
	18 inches	
		1256.00 m²
2.7 cm		
		907.46 cm²

14 Work out the area of each of these shapes.

a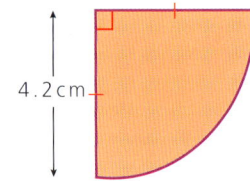

b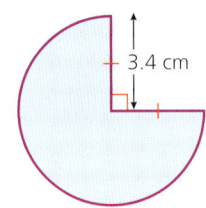

15 Arjun says that you can use the formula $Area = \dfrac{\pi d^2}{4}$ to work out the area of a circle with diameter d.

Jasmine says that Arjun is wrong and the formula is $Area = \dfrac{\pi d^2}{2}$.

 a Who is right? Explain your answer fully.
 b Use the correct formula to work out the area of a circle with diameter 7.4 metres.

16 Calculate the area of the shaded region.

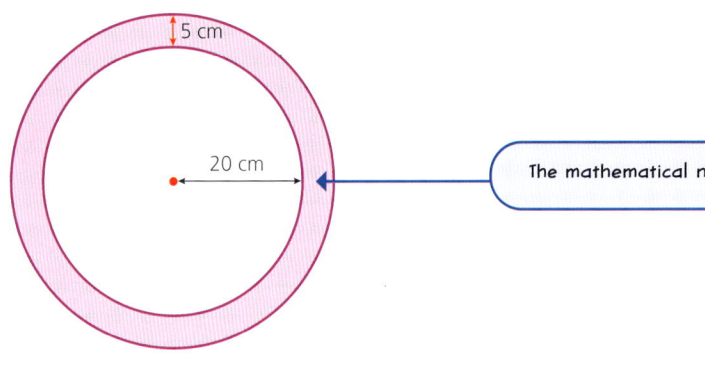

The mathematical name of this shape is an **annulus**.

17 A circular pond has an area of 15 m².

Calculate the diameter of the pond. Give your answer correct to the nearest centimetre.

18 The diagram shows a small courtyard garden.

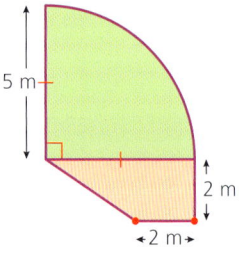

Calculate the area of the garden.

19 The area of a circle is 95 cm².

Calculate the circumference of the circle.

20 Blodwen has a piece of card in the shape of a quarter circle.

She cuts out a semicircle from the card.

 a Calculate the area of the card remaining.
 b What fraction of the card is left?

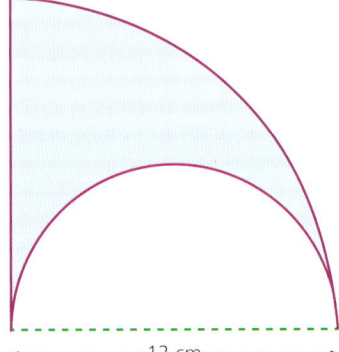

Key words

Here is a list of the key words you met in this chapter.

| Area | Circle | Circumference | Concentric | Diameter |
| Pi | Quadrant | Radius (radii) | Sector | Semicircle |

Use the glossary at the back of this book to check any you are unsure about.

15 Circles

Review exercise: circles

Round your answers to one decimal place, unless the question says otherwise.

Band 1 questions

1. For each of these circles, calculate:
   i the circumference
   ii the area.

   a 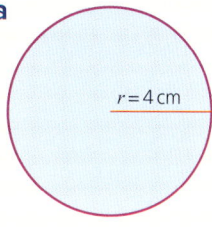   b 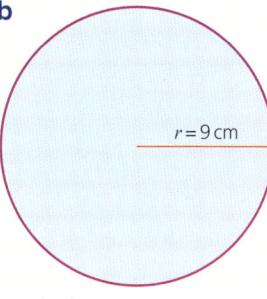   c 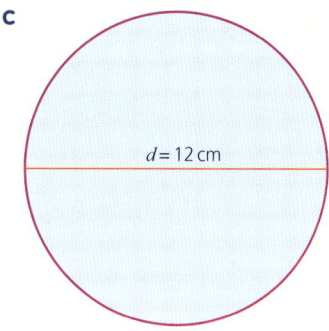   d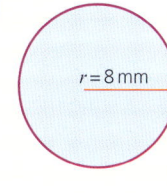

2. For each of these dimensions, calculate:
   i the circumference
   ii the area of the circles.

   a Diameter = 9 cm
   b Diameter = 23.6 km
   c Radius = 3.8 m

3. A circular table has a diameter of 10 metres.
   Twenty politicians are invited to an international conference.
   Each politician needs 1.5 m around the circumference of the table.
   Can they all sit around the table together?
   Explain your answer fully.

4. The picture shows a flowerpot.
   The rim of the pot is a circle of radius 12 cm.
   There is a leaf design pattern around the top made from exactly 16 leaves.
   What is the length of each leaf?

Band 2 questions

5. Marc measured the radius of this cartwheel to be 80 cm.

   a Calculate its circumference.
   b How far does the cart travel when the wheel turns 100 times?
      Give your answer in:
      i   centimetres
      ii metres
      iii kilometres.

275

6 Work out the area and perimeter of these semicircles.

a 6.9 m

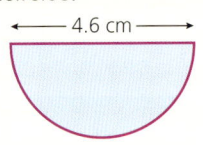

b 4.6 cm

7 A cylindrical bin has a circumference of 100 cm.

What is its diameter?

8 A cricket club has a circular field with a circumference of 110 metres.

Turf costs £9 per square metre and is only available in 1 m × 1 m square pieces.

How much would it cost the club to turf the whole field?

9 A circular table has a radius of 1.8 m.

Geraint has a circular tablecloth with circumference 12 m.

a Show that the tablecloth is large enough for the table.

b By how much will it overhang the table?

Band 3 questions

10 Calculate the area of the shaded region.

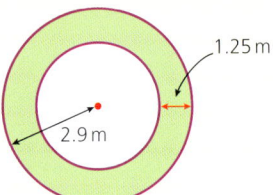

1.25 m
2.9 m

11 A circle is cut out from a square piece of card, of side 10 cm.

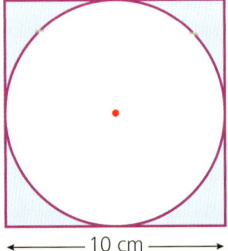

10 cm

What percentage of the card is left over?

12 A hotel swimming pool is in the shape of the letter P.

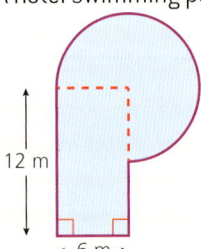

12 m
6 m

For health and safety reasons, each swimmer must have an area of at least 4 m².

How many swimmers can safely use the pool at the same time? You must show all your reasoning.

15 Circles

13 Work out the area and perimeter of these shapes.

a 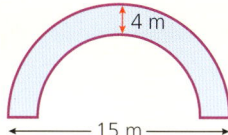   b 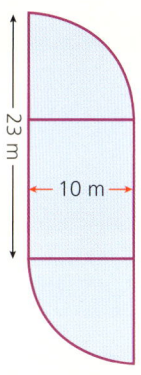   c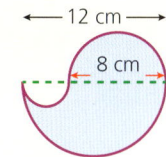

14 Work out the value of x, y and z.

Give your answers correct to two decimal places.

a 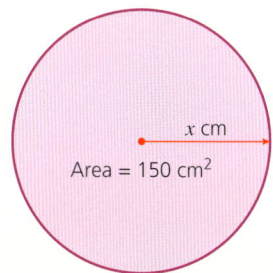   b 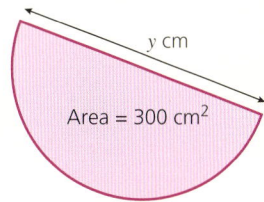   c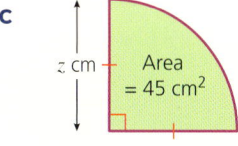

Cross-curricular activity

The Large Hadron Collider at CERN in Geneva, Switzerland, is a particle accelerator and one of the largest ever constructed. It has a circumference of 27 km and is situated deep underground. Can you calculate its diameter?

Subatomic particles, travelling close to the speed of light, are fired through the accelerator tube at each other to smash them apart. The fragments that are left are used to answer questions about the universe and where everything came from.

Find out more about this in a Science lesson.

16 Pythagoras' theorem

Coming up...
- ▶ Problem solving involving triangles
- ▶ Using Pythagoras' theorem to find an unknown side in a right-angled triangle

Conceptual understanding

Squares and triangles

You will need scissors, a ruler and 1-centimetre-squared paper for this activity.

Step 1: On squared paper, draw two squares with their corners touching, like this:

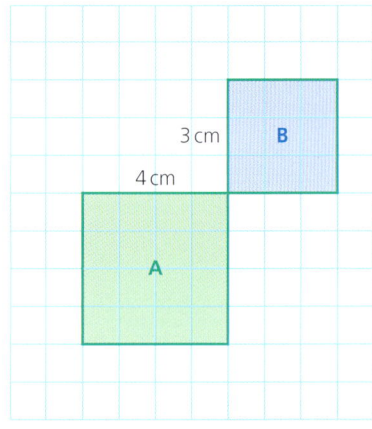

Step 2: Add four congruent right-angled triangles to your diagram, like this:

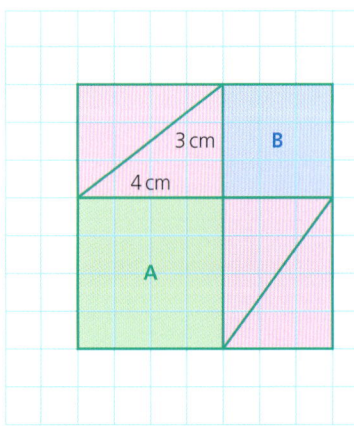
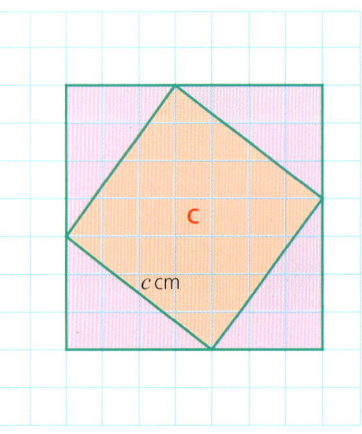

Remember
Congruent means 'exactly the same shape and size'.

Step 3: Cut out your squares and the four triangles. Arrange the four triangles to form the sides of a square.

What shape is **C**?

Explain why **steps 2** and **3** show that the sum of the area of squares **A** and **B** is the same as the area of square **C**.

16 Pythagoras' theorem

Step 4: Show that you can arrange the three squares (**A**, **B** and **C**) along the sides of one of the right-angled triangles.

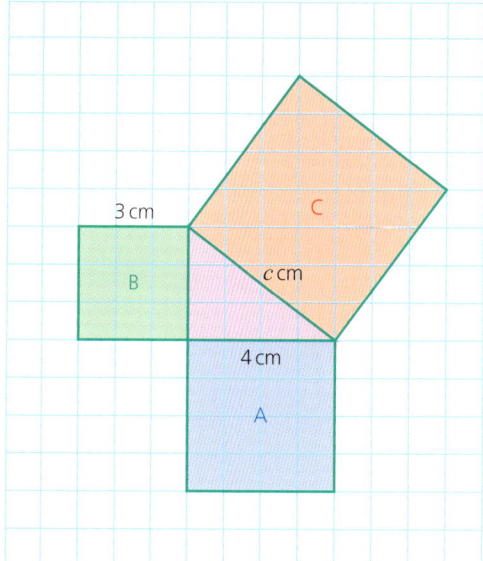

How can you use the area of the squares to work out the side lengths of the triangle? Measure them to check.

Copy and complete the table below.

Repeat the activity using different sizes of square in **Step 1**.

What do you notice? Does it always work?

Area of square A	Area of square B	Sum of areas A and B	Area of square C	Side lengths of the right-angled triangle
16 cm²	9 cm²	25 cm²	25 cm²	3 cm, 4 cm and 5 cm

16.1 Investigating triangles

Skill checker

1. Solve these equations.

   a $4.3 + a = 12.2$ b $b + 8.35 = 17.6$ c $15.3 + c = 37.24$

2. Find the side length of a square with an area of:

   a 100 cm^2 b 200 cm^2 c 300 cm^2.

3. For each of these triangles, write down the name of **i** each side and **ii** the longest side.

   a 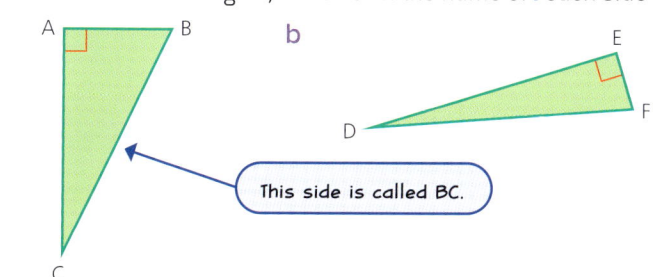   c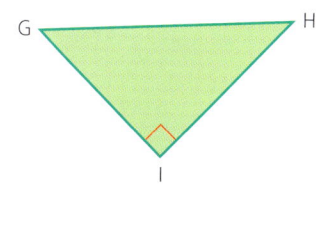

   This side is called BC.

▶ Right-angled triangles

The longest side of a right-angled triangle is called the **hypotenuse**.

The hypotenuse is always the side **opposite the right angle**.

In the starter activity you found that the area of the square on the hypotenuse of a right-angled triangle is equal to the sum of the squares on the other two sides.

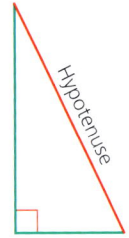

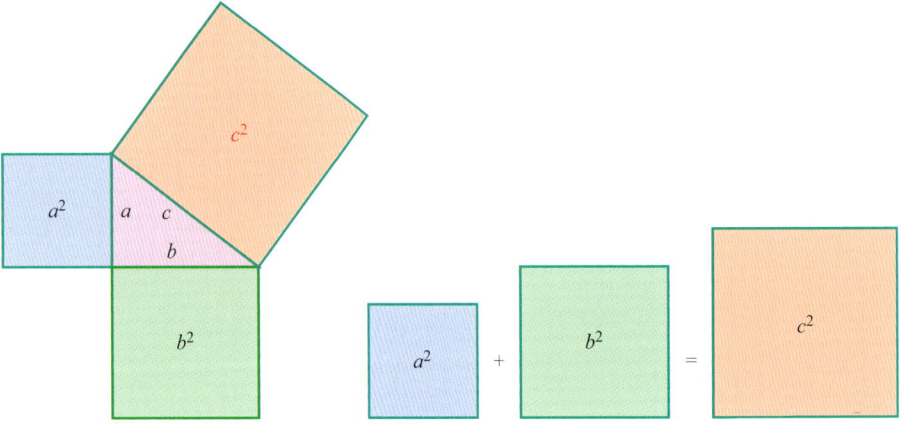

This is called Pythagoras' theorem and it is written as:

$$a^2 + b^2 = c^2$$

Pythagoras was an ancient Greek mathematician who lived from around 570 BC to 495 BC.

Worked example

Work out the length of the hypotenuse of this triangle.

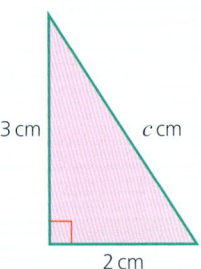

Solution

Draw a square on each side of the triangle. The area of the square on the hypotenuse is the same as the area of the other two squares.

$$2^2 + 3^3 = c^2$$
$$4 + 9 = c^2$$
$$c^2 = 13$$

So the area of the largest square is 13 cm²

$$c^2 = 13$$

square root ⟳ square root

$$c = 3.61$$

The hypotenuse is 3.61 cm (correct to two decimal places).

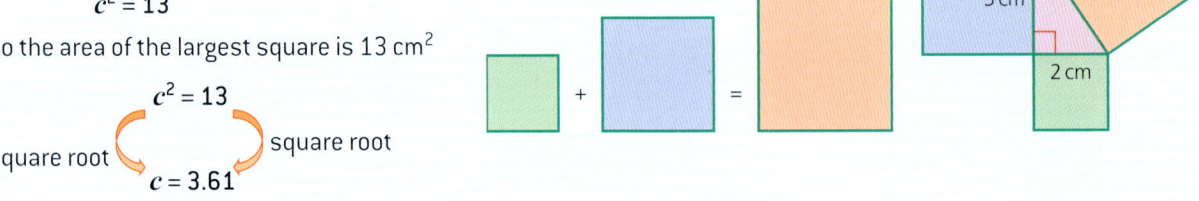

16 Pythagoras' theorem

You can also use Pythagoras' theorem to check whether a triangle is right-angled.

Worked example

Show that this triangle is not right-angled.

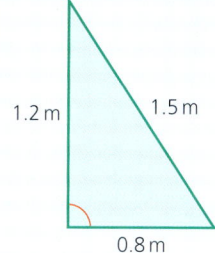

Solution

Pythagoras' theorem says that if a triangle is right-angled then $a^2 + b^2 = c^2$ where c is the length of the hypotenuse (longest side).

Substitute $a = 1.2$, $b = 0.8$ into $a^2 + b^2$ and $c = 1.5$ into c^2.

$a^2 + b^2$
$= 1.2^2 + 0.8^2$
$= 1.44 + 0.64$
$= 2.08$

c^2
$= 1.5^2$
$= 2.25$

> a and b are the two shorter sides – it doesn't matter which side you call a and which side you call b.

> $\neq$ means 'is not equal to'.

Since $2.08 \neq 2.25$ then $a^2 + b^2 \neq c^2$, which means the triangle is **not** right-angled.

16.1 Now try these

Band 1 questions

1 Work out these.

a 6^2
b 3^2
c 7^2
d 13^2
e 1.1^2
f 0.8^2
g 3.5^2
h 4.9^2

2 Work out these.

a $\sqrt{81}$
b $\sqrt{100}$
c $\sqrt{121}$
d $\sqrt{400}$
e $\sqrt{31.36}$
f $\sqrt{0.49}$
g $\sqrt{1.44}$
h $\sqrt{6.25}$

3 a Work out the area of each of these squares.

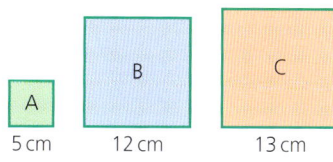

b Show that the sum of the areas of squares A and B is equal to the area of square C.

Band 2 questions

4 **a** Find the area of square A and of square B.

The area of the third square, C, is the same as the total area of squares A and B.

  b **i** Work out the area of square C.

  ii Work out the length of one side of square C.

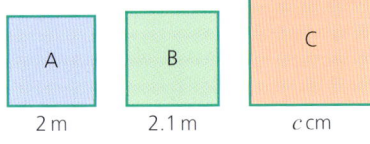

5 **a** Find the area of square A and of square C.

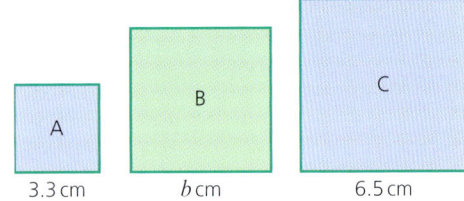

The area of square C is the same as the total area of squares A and B.

  b **i** Work out the area of square B.

  ii Work out the length of one side of square B.

6 For each of these right-angled triangles:

  i find the sum of the squares on the two shorter sides

  ii write down the area of the square on the hypotenuse

  iii work out the length of the hypotenuse.

  a 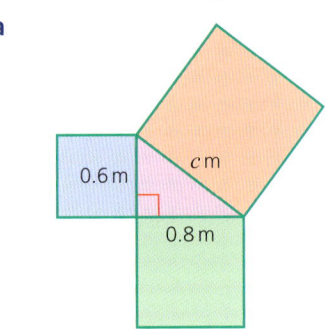   **b**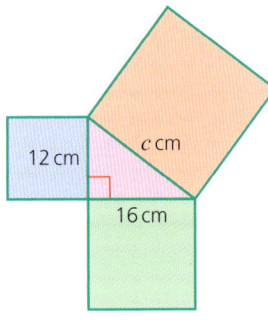

  Longest side.

7 For each of these right-angled triangles:

  i work out the area of the square on the side marked with a letter

  ii work out the length of the side marked with a letter.

  a 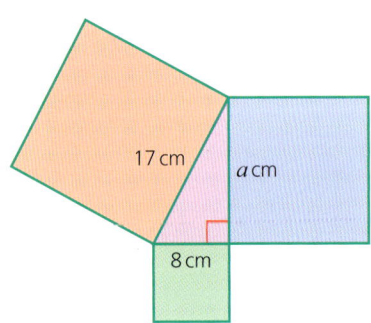   **b**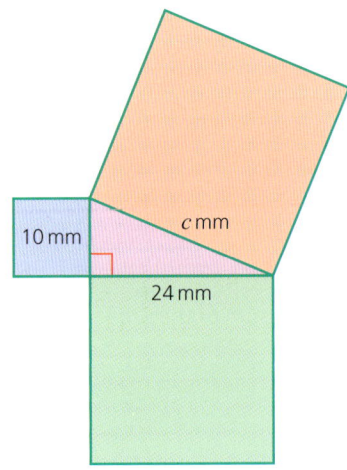

16 Pythagoras' theorem

Band 3 questions

8 Cai has drawn some triangles.

Sort Cai's triangles into right-angled triangles and non-right-angled triangles.

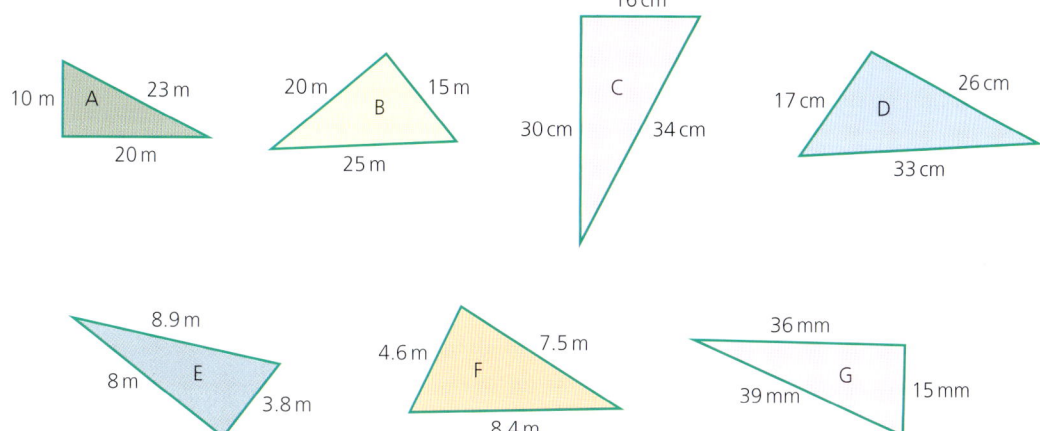

9 Erin has drawn some right-angled triangles, but she has forgotten to label the sides with their lengths.

Here, in no particular order, are the lengths of the sides of Erin's triangles.

50 cm	3 cm	30 cm	21 cm	20 cm	10 cm	15 cm	12 cm	4 cm
34 cm	9 cm	14 cm	26 cm	5 cm	48 cm	24 cm	16 cm	29 cm

Sketch Erin's triangles and use these lengths to label her sides.

Show that Pythagoras' theorem works for each triangle.

10 The large red square of side c cm has the same area as the sum of the two smaller squares.

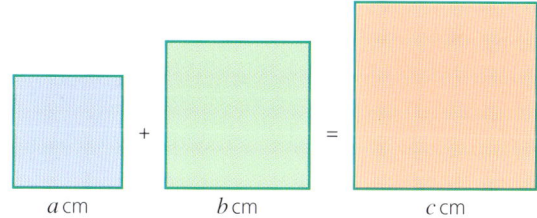

a cm b cm c cm

a Copy and complete the following:

$\square + \square = c^2$

a, b and c are all whole numbers.

The area of the large red square is less than 401 cm².

b Find possible values for the area of each square.
How many different answers can you find?

c Find possible values for a, b and c.

Hint

Find pairs of square numbers which add to give another square number.

Note

The numbers you have found in part **c** are examples of **Pythagorean triples**.

They are the solutions to the equation $a^2 + b^2 = c^2$ where a, b and c are integers.

There is an infinite number of Pythagorean triples.

The oldest record of a Pythagorean triple was on a Babylonian clay tablet dating from 1800 BC.

16.2 Using Pythagoras' theorem

Skill checker

Play a game of *Power Bingo* with a partner.

Each of you should draw a four-by-four grid like this one.

- Write the numbers from 1 to 16 at random in your grid.
- Take it in turns to choose a calculation from the list below.
- Answer the calculation and cross that number from your grid.

The winner is the first person to cross out a line of four numbers.

1	9	15	6
16	2	3	13
11	8	12	4
5	14	7	10

The winning line could be horizontal, vertical or diagonal.

$\sqrt{3^2 + 4^2}$	3^2	$\sqrt{6^2 + 8^2}$	$3 + 3^2$
$5^2 - (4^2 + 3)$	$7^2 + 6^2$	$\sqrt{196}$	$\dfrac{\sqrt{36}}{2}$
$2 \times \sqrt{16}$	$\sqrt{4}$	$6^2 - 5^2$	$\sqrt{256}$
$\sqrt{5^2 - 3^2}$	$\sqrt{25} - \sqrt{16}$	$3 \times 2^2 - 5$	$8^2 - 7^2$

▶ Problem solving

You can use Pythagoras' theorem to solve problems involving right-angled triangles without having to actually draw a square on each side.

This example shows you how to find the length of the hypotenuse.

Worked example

Find the length of side c in this triangle.

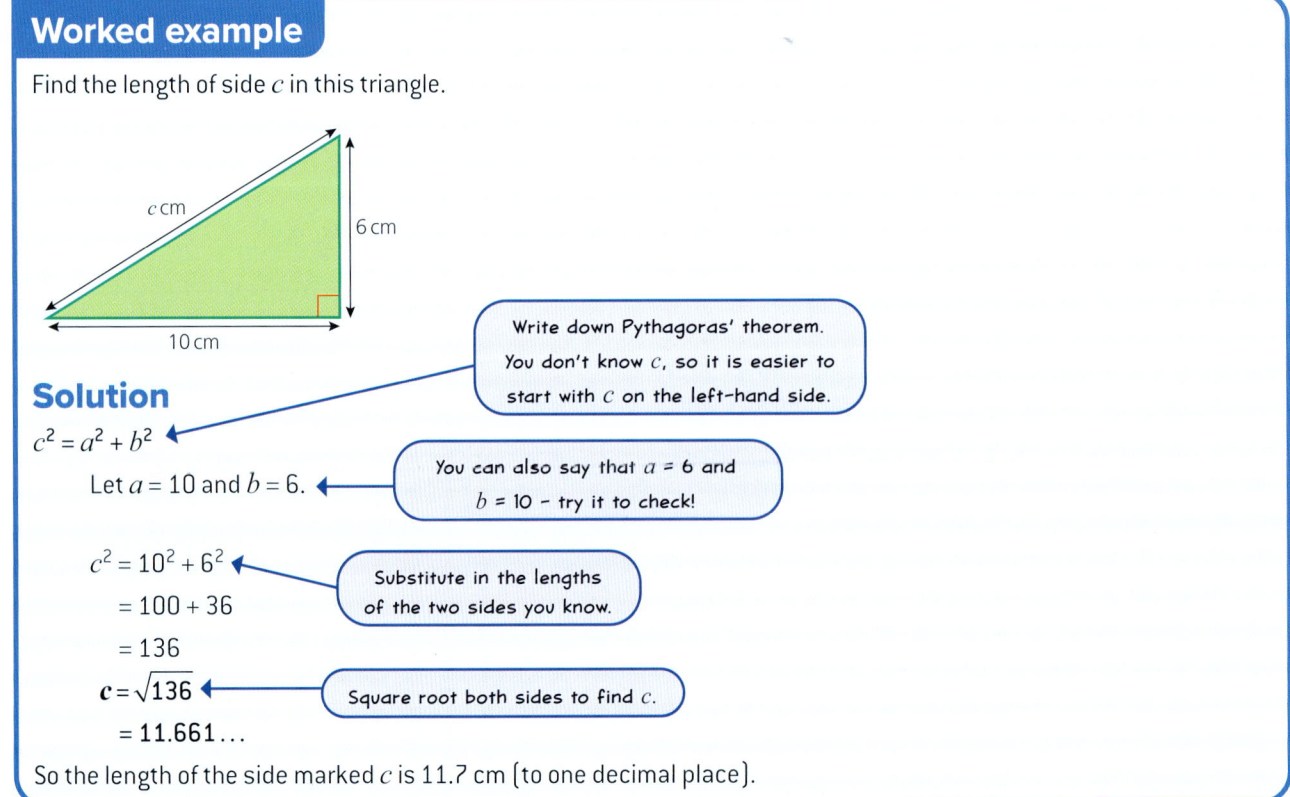

Solution

$c^2 = a^2 + b^2$

Let $a = 10$ and $b = 6$.

Write down Pythagoras' theorem. You don't know c, so it is easier to start with c on the left-hand side.

You can also say that $a = 6$ and $b = 10$ – try it to check!

$c^2 = 10^2 + 6^2$

$ = 100 + 36$

$ = 136$

Substitute in the lengths of the two sides you know.

$c = \sqrt{136}$

$ = 11.661\ldots$

Square root both sides to find c.

So the length of the side marked c is 11.7 cm (to one decimal place).

The next example shows you how to find the length of one of the two shorter sides.

Worked example

Lowri is a decorator.

She uses a ladder 6 m long.

The safety instructions on the ladder say that the base of the ladder must be at least 1.75 m from the wall on horizontal ground.

What is the maximum height her ladder will reach up a vertical wall?

Solution

$1.75^2 + h^2 = 6^2$

$3.0625 + h^2 = 36$

$h^2 = 32.9375$

square root ⟶ ⟵ square root

$h = \sqrt{32.9375}$

$h = 5.739...$

Maximum height reached by the ladder is 5.74 m (to the nearest centimetre).

> Write down Pythagoras' theorem. The longest side is 6 m and the two shorter sides are h m and 1.75 m.

> Draw a sketch and label it. h stands for the height reached by the ladder.

6 m, h m, 1.75 m

Remember

▶ Pythagoras' theorem only works for **right-angled triangles**.
▶ You need to know the length of **two of the sides** in order to use Pythagoras' theorem.

> To find the length of the **HYPOTENUSE** (longest side):
> - **SQUARE** the two given sides.
> - **ADD** the squares together.
> - **SQUARE ROOT** your answer.
>
> To find the length of one of the **SHORTER** sides:
> - **SQUARE** the two given sides.
> - **SUBTRACT** the smaller square from the larger.
> - **SQUARE ROOT** your answer.

Cross-curricular activity

In Physics, you can use Pythagoras' theorem to calculate resultant force.

An object experiences a force of 2.8 N to the right and 8.4 N downwards.

Calculate the size of the resultant force. Give your answer to one decimal place.

Resultant force

Curriculum for Wales Mastering Mathematics: Book 2

16.2 Now try these

Band 1 questions

1 Work out the value of $a^2 + b^2$ when:
   a $a = 5$ and $b = 4$
   b $a = 4.5$ and $b = 7.5$
   c $a = 2.6$ and $b = 1.3$.

2 The length of one side of a triangle is c cm. Work out the value of c when:
   a $c^2 = 25$
   b $c^2 = 100$
   c $c^2 = 64$.

3 Gwen is trying to find the length of the hypotenuse of this triangle.
Copy and complete her calculation.
Give your answer correct to two decimal places.

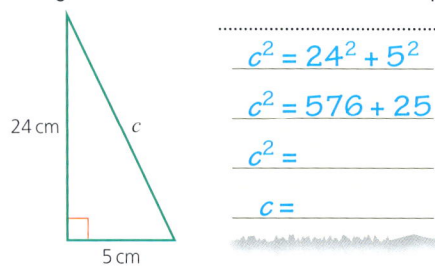

$c^2 = 24^2 + 5^2$
$c^2 = 576 + 25$
$c^2 =$
$c =$

4 Use $c^2 = a^2 + b^2$ to work out the length of the hypotenuse, c cm, of each of these triangles.

a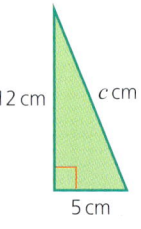

b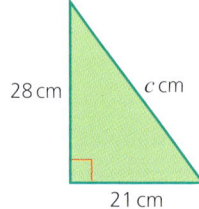

c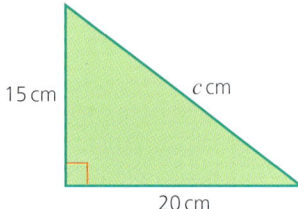

5 A right-angled triangle, ABC, has a base AB and height BC.
Find the length of the hypotenuse, AC, given that the base is 6 m and the height is 6.3 m.

Hint Draw a sketch of the triangle first.

Band 2 questions

6 Find the length of the side marked with a letter in each of these triangles.
Give your answers correct to one decimal place.

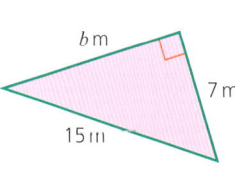

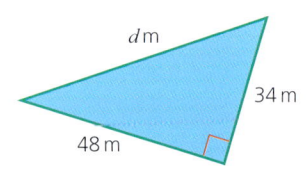

7 Work out the perimeter of this triangle.

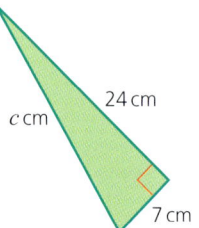

286

8. A right-angled triangle, ABC, has a right angle at C.
   AC = 6.5 cm. BC = 7.2 cm.
   Find the length of AB.

9. Calculate the length of the diagonal of this rectangle.

   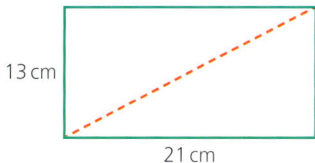

   Give your answer in centimetres correct to one decimal place.

10. Calculate the area of this triangle.

    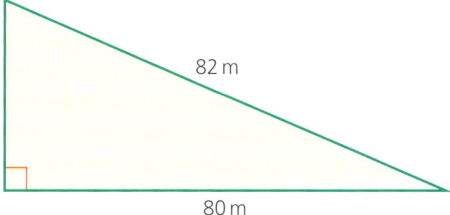

Band 3 questions

11. Calculate the value of d.

    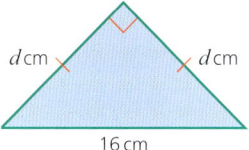

    Give your answer correct to one decimal place.

12. Sara makes earrings.
    Here is one of her designs, called 'Isosceles'.

    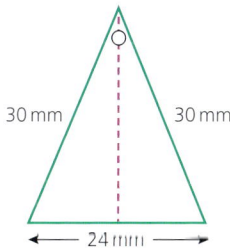

    a   Calculate the perpendicular height of the triangle.
    b   Calculate the area of the earring. Give your answers correct to one decimal place.

13. The diagram shows a ball touching a floor and a wall.

    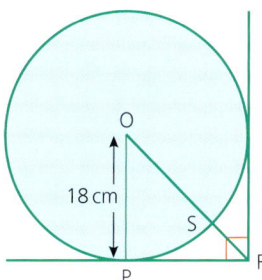

    Calculate the distance SR correct to the nearest millimetre.

Curriculum for Wales Mastering Mathematics: Book 2

Strategic competence

14. The quadrilateral ABCD is a diagram of a children's play park.

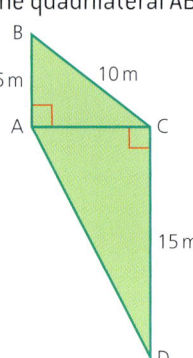

The town council wants to put up a fence around the park.

Fencing costs £32 per metre.

How much should it cost to put up a fence around the play park?

15. The points A(1, 2) and B(5, 4) are plotted on a coordinate grid.

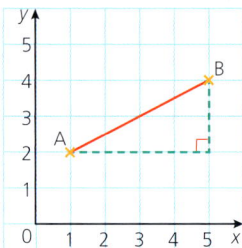

Calculate the distance AB.

Give your answer correct to two decimal places.

16. The perimeter of a square is 72 cm.

Calculate the length of its diagonal.

17. In the diagram below, the diagonal of the rectangle is equal to the side length of the square.

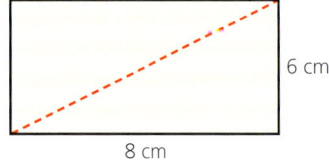 

Calculate the area of the square.

Key words

Here is a list of the key words you met in this chapter.

Hypotenuse   Pythagoras' theorem   Right-angled triangle   Square
Square root   Triangle

Use the glossary at the back of this book to check any you are unsure about.

16 Pythagoras' theorem

Review exercise: Pythagoras' theorem

Band 1 questions

1 Work out these.
  a 9^2
  b 5.3^2
  c $\sqrt{169}$
  d $\sqrt{20.25}$

2 For each of these right-angled triangles:
  i   find the area of the two squares with the known side lengths
  ii work out the area of the third square
  iii work out the side length of the third square.

  a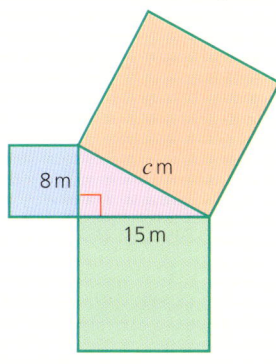
  b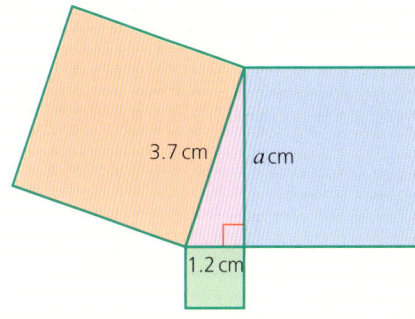

3 Use $c^2 = a^2 + b^2$ to work out the length of the hypotenuse, c m, of each of these triangles.

  a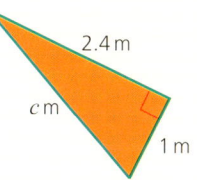
  b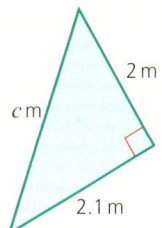

4 Find the length of the hypotenuse of right-angled triangles with these dimensions.
  a Base = 27 cm, height = 120 cm
  b Base = 65 cm, height = 72 cm

Band 2 questions

5 Calculate the length of the unknown sides in these triangles.
Give your answers correct to two decimal places.

  a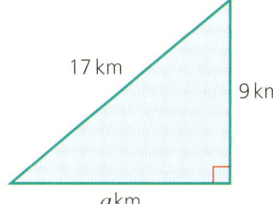
  b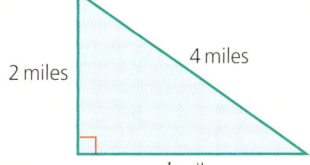
  c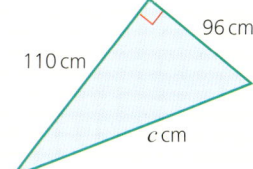

6. The diagram shows the flight path of an aircraft.

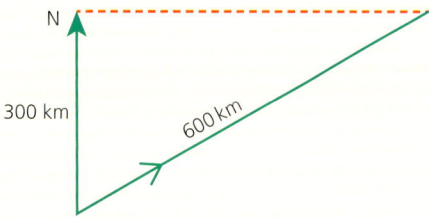

It has travelled 600 km.

By this time, it has gone 300 km North.

Find how far East it has travelled.

Give your answer correct to the nearest kilometre.

7. A ship leaves harbour and sails 53 miles North then 78 miles West.

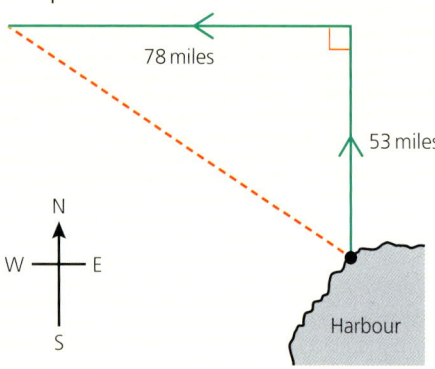

Calculate the shortest distance back to the harbour.

Give your answer in miles correct to one decimal place.

8. Here is one of Dylan's homework problems.

> The triangle ABC has a right angle at B.
> AC = 22 cm. BC = 17 cm.
> Find the length of AB.
> Solution
> $a^2 + b^2 = h^2$
> $22^2 + 17^2 = h^2$
> $484 + 289 = h^2$
> $773 = h^2$
> $h = \sqrt{773} = 27.8028\ldots$
> $= 27.8$ cm (to one decimal place)

Dylan's solution is wrong.

Find the mistake and write a correct solution.

16 Pythagoras' theorem

9 Adil draws a triangle with sides 7 cm, 9 cm and 11 cm.

Has Adil drawn a right-angled triangle? Explain your reasoning fully.

10 A ladder 5.6 m long rests against a wall. The bottom of the ladder is 1.5 m from the wall.

Calculate how far the ladder reaches up the wall.

Give your answer correct to the nearest centimetre.

Band 3 questions

11 ABCD is a square.

AC is 15 cm.

Calculate the area and perimeter of ABCD.

Give your answers correct to one decimal place.

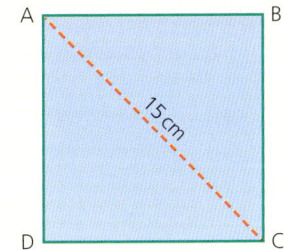

12 Work out the area of an equilateral triangle of side 6 cm.

Give your answer correct to one decimal place.

13 In the diagram, object B is hanging between two walls.

It is suspended by two strings, AB and CB.

A and C are at the same level.

B is 12 m below them.

B is 5 m from the nearer wall.

The total length of the strings is 33 m.

  a Find the distance between the two walls.

  b Prove that the angle ABC is not 90°.

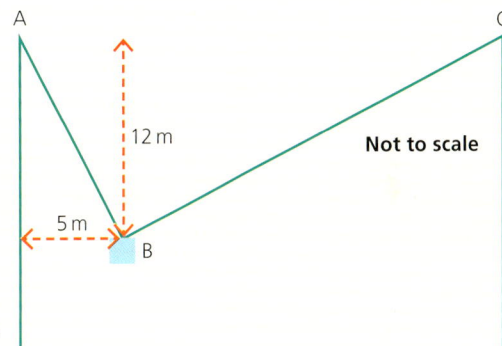

14 The diagram shows a flagpole, VT, tethered by two ropes, ST and TU.

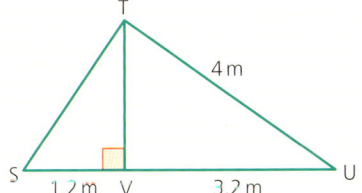

Calculate the length of rope ST.

Give your answer correct to the nearest centimetre.

15 The diagram shows the pentagon ABCDE.

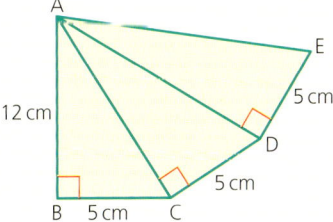

Calculate the length of AE.

Give your answer correct to the nearest millimetre.

Consolidation 6: Chapters 14–16

Band 1 questions

1 The radius of a circle is 7.2 cm.
  a What is the diameter of the circle?
  b Work out the circumference of the circle.
  Give your answer correct to one decimal place.

2 For each of these circles, find:
  i the circumference
  ii the area.
  Give your answers to one decimal place.

  a 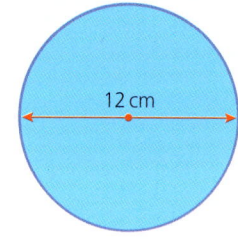   b 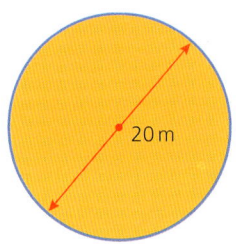   c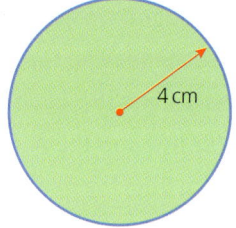

3 Seren is testing a special plant compost. She grows some seeds in the compost.

She measures and records the heights of the plants after six weeks. Here are her results:

3 cm	7 cm	4 cm	6 cm	11 cm	9 cm	17 cm
15 cm	8 cm	11 cm	16 cm	12 cm	4 cm	7 cm
8 cm	9 cm	10 cm	9 cm	13 cm	12 cm	

  a Construct a frequency table.
  b Work out:
    i   the range
    ii the mode
    iii the median.

4 a Use Pythagoras' theorem, $c^2 = a^2 + b^2$, to find the length of the hypotenuse of this triangle.

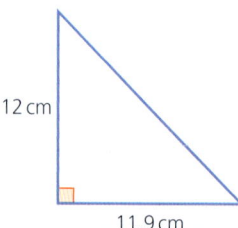

  b Work out the perimeter of the triangle.

5 Omar draws a pie chart to show which instruments the students in his Music class play.
  a Which instrument is the most popular?
  b What fraction of the students play the violin?
  c Three students play the piano.
     How many students are there in Omar's class?

This is an exercise you could try in your Music lesson with your class!

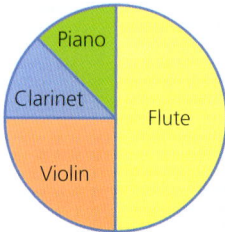

6 Work out the length of the missing side in each of these triangles.

a 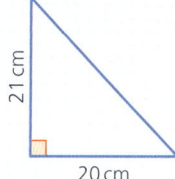   b 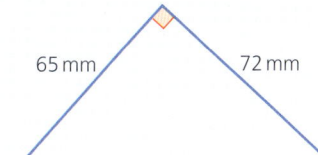   c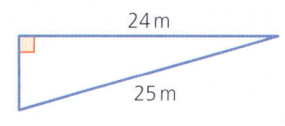

Band 2 questions

7 The radius of a pizza is 18 cm.

Calculate the area of:

a   the whole pizza

b   half the pizza.

Iwan cuts a piece of pizza for himself.

It is a quarter of the pizza.

c   What is the area of the piece he cuts?

Give your answers correct to the nearest square centimetre.

8 A bird builds its nest at the very top of an oak tree.

The height of the tree is 36 m.

The bird rests on level ground 77 m from the foot of the tree.

Work out the distance between the bird and its nest.

9 Calculate the area of this triangle.

10 Amir records how students in his year group travel to school.

Amir decides to draw a pie chart of his results.

How students get to school	Frequency	Angle of sector
Walk	125	
Cycle		72°
Bus	25	
Car	10	18°

a   Copy and complete Amir's table.

b   Draw a pie chart to show this information.

Remember

Spreadsheets can be used to quickly draw pie charts.

11 The table shows information about the ages of the members of a drama club.

Age of student	Frequency
13	6
14	8
15	5
16	10

> **Remember**
> The mean from a frequency table can be worked out from a spreadsheet very quickly.

  a What is the modal age?
  b Work out the range.
  c Work out the median of these ages.
  d Work out the mean of these ages.
     Give your answer correct to one decimal place.
  e Jabir, aged 13, joins the drama club.
     Without calculating the new mean, state whether the mean will increase, decrease or stay the same.

12 a Make d the subject of $C = \pi d$.
   b Make r the subject of $C = 2\pi r$.

Band 3 questions

13 Buddug says that a circle of area 200 cm² will have twice the radius of a circle of area 100 cm².

Is Buddug right?

Explain your answer fully.

14 The Babylonians, nearly 4000 years ago, knew that a rough approximation for π is 3. Your calculator can give you a much more accurate value.

  a i   Write down the value of π that your calculator gives you.
     ii Caron says 'π is 3.142'. Is Caron right?
  b Other approximations for π are $\frac{22}{7}$ and $\sqrt{10}$.
     i   Use a calculator to change $\frac{22}{7}$ to a decimal.
     ii What sort of decimal is your answer?
     iii Compare $\frac{22}{7}$ and the value of π on your calculator.
         How many decimal places are the same?
     iv Percentage difference is the percentage that the difference is of the true value.
         Find the percentage difference between $\frac{22}{7}$ and π.
  c i   Use a calculator to change $\sqrt{10}$ to a decimal.
     ii Find the percentage difference between your answer and the value given by the π button.

15 An aircraft leaves an airport and flies South-East for 200 km.
  a On what bearing is the aircraft flying?
  b Draw a diagram of the flight path.
  c Calculate how far East the aircraft has flown.
  d How far South has the aircraft flown?

Give your answers correct to the nearest kilometre.

16. A stepladder is 2.2 metres long.

    The stepladder is opened so that its feet are 60 centimetres apart.

    How high is the top of the ladder?

    Give your answer correct to the nearest centimetre.

17. The diagram shows a square inside a circle.

    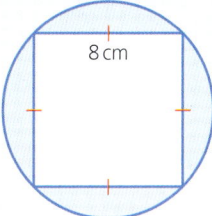

    Calculate the area of the shaded region.

    Give your answer correct to one decimal place.

18. The points A(3, 4) and B(6, 8) are plotted on a coordinate grid.

    Calculate the distance AB.

Glossary

3D shape A solid shape with three dimensions: length, width and depth.

Acute An acute angle lies between 0° and 90°.

Addition (add) Addition or adding is finding the total of two or more amounts. For example, $3 + 4 = 7$.

Allied angles A pair of angles on the inside of a pair of parallel lines and on the same side of a transversal. Allied angles are sometimes referred to as C angles because of the shape they make.

Alternate angles A pair of angles on the inside of a pair of parallel lines, but on opposite sides of a transversal.

Angle An angle is a measure of turning. Angles are measured in degrees. For example, a full turn is 360°.

Area The area of a shape is the amount of surface that it covers. Area is measured in square units such as mm^2, cm^2, m^2 and km^2.

Arithmetic sequence An arithmetic sequence is a sequence where the difference between consecutive terms is a constant. For example, 7, 11, 15, 19, …

Arrowhead An arrowhead is a kite with a reflex angle.

Average An average is a measure of the typical value in a data set. Common averages are the mean, the mode and the median.

Axes Axes is the plural of axis. An axis is a fixed reference line for the measurement of coordinates. The x-axis is horizontal and the y-axis is vertical.

Bar chart A bar chart is a chart that uses rectangular bars to display data. The height of each bar represents the frequency.

Base The base of a shape is the line or surface on which it appears to be standing.

Bearing An angle measured clockwise from North.

Best buy The cheaper or cheapest option when a decision must be made between two or more similar items for sale. The best buy can often be decided using the unitary method.

Biased Data is biased if some outcomes occur more or less than would be expected. For example, a dice is biased if the number 6 hardly ever comes up.

BIDMAS BIDMAS is a way of remembering the order in which you carry out the different operations in a calculation: Brackets, Indices, Division, Multiplication, Addition and Subtraction.

Brackets Brackets are a way of grouping numbers or algebraic terms together. For example, $(3 + 5) \div 2$ means the same as $8 \div 2$.

Cancel To cancel a fraction is to simplify it by dividing the numerator and denominator by a common factor.

Capacity The capacity of a 3D shape is the volume it can hold. It is measured in cubic units such as mm^3 and cm^3, or for liquids litre (l) and millilitre (ml).

Categorical Categorical data is non-numerical data that can be put into groups such as favourite colour or types of pet.

Centilitre 1 centilitre (1 cl) is one hundredth of a litre.

Centimetre 1 centimetre (1 cm) is one hundredth of a metre.

Centre of rotation The centre of rotation is the 'pivot' point about which an object is rotated.

Certain If an event is certain, its probability is 1. For example, it is certain that the Sun will rise tomorrow.

Chance The chance of something happening is how likely it is to happen. Words can be used – it is unlikely that Sam will get up before midday. Or numbers can be used – the probability of Sam getting up before midday is about 0.09.

Circle A circle is a shape made up of all the points that are a specific distance from the centre.

Circumference The circumference is the perimeter of a circle.

Clockwise The direction followed by the hands of a clock.

Coefficient The number written in front of a letter variable.

Co-interior angles See allied angles.

Column method Digits are written with units in the same column, tens in the same column, hundreds in the same column, etc.

Common denominator The common denominator of two or more fractions is a common multiple of all of the denominators. It is used when adding or subtracting fractions.

Common factor A common factor of two or more numbers is a number that divides exactly into all the numbers. You simplify fractions by cancelling by a common factor.

Commutative An operation is commutative if the order in which you do it doesn't matter. Addition and multiplication are commutative operations.

Concave All sides are oriented inwards.

Concentric Concentric circles have the same centre but different radii (sizes). They look like ripples after a stone is dropped into water.

Cone A cone is a 3D shape that has two faces: a flat circular base and a single curved face forming a single vertex. It could be described as a circular based pyramid.

Congruent Congruent shapes are exactly the same shape and size – they are identical.

Construct Draw accurately using only a pair of compasses and a straight edge.

Conversion graph A conversion graph is used to change one unit into another. This could be changing between

Glossary

miles and kilometres, pounds to a foreign currency, or the cost of a journey based on the number of miles travelled.

Convex At least one side is oriented outwards.

Coordinates Coordinates are a way of showing position on a pair of axes or graph. For example, the point $(3, -5)$ is 3 to the right and 5 down.

Corresponding angles A pair of angles in identical positions relative to a transversal.

Cube (number) To find the cube of a number you multiply the number by itself twice. For example, the cube of 5.6 (5.6^3) is $5.6 \times 5.6 \times 5.6$ or 175.616.

Cube (shape) A cube is a 3D shape with six identical square faces.

Cube root The cube root of a number is the number that, when multiplied by itself twice, gives the original number. The inverse of cubing is cube rooting. For example, the cube root of 27 is 3 (as $3 \times 3 \times 3 = 27$). The symbol $\sqrt[3]{}$ is used for the cube root of a number, so $\sqrt[3]{27} = 3$.

Cuboid A 3D shape whose six faces are all rectangular or square. A rectangular prism.

Currency The money used in a country, for example the currency of the UK is pounds.

Currency conversion The process of changing one currency to another, for example pounds to euros. A currency conversion can be done using a conversion graph.

Curve For example, the circumference of a circle. A curve can also be the shape of a graph described by the relationship between two variables, for example $y = x^2$.

Data A group of facts or statistics, it can be numerical (for example heights) or non-numerical (for example favourite crisp flavour).

Decagon A ten-sided polygon.

Decimal A decimal is a number written using a decimal point. For example, 82.17. The digits after the decimal point represent a value less than one.

Decimal place The number of decimal places in a decimal is the number of digits after the decimal point. For example, 3.2 is written to one decimal place and 5.678 is written to three decimal places.

Decimal point The decimal point is the dot in a decimal number. The digits before the decimal point represent whole numbers; the digits after the decimal point represent fractions. For example, the number 4.37 means four units plus three tenths plus seven hundredths.

Denominator The denominator is the bottom integer of a fraction. It tells you how many equal parts the whole is divided into. For example, $\frac{3}{8}$ has a denominator of 8, so the 'whole' has been divided into eight equal parts.

Diameter A diameter is a line that passes through the centre of a circle and joins two points on the circumference.

Difference The difference between two numbers is the result of subtracting the smaller from the larger. For example, the difference between 15 and 6 is $15 - 6 = 9$.

Digit A digit is one of the symbols 0, 1, 2, 3, 4, 5, 6, 7, 8 or 9. Digits are used to write numbers.

Directed A number with a positive or a negative sign in front of it. If a sign is missing, then we assume it is positive.

Divide To divide two numbers you share the first number into the number of equal parts given by the second number. For example, 24 divided by 3 is 8 ($24 \div 3 = 8$).

Division Division is the process of dividing two numbers.

Edge In a 3D shape, an edge forms the boundary between two faces. In a cube, for example, there are 12 edges that form the boundaries between the 6 faces.

Equal If two quantities or expressions are equal, they have the same value. The symbol = is used to show equality. For example, $14 + 8 = 22$.

Equally likely In probability, if two events are equally likely they have the same probability of happening. For example, when you roll a fair six-sided dice the possible outcomes of 1, 2, 3, 4, 5 or 6, are equally likely. They each have a probability of $\frac{1}{6}$.

Equation An equation is where one expression is equal to another. An equation always contains an equal sign. For example, $6 + 4 = 16 - 6$. When an equation contains an unknown, it can be solved. For example, the solution to the equation $x + 4 = 16 - x$ is $x = 6$.

Equilateral An equilateral triangle has three equal angles (all 60°) and three sides of equal length.

Equivalent Equivalent fractions represent the same value. For example, $\frac{3}{5}$ and $\frac{9}{15}$ are equivalent.

Evaluate Evaluate means 'work out the value of'.

Even An even number is an integer that is a multiple of 2. For example, 2, 4, 6, ..., 48, ..., etc.

Event An event is any of the possible outcomes from an experiment. For example, I roll two dice and add the numbers together. What is the probability of the event 'my total is 10'?

Expand To multiply out or remove the brackets from an expression.

Experiment An experiment or trial is a procedure in probability that can be repeated over and over again and where we know what the possible outcomes are. For example, rolling two dice and recording the total of the two numbers.

Expression An expression is numbers, symbols and operators grouped together. For example, $3x - 8$. An expression does not contain an equal sign.

Exterior The exterior angles of a polygon add up to 360°.

Face In a 3D shape, a face is a surface that forms a part of the boundary of the shape. For example, a cube has six faces.

Factor A factor of a number divides into that number exactly. For example, the factors of 18 are 1, 2, 3, 6, 9 and 18.

Factorise Opposite of expand.

Fair In probability, a dice or coin is fair if all the possible outcomes are equally likely.

Fibonacci sequence The Fibonacci sequence begins 1, 1, 2, 3, 5, 8, … Each term is the sum of the previous two terms.

Fibonacci-type sequence Any sequence in which each term is the sum of the previous two terms, for example 3, 4, 7, 11, 18, …

Formula A formula is a rule or relationship connecting two or more variables. For example, the formula for the area of a triangle is $A = \frac{1}{2}bh$.

Fraction A fraction, for example $\frac{3}{4}$, consists of a numerator (3 in this example) and a denominator (4 in this example). A fraction can represent part of a whole (3 parts out of 4) or represent a decimal number (0.75).

Frequency The frequency of a data value is the number of times it occurs.

Frequency table A frequency table shows the frequency of each data value in a data set.

Function machine A flow diagram that takes an input value, applies a set of mathematical operations and outputs the answer.

Geometric sequence A geometric sequence is a sequence where the ratio between any two consecutive terms is a constant. For example 5, 10, 20, 40, … (in this sequence the ratio is 2).

Gram Gram (g) is a metric unit of mass. 25 g is approximately the mass of a regular size bag of crisps.

Graph A graph is the depiction of a relationship between (usually) two variables. For example, it is possible to draw a graph of the relationship $y = x^2$.

Greater than The symbol > indicates the value to the left is bigger than the one to the right.

Grid method A method for multiplying numbers which involves dealing with units, tens, hundreds, etc. separately.

Height The height is the distance from the base to the top.

Heptagon A heptagon is a seven-sided polygon.

Hexagon A hexagon is a six-sided polygon.

Highest common factor (HCF) The highest common factor (HCF) of two numbers is the greatest integer that divides exactly into both numbers. For example, the highest common factor of 6 and 15 is 3.

Horizontal A horizontal line is parallel to the horizon. It runs from left to right.

Hundreds In place value, hundreds is the place value of the first digit in a three-digit number. In the number 714, the 7 represents seven hundreds.

Hundredths In place value, hundredths is the place value of the second digit after the decimal point. In the number 7.14, the 4 represents four hundredths. In the number 7246, the 2 represents two hundreds.

Hypotenuse The hypotenuse is the longest side of a right-angled triangle.

Image When a shape undergoes a transformation, the resulting shape is the image.

Impossible An event with probability 0 is impossible.

Included angle The angle between two given sides.

Included side The side between two given angles.

Indices The index is the power to which a number is raised. For example, in 4^3 the power (or index) is 3 and so $4^3 = 4 \times 4 \times 4$. The plural of index is indices.

Inequality Compares the size of two values, showing if one is less than , or greater than, the other.

Input A starting value in a function machine.

Integer An integer is a positive or negative whole number (including zero).

Interest Interest can be: 1) a payment made to an investor or 2) money charged for borrowing. See also **simple interest**.

Interior Inside a shape.

Internal The sum of the interior or internal angles of an n-sided polygon $= 180° \times (n - 2)$.

Inverse Inverse means opposite. Subtraction is the inverse of addition. Division is the inverse of multiplication.

Irregular (shape) A shape where all angles are not an equal size or all sides are not an equal length.

Isosceles An isosceles triangle has two equal angles and two sides of equal length.

Glossary

Key A key is used on some diagrams such as pictograms to show what a symbol represents. For example, Key: ▢ represents four people.

Kilogram Kilogram (kg) is a metric unit of mass equal to 1000 grams. 1 kg is approximately the mass of a bag of sugar.

Kilometre Kilometre (km) is a metric unit of distance equal to 1000 metres. 1 km is $2\frac{1}{2}$ times round an ordinary athletics track.

Kite A kite is a quadrilateral which has two pairs of adjacent, equal sides. Other properties are one pair of equal angles and diagonals that cross at right angles.

Length The length of an object is a way of measuring its size. For example, the length of this rectangle is 13 cm.

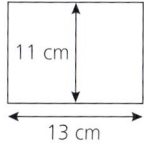

Less than The symbol < indicates the value to the left is less than the one to the right.

Likely An event with probability greater than 0.5 but less than 1 is likely.

Line A line is a straight one-dimensional figure with no thickness and extending infinitely in both directions. It is often called a straight line.

Line segment A line segment is part of a line. It has a beginning and an end.

Linear Relating to a straight line. An arithmetic sequence can be described as linear, since on a graph of the terms of the sequence, the points would lie on a straight line.

Litre Litre (l) is a metric unit of capacity equal to 1000 millilitres. 1 litre is usually the capacity of a carton of fruit juice.

Long division A method for dividing numbers which clearly shows how each remainder is calculated.

Long multiplication A method for multiplying two numbers.

Loss If an item is sold for less than the buying price, the **loss** is the difference between these two prices. If the selling price is greater than the buying price, there is a **profit**.

Lowest common multiple (LCM) The lowest common multiple of two numbers is the smallest integer that is a multiple of both numbers. For example, the lowest common multiple of 6 and 15 is 30.

Mass The mass of an object is a measure of how much matter it contains. Mass is measured, for example, in kilograms. In Maths at this level, mass and weight are considered to be the same thing.

Mean The mean is found by adding together all of the data values and then dividing this total by the number of data values. The mean is one of the three main ways to measure an average.

Median Median is the middle value when the data set is organised in order of size. The median is one of the three main ways to measure an average.

Metre Metre (m) is a metric unit of length equal to 1000 millimetres. 2 m is roughly the height of a doorway.

Milligram Milligram (mg) is a metric unit of mass equal to one thousandth of a gram. Milligrams are used to measure out doses of drugs. For example, a small aspirin might contain 75 mg.

Millilitre Millilitre (ml) is a metric unit of capacity equal to one thousandth of a litre. For example, a teaspoon holds 5 ml.

Millimetre Millimetre (mm) is a metric unit of length equal to one thousandth of a metre. For example, a small ruler is 150 mm long.

Millions In place value, millions is the place value of the first digit in a seven-digit number. In the number 2 714 806, the 2 represents two millions.

Mirror line The line in which a shape is reflected is known as the mirror line.

Modal The mode of a set of data can be called the modal value.

Mode The mode is the value occurring most often in a data set. The mode is one of the three main ways to measure an average.

Multiple The multiple of a number is the result when you multiply that number by a positive integer. For example, the multiples of 6 are 6, 12, 18, 24, 30, …

Multiplication (multiply) Multiplication comes from repeated addition, so 7 × 3 is the same as 7 + 7 + 7 or 21.

Multiplier The multiplier is the number you are multiplying by. For example, in 3 × 7 the multiplier is 7.

Mutually exclusive events Two events are mutually exclusive if they cannot happen at the same time. For example, when rolling a dice, getting a 3 and getting a 5 are mutually exclusive as you can't get both a 3 and a 5 from rolling one dice.

Negative A value which is less than zero.

Nonagon A nine-sided shape.

Number line A line with numbers indicating their size relative to each other.

Numerator The numerator is the top number in a fraction. For example, $\frac{3}{8}$ has a numerator of 3.

Object The object is the shape that a transformation is applied to.

Obtuse An obtuse angle lies between 90° and 180°.

Octagon An octagon is an eight-sided polygon.

Odd An odd number is an integer that is not a multiple of 2. For example, 1, 3, 5, ..., 57, ..., etc.

Ones In place value, ones is the place value of the digit in a single-digit number. In the number 8, the 8 represents eight ones. In the number 7246, the 6 represents six ones. Ones are also known as units.

Operations An operation is a procedure that is carried out to a number, such as dividing or raising to a power. See *BIDMAS* for the order in which to carry out operations.

Outlier In a set of data, an outlier is a value that lies outside the range of the rest of the data – it is much larger or much smaller. It also refers to a value that does not fit the same pattern as the rest of the data.

Output The final value, once the operations of a function machine have been applied.

Pair of compasses An instrument for drawing a circle, or an arc of a circle. Sometimes incorrectly referred to as a compass.

Parallel A pair of parallel lines can be continued to infinity in either direction without meeting.

Parallelogram A parallelogram is a quadrilateral that has two pairs of parallel sides. Its other properties are that opposite sides are equal and opposite angles are equal.

Partitioning Breaking numbers into smaller parts to make them easier to work with, for example into units, tens, hundreds, and so on.

Pentagon A pentagon is a five-sided polygon.

Percentage A percentage (%) is a number expressed as a fraction of 100. For example, 38% is equivalent to $\frac{38}{100}$ or 0.38.

Percentage change A change in a quantity expressed as a percentage of the original quantity. For example, if the price of an item falls from £4 to £3, the decrease is £1 and the percentage decrease is 25% (since $\frac{1}{4} = 25\%$).

Percentage error The error in the measurement of a quantity expressed as a percentage of the true quantity. For example, if the length of an object is measured as 49 cm but its true length is 50 cm, the error is 1 cm and the percentage error is 2% (since $\frac{1}{50} = 2\%$).

Perimeter The perimeter of a shape is the distance around the outside edge of the shape. Perimeter can be measured in mm, cm, m or km.

Perpendicular Two lines are perpendicular if they meet at right angles.

Pi, π The Greek letter pi is the number 3.14159..., it is equal to the circumference of a circle divided by its diameter.

Pictogram A pictogram is a chart that uses pictures or symbols to display data.

Pie chart A circular chart that is divided into sectors to represent different groups.

Place value The place value of a digit is the number that a particular digit in a number represents. For example, in the number 4137.59, the 4 represents four thousands and the 9 represents nine hundredths.

Polygon A polygon is a closed 2D shape made up of straight lines.

Position-to-term rule The position-to-term rule (or nth term) is a way of describing the terms of a sequence using their position in the sequence. For example, the position-to-term rule for the sequence 7, 13, 19, 25, 31, ... is $6n + 1$.

Positive A value which is greater than zero.

Powers For example, in 4^3 the power is 3 and so $4^3 = 4 \times 4 \times 4$.

Power of 10 10 raised to an integer power. For example, $10^3 = 1000$.

Prime A prime number is a number with exactly two factors: one and itself. The prime numbers are 2, 3, 5, 7, 11, ...
Note: 1 is not a prime number as it has only one factor.

Prism A 3D shape that has a constant cross-section and that does not have any curved faces.

Probability Probability is the study of chance. The probability of an event happening is a measure of how likely that event is to happen.

$$\text{Probability} = \frac{\text{number of favourable outcomes}}{\text{total number of equally likely outcomes}}$$

Probability scale The probability scale goes from 0 (an impossible event) to 1 (a certain event).

Product The product of two numbers is the result of multiplying them together. For example, the product of 8 and 3 is 24 ($8 \times 3 = 24$).

Profit The profit is the difference between the selling price and buying price of an item. If the selling price is lower than the buying price there is a loss.

Proportion A proportion is a part of a whole. Proportions can be given as fractions, decimals or ratios.

Pythagoras' theorem A theorem that says that the square of the hypotenuse (longest side) of a right-angled triangle is equal to the sum of the squares of the two shorter sides. It is written as $a^2 + b^2 = c^2$.

Quadrant A quadrant is any of the four regions of a graph divided by the x and y axes.

Quadrilateral A quadrilateral is a four-sided polygon (a four-sided plane shape).

Radius A radius is a line that joins the centre of a circle to a point on the circumference. The plural is radii.

Random An event is random if it is done without method or conscious decision.

Glossary

Range Range is a measure of the spread of a data set. The range is the difference between the highest and lowest data values.

Ratio A ratio compares two quantities.

Real-life graph Any graph that arises from a real-life situation rather than just an equation. Real-life graphs include travel graphs and conversion graphs.

Reciprocal The product of a number and its reciprocal is 1. So $4 \times \frac{1}{4} = 1$ and $5 \times \frac{1}{5} = 1$.

Rectangle A rectangle is a quadrilateral with the following properties: two pairs of parallel sides and four equal angles (90°). It also has opposite sides of equal length.

Reflection A reflection is a 'flip' movement about a mirror line. The mirror line is the line of symmetry between the object and its image.

Reflex A reflex angle is an angle between 180° and 360°.

Regular polygon A shape where all angles are equal sizes and all sides are an equal length.

Rhombus A rhombus is a quadrilateral with four equal sides. It also has two pairs of parallel sides, equal opposite angles and diagonals that cross at right angles.

Right angle A right angle is 90°.

Right-angled triangle A triangle where one of the internal angles is 90°.

Rotation A rotation is a 'turning' movement about a specific point known as the centre of rotation.

Scale In a scale drawing, the scale is the ratio or multiplier that tells you how much bigger (or smaller) the image is compared to the original drawing.

Scalene A scalene triangle has three sides of different lengths and three different angles.

Sector A sector is part of a circle that looks like a piece of pie, it is enclosed by two radii and part of the circumference.

Semi circle A semi circle is half a circle.

Sequence A sequence is a collection of terms arranged in a specific order.

Short division An abbreviated version of long division where the remainders are evaluated without writing down the calculation.

Sign A symbol written in front of a number to indicate whether it is positive or negative.

Simplify To simplify an expression you write it in an equivalent way but using smaller numbers or fewer terms. You can simplify a fraction, ratio or an algebraic expression. For example, you can simplify $\frac{9}{15}$ to $\frac{3}{5}$ or $x + x + y + y + y$ to $2x + 3y$ or $21:35$ to $3:5$.

Solid Another word for a 3D shape.

Solve To solve an equation you find the value of the unknown that satisfies the equation. For example, the solution of $7x = 35$ is $x = 5$.

Sphere A 3D shape that is perfectly round, like a ball.

Square (shape) A square is a quadrilateral with four equal sides and four equal angles (90°). It also has two pairs of parallel sides and diagonals that cross at right angles.

Square metre A square metre (m^2) is a unit of area. 1 m^2 is equivalent to the area of a square of side 1 m.

Square number A square number is the result when an integer is multiplied by itself. The square numbers are 1, 4, 9, 16, 25, ...

Square root The square root of a number is the number that, when multiplied by itself, gives the original number. The reverse of squaring is square rooting. Every positive number has two square roots. For example, the square root of 9 is 3 (as $3 \times 3 = 9$) or -3 (as $-3 \times -3 = 9$). The symbol $\sqrt{}$ is used for the positive square root of a number, so $\sqrt{9} = 3$.

Square-based pyramid A 3D shape whose base is a square. It has four other faces, all of which are triangles, meeting at a single point.

Straight line graph A straight line graph, or linear graph, is a graph in which the points are arranged in a straight line.

Subject The subject of a formula is the variable (letter) which appears on its own on one side of the formula.

Substitute To substitute is to replace the variables (letter symbols) in an expression or formula with numbers.

Subtraction (subtract) Subtraction or taking away is finding the difference between two or more amounts. For example, $7 - 4 = 3$.

Sum The sum of two numbers is the result of adding them together. For example, the sum of 8 and 3 is 11 ($8 + 3 = 11$).

Symmetry A shape has symmetry if it does not change under a transformation such as reflection or rotation.

Tally chart A tally chart is a way of collecting or representing data using tally marks. For example:

Score	Tally	Frequency
0		
1		
2	⅃⅃⅃⅃⅃	5
3	⅃⅃⅃⅃⅃ ⅃⅃⅃⅃⅃	10
4	⅃⅃⅃⅃⅃ ⅃⅃⅃⅃⅃ I	11
5	III	3
Total:		

301

Tens In place value, tens is the place value of the first digit in a two-digit number. In the number 57, the 5 represents five tens. In the number 7246, the 4 represents four tens.

Tenths In place value, tenths is the place value of the first digit after the decimal point. In the number 0.83, the 8 represents eight tenths.

Term Each number in a sequence is a called a term.

Term number The term number gives the position of a term in a sequence. For example, in the sequence 3, 13, 23, 33, 43, …, the term number of 13 is 2 as it is second in the sequence. The term number is often referred to as n.

Term-to-term rule A term-to-term rule describes how to use one term in a sequence to find the next term.

Terminating decimal A terminating decimal has digits that do not continue forever. For example, 0.123 and 0.987 654 321.

Thousands In place value, thousands is the place value of the first digit in a four-digit number. In the number 7246, the 7 represents seven thousands.

Thousandths In place value, thousandths is the place value of the third digit after the decimal point. In the number 0.691, the 1 represents one thousandth.

Tonne Tonne (t) is a metric unit of mass equal to 1000 kilograms. 1 t is approximately the mass of a small car.

Total The total is the result of adding numbers together.

Transformation A transformation changes the size, shape, position or orientation of an object. Examples of transformations are reflection, rotation, enlargement and translation.

Translation A translation is a sliding movement.

Trapezium A trapezium is a quadrilateral with one pair of parallel sides.

Triangle A triangle is a three-sided polygon.

Triangle numbers The sequence of numbers 1, 3, 6, 10, 15, …

Turn To turn is to rotate about a point. A full turn is a rotation of 360°.

Unitary method A method used to determine a best buy. The unitary method finds the cost of one item, when two or more purchase options are available. For example, if 6 eggs cost £1.20 the unit price is 20p; if 10 eggs cost £1.50 the unit price is 15p and the pack of 10 eggs is the best buy.

Units (number) In place value, units is the place value of the digit in a single-digit number. In the number 8, the 8 represents eight units. In the number 7246, the 6 represents six units. Units are also known as ones.

Units (shape) The standard units used for measurement, such as kilogram, metre and second.

Unknown The unknown is the letter you are trying to find a value for in an equation. The unknown is most commonly given the symbol x.

Unlikely An event with probability greater than 0 but less than 0.5 is unlikely.

Variable A variable is a quantity that may change in maths problems, such as the area of a triangle. A variable is usually represented by a letter.

VAT VAT stands for 'value added tax'. This is a tax that is added to most goods that we buy. The Government sets the percentage VAT that is charged.

Vertex A vertex of a shape is a point where two sides meet.

Vertical A vertical line is perpendicular to the horizon. It runs up and down.

Vertical line graph A vertical line graph is a graph that uses vertical lines to display data. The height of each line represents the frequency.

Vertically opposite Vertically opposite angles are formed when two lines cross. Vertically opposite angles are equal.

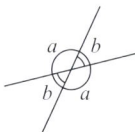

Volume Volume is the amount of space that a 3D shape takes up.

Weight Weight is an everyday term that means the same as mass. When Science and Maths get more precise, a different definition is used.

Width The width of an object is a way of measuring its size. For example, the width of this rectangle is 11 cm.

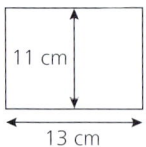

x-coordinate The x-coordinate is the horizontal value in a pair of coordinates. It is how far left or right the point is. In the coordinate (7, 2) the x-coordinate is 7.

y-coordinate The y-coordinate is the vertical value in a pair of coordinates. It is how far up or down the point is. In the coordinate (7, 2) the y-coordinate is 2.